The Rock-Eval Method

SCIENCES

Geoscience, Field Director – Yves Lagabrielle

The Surface of the Lithosphere and the Biosphere,
Subject Head – François Baudin

The Rock-Eval Method

Principles and Application

Coordinated by
François Baudin

iSTE WILEY

First published 2023 in Great Britain and the United States by ISTE Ltd and John Wiley & Sons, Inc.

ISTE Ltd
27-37 St George's Road
London SW19 4EU
UK

www.iste.co.uk

John Wiley & Sons, Inc.
111 River Street
Hoboken, NJ 07030
USA

www.wiley.com

Library of Congress Control Number: 2023942076

British Library Cataloguing-in-Publication Data
A CIP record for this book is available from the British Library
ISBN 978-1-78945-153-5

ERC code:
PE10 Earth System Science
PE10_5 Geology, tectonics, volcanology
PE10_7 Physics of earth's interior, seismology, volcanology
PE10_10 Mineralogy, petrology, igneous petrology, metamorphic petrology

Contents

Presentation of Units, Symbols and Acronyms

The units of measurement and symbols used in this book are those of the International System. They are generally presented as a power of 10 and as exponents (e.g. $kg.m^{-2}$). The common prefixes of the powers of 10 have been used:

– P (peta): 10^{15} = 1,000,000,000,000,000

– T (tera): 10^{12} = 1,000,000,000,000

– G (giga): 10^{9} = 1,000,000,000

– M (mega): 10^{6} = 1,000,000

– k (kilo): 10^{3} = 1,000

– h (hecto): 10^{2} = 100

– c (centi): 10^{-2} = 0.01

– m (milli): 10^{-3} = 0.001

– µ (micro): 10^{-6} = 0.000001

– n (nano): 10^{-9} = 0.000000001

– p (pico): 10^{-12} = 0.000000000001

Nevertheless, in order not to break away from well-established habits, certain data are presented in more common units. A conversion table is provided below:

SI Unit			Multiples and submultiples and SI value			Non-SI units and SI value		
Length	meter	m	angström	Å	10^{-10}			
Area	square meter	m^2						
Volume	cubic meter	m^3	liter	l	10^{-3}	barrel	bbl	0.159
Mass	kilogram	kg	ton	t	10^3			
Time	second	s	day	d	86,400			
			year	yr	31,557,600			
Pressure	pascal	Pa	bar	bar	10^5			
Heat-Work-Energy	joule	J				calorie	cal	4.185

The following abbreviations and acronyms have been used in this book:

A	Arrhenius constant
API	American Petroleum Institute
BP	Before Present
BSR	Bacterial Sulfate Reduction
Ca	Active Carbon
Cs	Stable Carbon
DCM	Dichloromethane
DOC	Dissolved Organic Carbon
E/Ea	Activation Energy
FID	Flame Ionization Detector
GC	Gas Chromatography
GC-MS	Gas Chromatography–Mass Spectrometry
GPI	Gas Production Index
HC	Hydrocarbons
HI	Hydrogen Index

IFPEN	*Institut Français du Pétrole Energies Nouvelles* = French Institute of Petroleum and New Energies
IR	Infrared
ka	Thousand years
Ma	Million years
MinC	Mineral Carbon
NMR	Nuclear Magnetic Resonance
NSO	Nitrogen–Sulfur–Oxygen
OI	Oxygen Index
OM	Organic Matter
OPI	Oil Production Index
OSA	Oil Show Analyzer
OSI	Oil Saturation Index
PAH	Polycyclic Aromatic Hydrocarbons
PC	Pyrolyzable Carbon
PI	Production Index
POC	Particulate Organic Carbon
POM	Pure Organic Matter
ppm	Parts per million
RC	Residual Carbon
RE	Rock-Eval®
RE6	Rock-Eval® 6
RE7S	Rock-Eval® 7 Sulfur
Ro	Vitrinite Reflectance
SI	Sulfur Index
SOC	Soil Organic Carbon
SOM	Soil Organic Matter

Sorg	Organic Sulfur
SPM	Suspended Particulate Matter
Spyr	Pyritic Sulfur
Stot	Total Sulfur
TC/Ctot	Total Carbon
TGA	Thermogravimetric Analysis
TIC	Total Inorganic Carbon
Tmax	Maximum Temperature
TOC	Total Organic Carbon
TOM	Total Organic Matter
TOS	Total Organic Sulfur
Tpeak	Temperature at the peak of the parameter
TR	Transformation Ratio
TSR	Thermal Sulfate Reduction
UV	Ultraviolet
XRD	X-Ray Diffractometry

List of Formulae

Useful Formulas and Parameters Derived from the Rock-Eval® Method and their Units

– **Ci index** = F3/(F1+F2+F3+F4+F5+F6) *unitless*

– **CL index** = (F1+F2)/(F1+F2+F3+F4+F5+F6) *unitless*

– **Coef c** = PyroOrgS/TotOrgS **Ratio of labile organic S to total organic S** *unitless*

– **CP index** = (F4+F5+F6)/(F1+F2+F3+F4+F5+F6) *unitless*

– **HCcont** = Sh0 + SH1 **Hydrocarbon Content Index** *in mg HC/g sample*

– **HI** = (S2/TOC) x 100 **Hydrogen index** *mg HC/g TOC*

– **HI/OI**= S2/S3 *mg HC/mg* CO_2

– $\mathbf{HI_{400}}$ = (R_{400}/TOC) x 100 *mg HC/g TOC*

– **HQI** = [Sh0/(Sh0+SH1)] x 100 **Hydrocarbon Quality Index** *%*

– **HS index** = (F4+F5+F6)/(F3+F4+F5+F6) *unitless*

– **I-index** = log[(F1 + F2)/F3] *unitless*

– **MinC** = PyroMinC + OxiMinC **Mineral carbon** *% by weight in sample*

– **OI** = (S3/TOC) x 100 **Oxygen index** *mg CO_2/g TOC*

– **OICO** = (S3CO/TOC) x 100 *mg CO/g* TOC

– **OIRE6** = [(16/28) x OICO + (32/44) x OI] **Oxygen index RE6** *mg O_2/g TOC*

– **OxiMinC** = (S5 x 12/44)/10 **Oxidation mineral carbon** *% by weight in sample*

– **OxyFeS** = PyroFeS x (1-Coef p)/Coef p **Residual pyritic sulfur** *% by weight*

– **OxyOrgS** = Residual S - OxyFeS **Residual organic sulfur** *% by weight of sample*

– **PC** = [(S1 + S2) x 0.83 + (S3 x 12/44) + [(S3CO) + (S3'CO/2)] x 12/28]/10 **Pyrolyzable carbon** *% by weight*

– **PC/RC Pyrolyzable organic carbon/Residual organic carbon** *unitless*

– **PC/TOC Pyrolyzable organic carbon/Total organic carbon** *unitless*

– **PI** = S1/(S1+S2)] x 100 **Production index** *unitless*

– **PIShale** = [(Sh0+SH1)/(Sh0+Sh1+Sh2)] × 100 **Production index** (Shale Play^{TM} method) *as a %*

– **PP** = S1 + S2 **Oil potential** *mg HC/g sample* or *kg HC/t*

– **PyroMinC** = [(S3 x 12/44)) + (S3'CO/2) x (12/28)]/10 **Pyrolysis mineral carbon** *% by weight in sample*

– **PyroOrgS** = S1OrgS + S2OrgS **Pyrolysis organic sulfur** *% by weight of sample*

– $\mathbf{R_{400}}$ = S2(before 400 °C)/S2 **Proportion of S2 before 400°C** *unitless*

– **RC** = [(S4CO x 12/28) + ($S4CO_2$ x 12/44)]/10 **Residual carbon** *% by weight in sample*

– **RC CO** = (S4CO x 12/28)/10 **Residual carbon in CO form** *% by weight in sample*

– **RC CO_2** = ($S4CO_2$ x 12/44)/10 **Residual carbon in CO_2 form** *% by weight in sample*

– **RC/TOC Pyrolyzable organic carbon/Residual organic carbon** *unitless*

– **R-index** = F1/F2 *unitless*

– **S'1** or **Sh1 Free liquid hydrocarbons** *mg HC/g sample*

– **S0** or **Sh0 Free gaseous hydrocarbons** *mg HC/g sample*

– **S1 Free hydrocarbons** *mg HC/g sample*

– **S1S Free sulfur** *mg SO_2/g sample*

– **S2 Pyrolyzable hydrocarbons** *mg HC/g sample*

– **S2S Pyrolyzable sulfur** *mg SO_2/g sample*

– **S3' CO_2 of pyrolysis after 400°C** *mg CO_2/g sample*

– **S3 CO_2 of pyrolysis before 400°C** *mg CO_2/g sample*

– **S3'CO CO of pyrolysis after 400°C** *mg CO/g sample*

– **S3CO CO of pyrolysis before 400°C** *mg CO/g sample*

– **S4** or **S4CO_2 CO_2 of oxidation before 650°C** *mg CO_2/g sample*

– **S4CO CO of oxidation before 650°C** *mg CO/g sample*

– **S5 CO_2 of oxidation after 650°C** *mg CO_2/g sample*

– **SI** = Total OrgS/TOC **Sulfur index** *unitless*

– **Tmax Temperature at the peak of the S2 parameter** *°C*

– **TOC** = PC + RC **Total organic carbon** *% by weight in sample*

– **Total S** = PyroOrg S + PyroFeS + Residual S **Total sulfur** *% by weight of sample*

– **TotFeS** = PyroFeS/Coef p **Pyrolyzable pyritic sulfur** *% by weight of sample*

– **TotOrgS** = Total S – TotFeS **Total organic sulfur** *% by weight of sample*

– **T*p*** or **Tpeak Temperature at the maximum of the parameter considered** *°C*

– **TpeakS2** = Tmax + 39°C **Temperature at the maximum of the S2 parameter** *°C*

1

Historical Perspectives

Violaine LAMOUREUX-VAR[1], Nicolas BOUTON[2] and François BAUDIN[3]

[1] *IFP Energies nouvelles, Rueil-Malmaison, France*

[2] *Vinci Technologies, Nanterre, France*

[3] *ISTeP, Sorbonne Université, Paris, France*

1.1. History of Rock-Eval®

Fifty years of research and innovation mark the history of Rock-Eval®. It was in the early 1970s that Jean Espitalié, then a young researcher at the French Institute of Petroleum (IFP, *Institut Français du Pétrole*, now IFPEN), sought to develop an analytical method to evaluate the potential of rocks to generate hydrocarbons. The scientific community had already been working on the subject for the past decade and publications on the characterization of oil source rocks were flourishing (Forsman and Hunt 1958; Bordenave et al. 1970; Douglas et al. 1970). The fundamental knowledge of that time enabled the estimation of the oil potential of insoluble organic matter contained in the rocks, also called kerogen, based on physico-chemical analyses. It was also known how to classify kerogens based on their origin and assess their degree of thermal evolution (McIver 1967; Durand et al. 1972; Durand and Espitalié 1973; Tissot et al. 1974). However, these analyses were costly, and Espitalié sought a way of evaluating these key characteristics for the oil industry, at a lower cost. He aimed to develop an industrial method meeting the requirements of oil exploration by screening a large number of samples in a minimum amount of time. With this goal in mind, he turned to pyrolysis, a technique consisting of heating rock under an inert atmosphere (Barker 1974a). By developing open-medium pyrolysis under temperature programming, he could see two major advantages in terms of the characterization of kerogens and their thermal evolution.

First, a conceptual advantage, since this thermal technique made it possible to simulate, in a reduced time at the laboratory scale, the natural thermal maturation of source rocks, which occurs between 50 and 200 million years at the scale of sedimentary basins. It also made it possible to simulate to some extent the expulsion of oil and gas from source rocks. He also saw an analytical advantage, since this technique thermally cracked the large organic molecules constituting kerogens into small molecules, which then became transportable by a carrier gas and could be analyzed by a flame ionization detector (FID). He ceased using molecular analysis detectors, despite them being more accurate than the FID, because they were much longer to implement, thus hindering the rapidity of the method. A further significant objective needed to be achieved: distinguishing the hydrocarbons already contained in the rock from those that could be generated by kerogen cracking. To do so, he defined a pyrolysis temperature program based on the scientific results of the time (Barker 1974a): a temperature plateau at 300°C would vaporize the hydrocarbons contained in the rock without cracking the kerogen, then a linear increase in temperature at a rate of 20°C per minute up to 550°C would lead to the cracking of kerogen (Espitalié et al. 1977a). Espitalié and his colleagues thus succeeded in quantifying the oil potential of a rock in under an hour, without even having to chemically isolate the kerogen from the rock.

Very rapidly, these researchers would attempt to improve this method to characterize the different types of pyrolyzed organic matter. To this end, they added a detector measuring the CO_2 resulting from the thermal cracking of oxygenated compounds bound to organic matter, such as carbonyl compounds, carboxylic acids and esters. They needed to ensure that CO_2 of organic origin was distinguishable from CO_2 of mineral origin, resulting from the thermal decomposition of carbonates found in sedimentary rocks. A study of the thermal behavior of clays and carbonates led them to note that the decomposition of carbonates started at a temperature between 400°C and 600°C and generated a significant quantity of CO_2 compared to that resulting from the cracking of organic matter. Therefore, only CO_2 released before 400°C could be considered to be definitely organic. They also showed that the quasi-continuous loss of clay water between 50°C and 650°C prevented the analysis of water lost by organic matter, but that it did not hinder the detection of hydrocarbons or CO_2 (Espitalié et al. 1977a).

Several patents describing the method and equipment were created in succession between 1974 and 1978, establishing the foundations of the Rock-Eval® method (Espitalié et al. 1974, 1977b, 1978). Fifty years later, the same foundations are still in place (see Chapter 2).

The application of the first Rock-Eval® to the study of oil source rocks provided a number of data, which Espitalié and his team interpreted. They deduced four descriptors of the source rocks: parameter S1 representing the hydrocarbons still trapped in the source rock; parameter S2 representing the oil potential of the source rock, i.e. the quantity of hydrocarbons generated by thermal cracking of the kerogen contained in the rock; parameter S3 representing the CO_2 released by the kerogen before 400°C under the same heating conditions as those providing S2; and finally the parameter Tmax, the temperature of the maximum of the S2 peak, indicating the degree of maturation of the organic matter (Espitalié et al. 1977a). These four parameters came to be reference descriptors, still commonly used today in oil exploration (see Chapters 6–8), and also more widely in organic geochemistry (see Chapters 9–12).

Afterwards, two new descriptors were proposed: the hydrogen index, HI, derived from S2, and the oxygen index, OI, derived from S3. These estimated the hydrogen and oxygen, respectively, released by the pyrolyzable fraction of kerogen, which is interpreted as being that capable of generating oil and gas under geological conditions. These parameters proved particularly useful, as it was already known at the time that the more hydrogenated a kerogen was, the greater its capacity to generate oil, and the higher the quality of this oil. It was also known that an immature, hydrogen-rich kerogen was generally derived from algae and microorganisms living in fresh water. Conversely, the more oxygenated a kerogen, the lesser its capacity to generate oil and the poorer the quality of this oil. It was also known that an immature, oxygen-rich kerogen was often derived from higher terrestrial plants. Finally, it was known that kerogens with intermediate hydrogen and oxygen contents were often derived from algae and marine microorganisms. Thus, it was possible to establish a schematic classification of kerogens according to their origin, based on the H/C-O/C diagram, called the van Krevelen diagram, from an extraction of kerogen followed by an analysis of the C, H and O elements (Tissot and Welte 1984). With the Rock-Eval® analysis supplemented by an elemental analysis of the carbon in the rock, an HI-OI diagram comparable to the van Krevelen diagram could be constructed, for the first time offering access to this crucial information, the type of kerogen, without having to isolate it (Espitalié et al. 1977a).

A seventh descriptor was developed from the work of Barker (1974b) to characterize the state of evolution of organic matter in rocks: the petroleum index, PI = S1/(S1+S2). It provided information on the quantity of hydrocarbons present in the source rock (S1), relative to their source (S2). Yet, since S1 came from S2, if the system remained closed, then S1 increased as S2 decreased. Based on this principle, the PI index also provided information on the layers of the source rock from which oil had been expelled to carrier rocks. This PI parameter continues to be used in this sense (see Chapter 6).

Rock-Eval® I[1] was marketed in 1977, in partnership with GIRDEL. Two years later, supported by a new patent that defended two major innovations, Rock-Eval® II was marketed in turn, in partnership with DELSI (Espitalié et al. 1979, 1985a). The instrument had gained in autonomy through its automation thanks to a microprocessor and an autosampler of 50 samples. The second innovation came from the addition of a combustion furnace combined with a thermal conductivity detector (katharometer). This system made it possible to characterize how much carbon remained in the rock after pyrolysis, thus giving access to the most thermally refractory organic and mineral carbon compounds. The total organic carbon, TOC and the HI and OI indices could then be quantified from a single Rock-Eval® analysis (Espitalié et al. 1985a). This was a major step forward: first, because this TOC was a fundamental parameter, being the first to be quantified when evaluating the organic-matter richness of a sediment, and second, because it was now possible to replace the conventional van Krevelen diagram with an equivalent HI-OI diagram, also called the van Krevelen pseudo-diagram, obtained much more rapidly (Espitalié et al. 1985b, 1986). Furthermore, an HI-Tmax diagram gave access to the kerogen maturity according to its class. Finally, an S2-TOC diagram added information on the kerogen families within the same class (Langford and Blanc-Valleron 1990). Today, the Rock-Eval® method remains a reference method for quantifying TOC in sediments, and these three diagrams continue to be widely used in oil exploration (see Chapter 6) and more generally in sedimentary and soil organic matter characterization (see Chapters 9 and 10).

In 1979, Rock-Eval® III left the laboratory to be installed on oil drilling sites. Protected by a patent (Espitalié et al. 1980a), it was industrialized and marketed by DELSI. It met the need for real-time analysis of cuttings and cores extracted from the subsurface during drilling (Espitalié and Pages 1981, Espitalié et al. 1984b). The state and evolution of the organic matter quantified as the drilling process progressed, provided decision support for coring, casing and sometimes further drilling. This Rock-Eval® III, also known as Rock-Eval® OSA (for "Oil Show Analyzer"), differed from Rock-Eval® II in that it was able to quantify gas and oil separately. For this, an ingenious system heated the rock to 90°C for two minutes, even before it entered the pyrolysis furnace. The lightest hydrocarbons were thus volatilized and detected (S0 parameter). Then, the rock sample was introduced into

1. Versions I–IV of Rock-Eval® were designated by Roman numerals. For subsequent versions, Arabic numerals were chosen for greater clarity.

the pyrolysis furnace at 300°C, where the oil was vaporized and detected (S1' parameter). From these parameters, the gas production index GPI = S0/(S0+S1'+S2) and the oil production index OPI = S1'/(S0+S1'+S2) were deduced, which are valuable indications of fossil resources encountered in the source- and reservoir-rock levels crossed by the drilling. Veritable geochemical logs were obtained, used to identify source-rock levels, locate and characterize oil and gas accumulation zones, and investigate migration pathways through well correlations (Espitalié et al. 1984a).

In the 1980s, different industrial versions of Rock-Eval® appeared one after another. Rock-Eval® II-Geocom, resulting from a joint venture between COMSIP and GEOMECANIQUE (IFP), was industrialized and marketed by the American company GEOCOM in the USA back in 1981. It differed from DELSI's Rock-Eval® II owing to the presence of a linear autosampler in place of a carousel, and especially in that it did not analyze residual organic carbon or carbonates. It did not encounter much success, with geochemists preferring DELSI's Rock-Eval® II, which was equipped with a TOC module. A prototype of version "IV" was later designed, but it did not leave the walls of the laboratory. Rock-Eval® 5 was then developed[2]. The instrument, small in size and without an autosampler, was mobile, intended to be transported in a mobile laboratory from site to site (Espitalié 1984). This version had little success, with it proving ultimately less efficient to convey a Rock-Eval® to the explored zones than to convey the samples to the laboratory. The Well Site Pyrologger was built by GEOSERVICES at the time of Rock-Eval® 5, in collaboration with IFP. Designed for the specific needs of GEOSERVICES drilling sites, it differed from DELSI's Rock-Eval® 5 mainly in that it did not require gas cylinders to operate, with gases being generated inside the device itself. As with DELSI's Rock-Eval® 5, the Pyrologger met with resistance from customers, who preferred to conduct their analyses in the laboratory.

All of these innovations and developments propelled the Rock-Eval® method to the status of reference geochemical method in oil exploration within 10 years (Tissot and Welte 1984; Peters 1986). The analysis method was simple, brief and moderately priced, and the interpretation methods were judiciously adapted and pragmatic.

This new analysis and interpretation method, based on the same tool, rapidly proved pertinent for identifying and characterizing reservoir rocks and oils (Espitalié et al. 1985b). These methods are still used in oil production today (see Chapter 6).

2. From Rock-Eval® 5 onwards, researchers designated the Rock-Eval® version with Arabic numerals.

In the 1990s, Rock-Eval® 6 was designed with the aim of quantifying the kinetics of the thermal cracking of kerogen. The pyrolysis temperature recorded now needed to be as close as possible to the actual temperature of the sample analyzed, and no longer to that of the furnace wall as in previous versions. For this purpose, the thermocouple in contact with the wall of the pyrolysis furnace was complemented by a second thermocouple placed as close as possible to the rock, under the crucible that contains it. Thus, Rock-Eval® pyrolysis reproduced at different heating rates on several aliquots of an immature source rock, now made it possible to define a kerogen cracking rate, of order one (Lafargue et al. 1998a) (see Chapter 7). A second innovation significantly improved the accuracy of TOC quantification, and at the same time made it possible to quantify the content of mineral carbon associated with carbonates. Unlike previous versions, this new instrumental version allowed CO_2 to be measured continuously during heating. In addition, thanks to an infrared spectrometer, Rock-Eval® 6 analyzed not only CO_2, but also CO. Since the CO detected during oxidation comes exclusively from the cracking of organic matter (S4CO peak), unlike CO_2, it was then possible, by comparing the CO and CO_2 oxidation thermograms, to distinguish the organic CO_2 (S4 peak) from the mineral CO_2 (S5 peak). This was of great interest since it was now possible to access not only the total carbon content of the rock, but also those of organic carbon (TOC) and mineral carbon (MinC) (Behar et al. 2001). Rock-Eval® had now become an instrument for precise and rapid carbonate quantification (see Chapter 13). The accuracy of the OI index and the TOC was also improved, with their values now incorporating the quantities of CO. These innovations were protected by a new patent (Espitalié et al. 1997) and Rock-Eval® 6 was industrialized and marketed by Vinci Technologies as of 1998.

That same year, Pollut-Eval®, a cousin of Rock-Eval® designed to rapidly recognize and quantify hydrocarbon pollutant classes in soils, saw the light of day, supported by a new patent (Lafargue et al. 1998b). In particular, it gave access to hydrocarbons composed of more than 40 carbon atoms, which was an advantage over molecular analysis techniques for diagnosing soil pollution from heavy fuel oils or bitumen (Ducreux et al. 1997; see Chapter 12). Intended for rapid screening, it offered the possibility of quickly developing a mapping of the contamination of a site, to estimate the volumes of polluted land, select samples for further analysis and assist in the choice of decontamination processes (Benoit et al. 2002; Tzovolou et al. 2009). A field version, Pollut-Eval® T, offered the added benefit that it could be conveyed and used directly on polluted sites, bringing a gain in responsiveness. Despite its undeniable benefit in meeting the need for diagnosis of soil pollution, Pollut-Eval® did not prove as commercially successful as expected. Its selling price and the skills needed to implement it certainly weighed in the balance; but the factor that no doubt ruled it out of the market was that it did not form part of the standardized analysis techniques for diagnosing polluted sites and soils.

Aside from carbon, hydrogen and oxygen, sulfur is the most significant element of organic matter contained in source rocks. It significantly conditions the biological and thermal evolution of organic matter, as well as the composition of the oil and gas derived from it. Sulfur, presenting a major issue in the oil industry, rapidly aroused the interest of Espitalié to improve Rock-Eval® to this end. Hence, a first Rock-Eval® device associated with a sulfur analysis module was patented in 1976 (Madec et al. 1978; Madec and Espitalié 1985), followed by a second device in 2010 (Espitalié et al. 2010). This second patent resulted in the designing of Rock-Eval® 7 Sulfur, which began to be industrialized and marketed by Vinci Technologies in 2019. This version presents the benefit of simultaneously detecting hydrocarbons, organo-oxygen compounds and organosulfur compounds of organic matter. The composition of the pyrolyzable fraction of a kerogen or an oil can then be more fully characterized. In addition, the sulfur associated with pyrite can be distinguished from that associated with organic matter, which is long and difficult to achieve with physico-chemical methods (Aboussou 2018). Finally, Rock-Eval® 7 Sulfur can be used to quantify the total and pyrolyzable sulfur contents of rocks, along with the carbon contents. This version has thus proved to be relevant in the oil industry, where sulfur is so significant that its role is scrutinized throughout the chain, from the diagenesis of organic matter to the refining of sulfur-containing crude oils, via the cracking of kerogen, and oil production and transportation (see Chapter 8).

With 50 years of research, innovation and development by dozens of researchers and engineers, the Rock-Eval® methodology is constantly evolving and is considered a cornerstone of global analytical techniques in organic geochemistry (Figure 1.1). Simple to implement, fast and pragmatic, it is a true method of geochemical screening, applicable not only to oil exploration and production (see Chapters 6–8), but also to sedimentary geology (see Chapters 9 and 13), pedology (see Chapter 10), the study of black carbon (see Chapter 11) and the study of polluted soils (see Chapter 12).

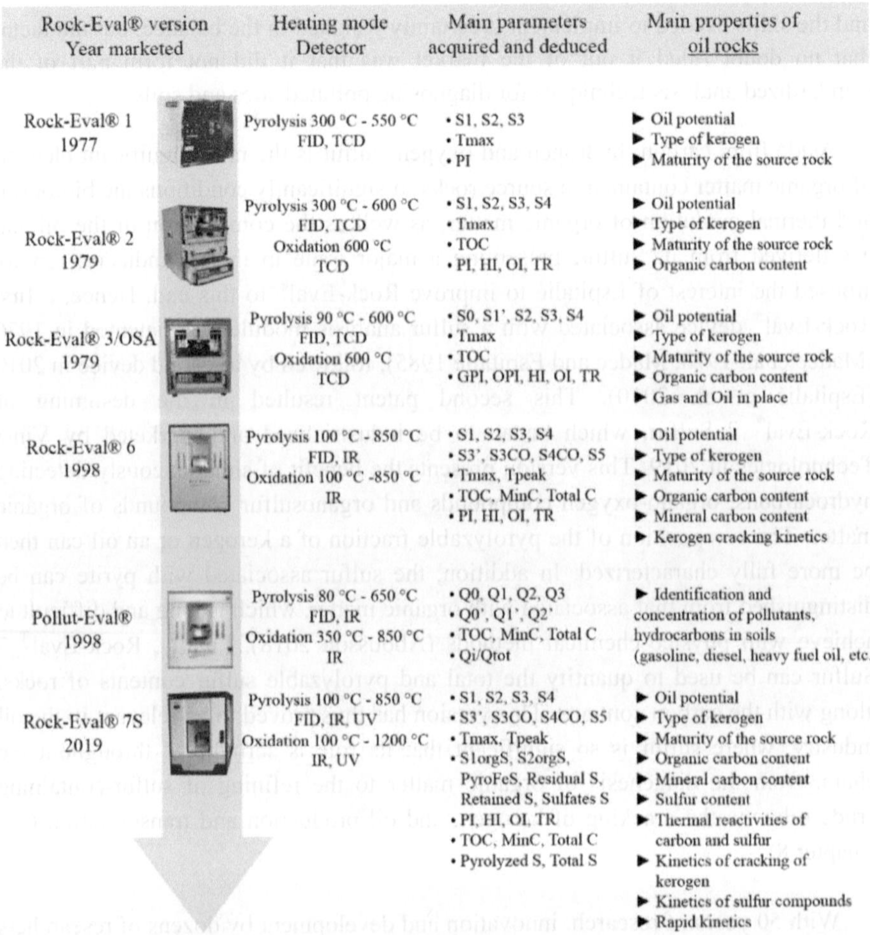

Figure 1.1. *Timeline of the main Rock-Eval® versions and their features. For a color version of this figure, see www.iste.co.uk/baudin/rock-eval.zip*

1.2. Geographical distribution of Rock-Eval® devices worldwide and areas of application of the method

By the end of 2020, there were more than 200 Rock-Eval® devices in service worldwide, mainly in countries where oil research was active (Figure 1.2). Half of these devices were distributed among oil and gas companies (~51 %), then equally between universities (18%) and public research institutes (18%), and finally, the remainder (13 %) were distributed in service laboratories (Figure 1.2). Most of these devices were Rock-Eval® 6 models, but there were still some Rock-Eval® II and III in operation. The Rock-Eval® 7 and 7S models, marketed since 2019, were operational in 10 countries.

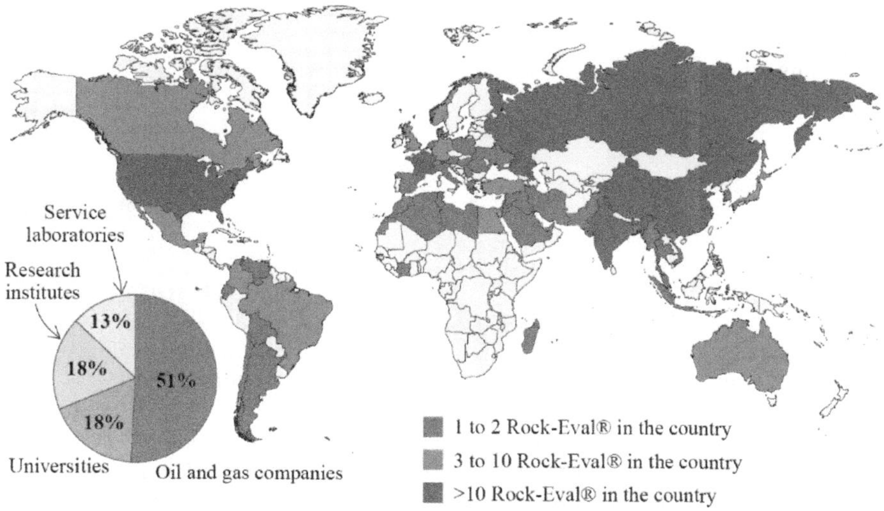

Figure 1.2. *Map of the worldwide distribution of Rock-Eval® devices in service (end of 2020) and their distribution according to the types of entity possessing them. For a color version of this figure, see www.iste.co.uk/baudin/rock-eval.zip*

While it is by far the oil sector that uses Rock-Eval® the most, the cement and petrochemical industries (catalyst recycling) also make use of it. Over time, the method has won over other geological disciplines, and then surface sciences in the broader sense. Its use for the study of recent sediments presents an interest in the fields of geomorphology, (paleo)oceanography and (paleo)climatology (Meyers 1997; Freudenthal et al. 2001; Copard et al. 2006; Carrie et al. 2012; Baudin et al. 2015) (see Chapters 9 and 11). In the field of pedology, research conducted since the 2000s has highlighted the benefit of Rock-Eval® analysis in quantifying organic and inorganic carbon concentrations in soils, and in characterizing organic matter (Disnar et al. 2003). These advances point to the device eventually counting among routine analyses to assess soil fertility and carbon sequestration, opening up prospects for application in the agronomic, forestry and climate sectors (Feller et al. 2010; Albrecht et al. 2015; Derenne and Quénéa 2015; Barré et al. 2016; Sebag et al. 2016; Cécillon et al. 2018; Soucémarianadin et al. 2018a, 2018b) (see Chapter 10).

The dissemination and expansion of the Rock-Eval® method can be appreciated through the publications referenced in web search engines or bibliographic databases. Thus, a bibliometric analysis on the *Web of Science* of publications containing "Rock-Eval" among the keywords listed more than 2,000 titles as of late 2021 (Figure 1.3).

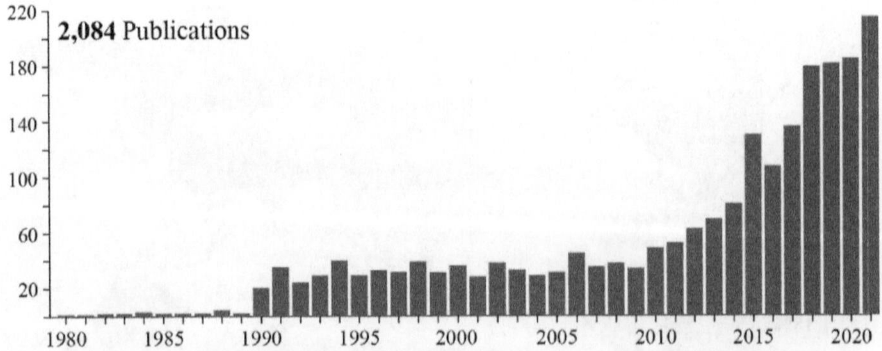

Figure 1.3. *Number of annual publications mentioning Rock-Eval® from 1980 to 2021 according to the Web of Science database. For a color version of this figure, see www.iste.co.uk/baudin/rock-eval.zip*

There has been an observable increase in the annual number of articles referring to Rock-Eval®, across three periods. In the 1980s, there were fewer than 10 articles per year, between 1990 and 2010, there were between 30 and 50, and since 2010, the number of publications has steadily increased to over 200 articles in 2021. Even taking into account the general increase in the number of scientific publications and the number of citations per article since the beginning of the 21st century, this trend nevertheless reflects the growing success of the Rock-Eval® method and the diversification of its sectors of application.

2

Principles and Methods

Violaine LAMOUREUX-VAR[1], Nicolas BOUTON[2], Jean ESPITALIÉ[2] and Yves BENOIT[1]

[1] *IFP Energies nouvelles, Rueil-Malmaison, France*

[2] *Vinci Technologies, Nanterre, France*

Rock-Eval® is a laboratory thermal analyzer that allows samples to be heated in a temperature-controllable oven, while continuously measuring the gases resulting from chemical reactions favored by the temperature rise of the sample (pyrolysis and oxidation). The quantity of hydrocarbons produced by the organic matter of the sample is thus determined.

2.1. Analysis process

Around 60 mg of the powdered sample[1] is weighed in a crucible using a spatula. By means of an autosampler, the crucible containing the sample is placed on a piston, which is then guided into a heated tube, called a pyrolysis oven. This oven, whose heating conditions are adaptable, is swept by an inert carrier gas flow (N_2 since Rock-Eval® 6 and helium in earlier versions). Molecules resulting from the thermal cracking reaction (pyrolysis) of organic matter – mainly hydrocarbons (gas and oil), heavy hydrocarbon compounds and other gases (CO, CO_2, H_2S, SO_2, etc.) – are driven by the flow of di-nitrogen to the detectors:

– the flame ionization detector (FID), to quantify the hydrocarbon molecules (HC);

1. See Chapter 4 for details on field sampling, transport conditions to the laboratory and their preparation for analysis.

– spectrophotometric detectors, infrared (IR) for CO, CO_2 and ultraviolet (UV) for SO_2, which record the concentration of these compounds in real time.

Other molecules can be formed during this pyrolysis reaction, and are often of high molecular masses: this organic residue is called coke or pyrobitumen.

At the end of this pyrolysis stage, the oven is cooled to its initial temperature. The pyrolysis piston returns downwards, and the crucible is transferred by the sampler to another piston, under the oxidation oven. This piston rises in the oven, driving the crucible in a flow of pure air (without CO_2 or water), to ensure the controlled combustion of the sample. Generally, oxidation starts at 300°C and ends at 850°C (1,200°C for Rock-Eval® 7S) with a temperature ramp of 20°C/min. The combustion of the remaining organic matter after pyrolysis produces water (H_2O), carbon oxides (CO and CO_2), as well as the other oxides of the heteroatoms present (SO_2). Water is removed using purifying traps, and CO and CO_2 are measured, as well as SO_2 in the case of Rock-Eval® 7S.

2.2. Schematic diagram of the Rock-Eval® 6 analyzer

The schematic diagram of Rock-Eval® 6 is reproduced in Figure 2.1.

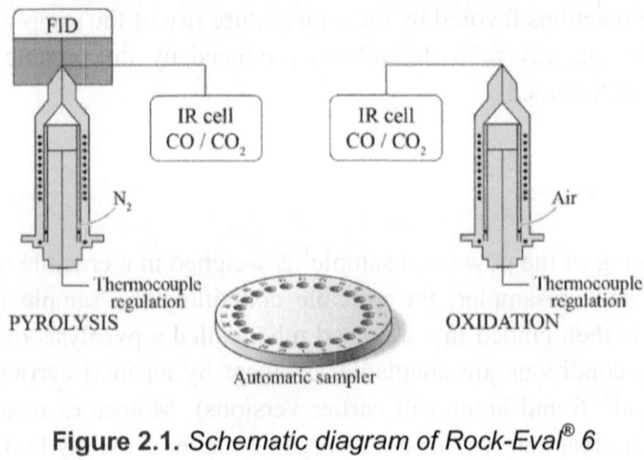

Figure 2.1. *Schematic diagram of Rock-Eval® 6*

Each oven is equipped with two thermocouples:

– a piston thermocouple, which is placed just below the crucible, and which measures the temperature as close as possible to the sample;

– a wall thermocouple, which is located outside the oven, and which touches the oven wall.

A flow rate of 100 mL.min^{-1} (150 mL.min^{-1} for Rock-Eval® 7 Sulfur) of di-nitrogen sweeps the pyrolysis oven. Di-nitrogen replaced helium, which was used in early versions of the instrument, for several reasons:

– it is more practical, and can be generated in situ from ambient air using generators;

– it is less expensive than helium, which is a fossil gas, therefore non-renewable, extracted from natural gas;

– as the helium atom, small in radius, tends to escape very easily, it was more difficult to achieve pyrolysis oven tightness with helium;

– and finally, di-nitrogen is a better carrier gas than helium for hydrocarbon molecules in general (Behar et al. 2001; see section 7.4 of this book).

A flow rate of 150 mL.min^{-1} of pure air (300 mL.min^{-1} for early versions of Rock-Eval® 6 using older IR detectors and 100 mL.min^{-1} for Rock-Eval® 7 and 7S) sweeps the oxidation oven for the combustion stage. A flow rate of 30 mL.min^{-1} of hydrogen and a flow rate of pure air 250–270 mL.min^{-1} are used, respectively, as the fuel and oxidizing agent for the flame of the FID detector.

2.3. Detectors

2.3.1. *Flame ionization detector (FID)*

The flame ionization detector (FID) was introduced in New Zealand by McWilliam and Dewar (1958). It was very quickly used as a detector in gas chromatography (GC) systems that were then beginning to be used on a large scale in laboratories to separate gaseous molecules constituting a complex mixture. The success of the FID thus followed in the footsteps of GC. It involves a flame produced by the combustion of hydrogen in air. When molecules containing C-H bonds are introduced into this flame, they are burned (oxidized) to produce CO_2 and H_2O. But the combustion process is a complex chemical process, which, although still poorly described (Dewar 1961; Holm 1999), goes through different chemical mechanisms. During this combustion, CHO^+ ions are produced in the core of the flame: about 1 ion per 10^6 carbon atoms introduced into the flame. An electrode polarized at +180 VDC placed above the flame attracts electrons, while the cathode (grounded) attracts the CHO^+ ions. An electric current, of very low intensity, is therefore generated each time hydrocarbon molecules enter the flame, by the ions and electrons that are produced proportionally to the flow of incoming molecules. It is this very low electric current (Coulombs/second) that is detected and amplified: its magnitude is therefore noted in pA (10^{-12} amperes) or "µV" in the case of Rock-Eval® 6.

The FID is a mass detector: it does not react to the hydrocarbon concentration in the carrier gas, but to the quantity of molecules transported. Its detection limit is therefore expressed in $pg.s^{-1}$. As a first approximation, this detector is considered to be universal to all hydrocarbons; it has a very low background noise level, is very sensitive and is linear between small quantities (ppm) up to high contents (%).

The Rock-Eval® FID signal is calibrated using one or several reference materials, natural rocks, which have been grounded, homogenized and characterized to this end. If a precise quantity of such a rock is weighed (Quant) and analyzed, an FID signal is obtained during pyrolysis. The surface area under the curve S2, noted Surf_12, is proportional to the quantity of hydrocarbons produced by this rock, StdS2, in mg $HC.g^{-1}$. This makes it possible to calculate the mass response factor (K) of the FID, called KFID: KFID = (StdS2 x Quant)/((Surf_12) x 100).

2.3.2. *CO–CO₂–SO₂ detectors*

The molecules of CO, CO_2 and SO_2 are measured according to the Beer–Lambert law by spectroscopy. The carrier gas passes into a measuring chamber, which is subject to IR radiation (a filament heated by an electric current) or UV radiation (using a gas lamp). The molecules adsorb part of this electromagnetic radiation, as a function of their concentration in the carrier gas. The analyzers therefore indicate a concentration in number of moles of CO, CO_2 or SO_2 per mole of total gas. The usual unit is ppm ($mol.mol^{-1}$) or parts per million (1 molecule of CO or CO_2 or SO_2 for 10^6 molecules present).

The ppm values measured by the IR and UV analyzers are converted to $ng.s^{-1}$ using K-factors (response factors). If we know the weight of the analyzed rock, the contents are expressed ultimately in mg/g (milligrams of CO and CO_2 and SO_2 per gram of sample). There is a mathematical, absolute way to calculate these K-factors. However, in practice, these K-values are calibrated using reference materials.

2.4. Thermograms (Figure 2.2)

While the experimental device has changed over the years, the analysis principle has remained the same and is made up of two stages: pyrolysis and oxidation.

2.4.1. *Pyrolysis: S1 parameter*

A conventional pyrolysis analysis (Bulk Rock method – Basic cycle) starts with an isotherm of 3 minutes at 300°C as soon as the sample is introduced into the oven. If there are hydrocarbons present in the sample, they are thermally desorbed and then driven by the neutral carrier gas flow to the FID located just above the oven. A hydrocarbon thermal-desorption signal is thus obtained, called the S1 parameter, whose area under the curve corresponds to the quantity of gas and oil present in the few mg of sample introduced into the crucible. This quantity of hydrocarbons in the rock, also referred to as free, is known as S1 by extension, and can be expressed in milligrams of hydrocarbons per gram of sample (mg $HC.g^{-1}$).

2.4.2. *Pyrolysis: S2 parameter*

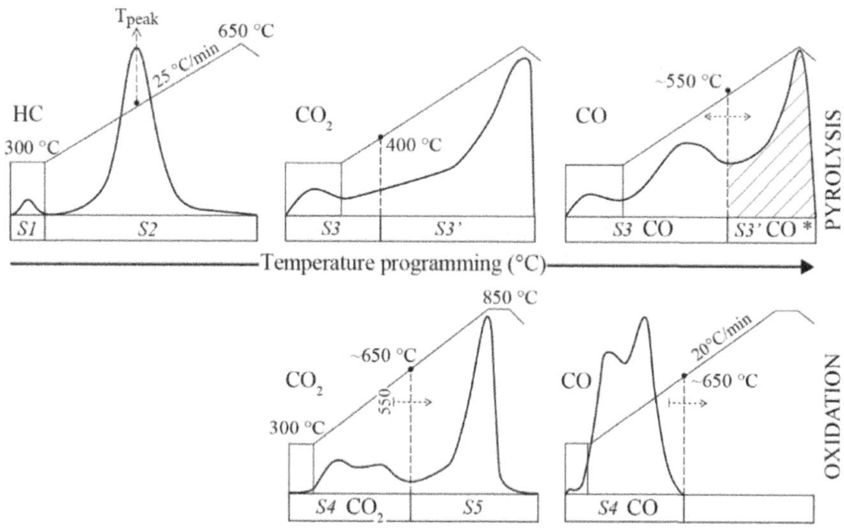

Figure 2.2. *Thermograms obtained during a Rock-Eval® 6 analysis according to the Bulk Rock method – Basic cycle. All thermograms are normalized to their maximum values, i.e. the surfaces/heights are not comparable from one thermogram to another (modified from Behar et al. (2001))*

Once the 3 minutes of this 300°C isotherm are completed, the oven is heated to 650°C in 14 minutes, i.e. at a gradient, or ramp, of 25°C.min^{-1}. The organic matter (e.g. the kerogen in petroleum source rocks) that may be present in the sample then undergoes a chemical reaction, called pyrolysis or thermal cracking. It depolymerizes to give molecules containing mainly carbon and hydrogen, i.e. hydrocarbons, commonly called natural gas or oil. A second hydrocarbon parameter, called the S2 parameter, is thus obtained. The detected quantity of hydrocarbons, in mg $HC.g^{-1}$, corresponds to the oil potential of the sample.

2.4.3. *Pyrolysis: Tpeak/Tmax*

The shape of the S2 parameter is often Gaussian: initially, the reaction rate increases with temperature, and then the reaction rate decreases as the kerogen is consumed by being converted into hydrocarbons. The temperature at the peak of the S2 parameter is called Tpeak. This Tpeak value is a direct index of the measurement of the thermal stability of the chemical bonds present in the organic matter (e.g. kerogen) that are at the origin of the production of oil and natural gas.

If the kerogen has not yet undergone a chemical reaction during geological time, and the weakest bonds have been preserved over time, this organic matter is little evolved or immature. If organic matter has already produced hydrocarbons over geological time (the weakest bonds have already reacted), only the strongest chemical bonds remain in the kerogen. The remainder to be produced needs to be done at higher temperatures. The organic matter is then said to be mature. The value of the temperature at the maximum of the peak, Tpeak, therefore gives a maturity index of the sample. The historical index of Rock-Eval® pyrolysis is not Tpeak, however, but Tmax.

The instrument's design changed significantly between the first versions of Rock-Eval® 1 and 2 in the late 1970s, and the marketing of Rock-Eval® 6 in the late 1990s, then recently with the introduction of Rock-Eval® 7. In the first versions of Rock-Eval®, the Tpeak was called Tmax, a parameter that the scientific and industrial community has used for decades as a source rock maturity index (see Chapter 6).

With the introduction of Rock-Eval® 6 in 1997, a technical advance made it possible to more accurately measure the temperature of the rock inside the crucible. While in earlier versions of Rock-Eval®, the temperature was read on the outside of the oven (on the wall), a thermocouple, placed directly in the piston in contact with the crucible, enabled the temperature to be read as close as possible to the sample during the pyrolysis reaction. A difference of +39°C was systematically observed between Tmax and Tpeak when analyzing the same samples with a Rock-Eval® 2 and a Rock-Eval® 6. The Rock-Eval® 6 Tpeak was higher, and also more accurate, than that of Rock-Eval® 2.

It was then decided to keep the Tpeak value as that observed at the peak of the S2 parameter with Rock-Eval® 6, and to recalculate from this Tpeak value the old Tmax value that would have been found if the rock had been analyzed with a Rock-Eval® 2. Tmax therefore remains as an organic matter maturity index.

Since Tpeak is the temperature at the peak of the S2 curve, it is an instrumental measurement. Its value can change as a function of the temperature ramp applied to analyze the same sample, precisely because of the kinetic laws of the pyrolysis reaction (Table 2.1). The slower the reaction, the more it can occur at low temperature (as in nature), and the higher the temperature ramp, the faster it will occur (as in the Rock-Eval®). The correlation between Tpeak and Tmax is therefore not always set, but it indeed corresponds to a net difference of 39°C if the sample is analyzed at 25°C/min.

Temperature ramp	Tmax (°C)	Tpeak (°C)
0.1°C/min	424	356
1°C/min	419	395
2°C/min	418	408
5°C/min	417	424
10°C/min	416	437
15°C/min	419	448
20°C/min	421	455
25°C/min	416	455
30°C/min	419	461
35°C/min	419	464
40°C/min	422	470

Table 2.1. *Evolution of Tpeak of analysis of reference material IFPEN-160000, as a function of the temperature ramp applied. The Tmax recalculated using an empirical equation remains relatively stable because it is an intrinsic, invariant index of kerogen*

2.4.4. *Specificities of the S2 parameter: pyrolysis in an open medium*

In addition to the time/temperature range, there are other obvious differences between the reaction at work in geological formations and in the Rock-Eval®:

– The Rock-Eval® oven is at ambient pressure, where geological formations can reach several hundred bars of pressure.

– The reaction in the pyrolysis oven takes place with a rock sample that is previously dried (see Chapter 4), at least roughly, where the nature organic matter is buried with the often-salty water of sediment.

– Hydrocarbons formed during Rock-Eval® pyrolysis are immediately transported out of the rock, where those in the sediments slowly migrate through the geological layers to a reservoir or are directly trapped in their source rocks in the case of "shale" gas or oil.

– In sedimentary basins, over time, the liquid hydrocarbons generated initially can in turn undergo a cracking reaction, referred to as secondary, to form gas and coke (pyrobitumen), whereas Rock-Eval® is an open-medium pyrolysis, simulating only the primary cracking of organic matter into oil and gas.

As the influence of pressure and water are in fact quite negligible, these known differences are not considered limiting factors to effectively simulate the processes at work in nature with Rock-Eval® (Tissot and Welte 1984).

2.4.5. *Pyrolysis: S3 and S3CO parameters*

In parallel with the detection of hydrocarbons, Rock-Eval® makes it possible to accurately measure, in real time, the CO and CO_2 molecules emitted by the sample during pyrolysis. For this, the flow from the pyrolysis oven is divided in two: one half goes to the FID and the other half is drawn by a pump through an infrared spectrophotometer, which records the CO and CO_2 content in the carrier gas. These signals are called S3 parameters: S3CO and $S3CO_2$.

The recorded signals correspond to the cracking of the C–O bonds of the organic and inorganic matter (see Chapters 3 and 14). S3 signals are divided into four categories:

– $S3CO_2$ (commonly called S3), which is due to the degradation of the bonds linking oxygen atoms to the kerogen skeleton: these are carbon and oxygen of organic origin;

– S3CO, which is also due to the same phenomenon: carbon and oxygen of organic origin;

– $S3'CO_2$ (commonly called S3'), which is mainly the result of the thermal cracking, above 400°C, of the carbonates contained in the sample (CO_2 of inorganic origin);

– S3'CO, half of which is due to organic carbon, and half to inorganic carbon, under the effect of the Boudouard reaction, which showed that at high temperature, carbon reacts with CO_2 to form CO, according to the reaction:

$$CO_2 + C \rightarrow 2\,CO$$

The difference between the signal S3CO and S3'CO is often separated by a minimum located around 550°C (Figure 2.2).

2.4.6. *Oxidation: S4 parameters*

Once pyrolysis is complete, the oven returns to its initial temperature, and the crucible is transferred to the oxidation oven using the sampler. In air, the sample residue undergoes oxidation. The combustion of organic matter is an oxidation reaction that produces carbon oxides (CO and CO_2), water (H_2O) and heteroatom oxides, SO_2 in the case of sulfur, for example. The carrier gas (air), which is also a reagent of this reaction (the oxidizer), carries these molecules to the infrared detectors. The peaks observed during the combustion of organic matter are called S4CO and S4CO_2 (often noted S4). They are always superposable, and the quantity of CO is typically less than that of CO_2. The carbon of the S4CO_2 and S4CO parameters is used to quantify the residual organic carbon (RC) remaining in the sample after pyrolysis.

2.4.7. *Oxidation: S5 parameter*

The decomposition of carbonates most commonly produces only CO_2 (e.g. $CaCO_3 \rightarrow CaO + CO_2$), usually at temperatures above 600°C for the carbonates most frequently found in natural rocks (see Chapter 14). This parameter is called S5, and it is counted as MinC (Mineral Carbon). Most often, a minimum value is observed in the thermogram between the organic part and the inorganic part of the carbon (the oxygen of the CO and CO_2 organic molecules comes from the oxidizing gas). Combustion most often precedes the decomposition of the carbonates.

Ultimately, Rock-Eval® provides several basic parameters, which are (Table 2.2):

– oil content (S1);

– pyrolyzable organic matter content (S2);

– maturity of the organic matter (Tpeak, Tmax);

– content of residual organic matter (S4);

– mineral carbon content (S5).

Indirectly, many other parameters can be calculated on their basis (see Chapter 3).

Signal	Detector/oven	Unit	Name
S1	FID/pyrolysis	mg(HC)/g(sample)	Free hydrocarbons
S2	FID/pyrolysis	mg(HC)/g(sample)	Oil potential
Tpeak	FID/pyrolysis	°C	Temperature at the S2 parameter maximum
S3	IR/pyrolysis	mg(CO_2)/g(sample)	CO_2 of organic origin
S3'	IR/pyrolysis	mg(CO_2)/g(sample)	CO_2 of mineral origin
S3CO	IR/pyrolysis	mg(CO)/g(sample)	CO of organic origin
S3'CO	IR/pyrolysis	mg(CO)/g(sample)	CO of organic and mineral origin
S4	IR/oxidation	mg(CO_2)/g(sample)	CO_2 of organic origin
S4CO	IR/oxidation	mg(CO)/g(sample)	CO of organic origin
S5	IR/oxidation	mg(CO_2)/g(sample)	CO_2 of mineral origin

Table 2.2. *Signals/parameters measured by the various Rock-Eval® detectors and measurement units*

2.5. Methods and cycles

Different analysis methods are available with the Rock-Eval® instrument:

– **Bulk Rock**, which is the basic method, called screening (rapid measurement);

– **Reservoir**, used for petroleum reservoir samples containing a lot of hydrocarbons;

– **Pure Organic Matter** or **Kerogen**, for samples containing only organic matter;

– **Multi-Heating Rates** or **Shale Play**, initially used for the Pollut-Eval® method (see Chapter 12), using several isothermal plateaus and temperature ramps within the same pyrolysis analysis;

– **Gas Shale Single Rate**, Rock-Eval® 7 only: modified Reservoir method with calculations for Shale Play;

– **Pure Oil**, Rock-Eval® 7 only: for the study of pure oils;

– **Coal**, Rock-Eval® 7 only: for coal analysis;

– **Standard**, used for instrument calibration.

Each method enables specific parameters to be calculated.

For each method, it is possible to create different analysis cycles, which can have initial/final temperatures, heating ramps and configurable analysis times. Table 2.3 gives three examples of cycles.

Cycle		Initial temperature (°C)	Initial time (min)	Ramp (°C/min)	Final temperature (°C)	Final time (min)
Bulk Rock/Basic	Pyrolysis	300	3	25	650	0
	Oxidation	300	1	20	850	5
Reservoir	Pyrolysis	180	10	25	650	0
	Oxidation	300	1	20	850	5
Soil	Pyrolysis	200	3	30	650	0
	Oxidation	300	1	20	850	5

Table 2.3. *Description of the analytical conditions of the Basic cycles of the Bulk Rock, Reservoir and Soil methods. Note that the oxidation conditions remain identical for all three cycles*

2.6. Rock-Eval® 7 Sulfur

Another instrumental design allows the analysis of the sample sulfur along with all of the above parameters (Figure 2.3). For this purpose, the carrier gas flow from the pyrolysis oven is divided into three:

– one-third is drawn to the CO–CO_2 detectors;

– one-third is drawn to an oven maintained at 850°C, and swept in air, in order to burn the sulfur-containing products resulting from pyrolysis (predominantly H_2S) into SO_2;

– the remaining third is directed to the FID.

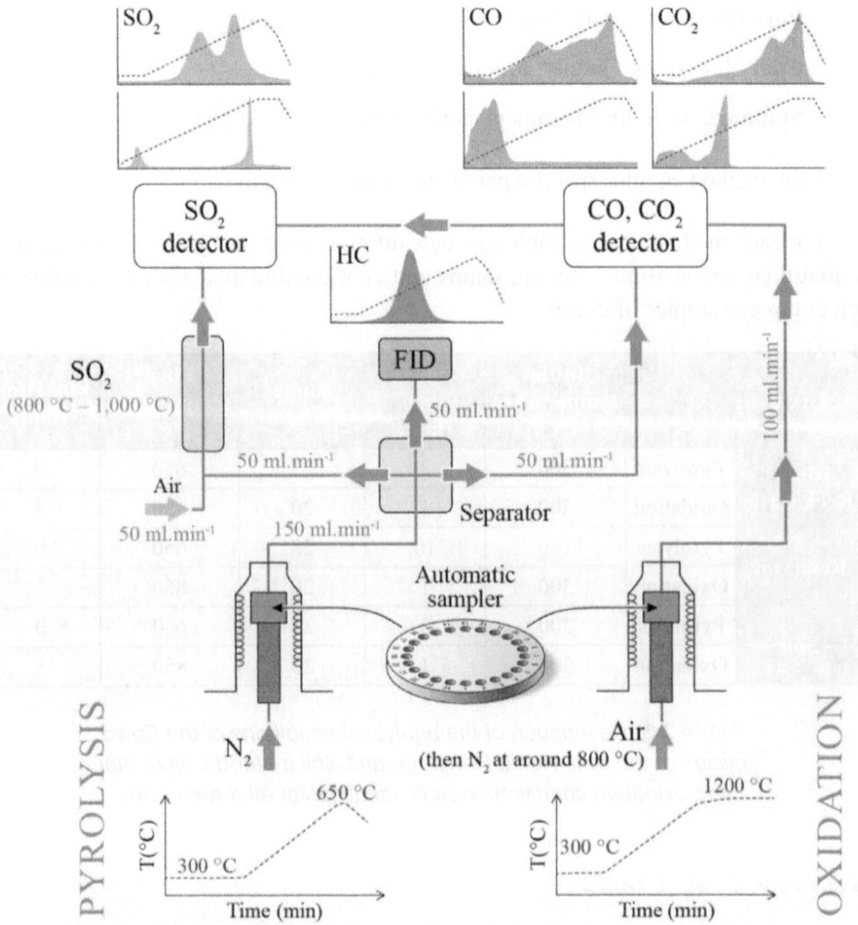

Figure 2.3. *Block diagram of Rock-Eval® 7S and thermograms. For a color version of this figure, see www.iste.co.uk/baudin/rock-eval.zip*

For chemical and thermal reasons, the materials chosen for Rock-Eval® 7S are different from those of Rock-Eval® 6. From the moment when the researchers tried to highlight the pyrolysis reactions of the sulfur compounds, it was necessary to use ceramic crucibles and two ceramic ovens, which are chemically inert and refractory (able to withstand 1,200°C). New signals and parameters are recorded:

– Native sulfur is detected at the start of pyrolysis during the initial isotherm at 300°C.

– The sulfur contained in the oil gives an S1S parameter.

– The sulfur contained in the organic matter that pyrolyzes is called the S2S parameter.

– Pyrite (FeS_2) decomposes partially at the end of pyrolysis, parameter FeS.

During the oxidation cycle, the pyrite ends up decomposing at the same time as the sulfur of the residual organic matter oxidizes, but knowing the FeS content, this parameter, called residual sulfur, can be divided into a part initially due to the pyrite and a residual organic part (see Chapter 8).

Once the organic matter has been burned, the carrier gas in the oxidation furnace is automatically changed at around 700°C, passing from air to di-nitrogen. A new parameter, called "retained sulfur", appears after 700°C; this parameter does not exist when pure organic matter is analyzed. This retained sulfur parameter is therefore the result of a chemical reaction taking place in the crucible between the sulfur and the mineral matter during pyrolysis and/or the oxidation stage.

The oxidation oven is then heated up to 1,200°C under di-nitrogen, and the sulfates then decompose (e.g. gypsum or anhydrite: $2\ CaSO_4 \rightarrow 2\ CaO + 2\ SO_2 + O_2$). Note that the sulfate parameter does not appear if the carrier gas remains air, because the presence of oxygen prevents the reduction of sulfur +VI of gypsum to sulfur +IV of SO_2.

Finally, these new parameters provide access to new qualitative and quantitative information regarding the different forms of sulfur, organic and inorganic, initially present in the rock:

– total sulfur;

– quantity of organic sulfur in the rock (labile and refractory sulfur);

– pyrite content of the rock;

– sulfate content of the rock;

– kinetics of cracking of organic sulfur.

It should be noted that no current laboratory technique allows direct measurement of organic sulfur in a bulk sample without prior preparation of the sample (Cohen-Sadon et al. 2021; see Chapter 8 of this book).

2.7. Pollut-Eval®

The device and the Pollut-Eval® method were developed by IFPEN in the second half of the 1990s for the analysis of soils polluted by hydrocarbon compounds.

2.7.1. *Instrumentation*

Two versions were developed by Vinci Technologies: one laboratory-based and the other for field work. The basic principle consists of heating a contaminated soil sample (100–300 mg wet) under conditions slightly different from those of the Bulk Rock – Basic cycle method.

The pyrolysis phase begins at 80°C and follows a temperature scheduling that is predefined and adjusted so as to progressively release the different petroleum cuts that may be present (gasoline, kerosene, diesel, lubricating oils, polyaromatic hydrocarbons and coal tars), which are still detected by the FID. The infrared (IR) detector makes it possible to quantify the release of CO and CO_2 resulting from the thermal decomposition of the natural organic matter that may be present in the analyzed soils. In fact, polluting hydrocarbons containing only carbon and hydrogen pyrolyze without producing CO and CO_2, which makes it possible to distinguish them from natural organic matter.

The oxidation phase is similar to that of the Rock-Eval® Bulk Rock method. For the analysis of polluted soils, this oxidation phase is necessary for the analysis of tars and coals, and makes it possible to complete the quantification of these compounds, whose detection begins at the end of the pyrolysis phase.

In order to minimize light-compound losses (case of soils contaminated by gasoline or jet-fuel petroleum cuts presenting high vapor pressures), the samples to be analyzed are deposited in closed crucibles and placed in a cooled autosampler.

2.7.2. *Pollut-Eval® method or multi-heating rate*

In the standard laboratory method, the initial temperature of the pyrolysis is 80°C. It is then raised to 120°C according to a gradient of 5°C/min, and then to 350°C at a rate of 15°C/min. The final temperature of 650°C is reached according to a gradient of 25°C/min (Table 2.4). Depending on the pollutants encountered and in particular for "light" cuts, the final temperature can be lowered, which reduces the analysis time accordingly.

Initial temperature (°C)	Gradient (°C/min)	Final temperature (°C)	Duration (min)
80	0	80	1
90	5	120	8
120	15	350	15.3
350	25	650	12
Total duration of the analysis cycle			**36.3**

Table 2.4. *Description of the analytical conditions and the duration of analysis for the Pollut-Eval® method*

As with the Rock-Eval® method, the thermogram of a contamination by a petroleum cut is divided into several zones using cursors positioned at **90°C, 120°C** and **350°C**, thus delimiting zones **Q0, Q1, Q2** and **Q3** (Figure 2.4). The integration of the signals in each of the zones makes it possible to quantify the fractions of light or volatile hydrocarbons for the first two (<12 carbon atoms), the semi-volatile hydrocarbons for the third (between 12 and 30 carbon atoms) and the heavy hydrocarbons for the last zone (>30 carbon atoms). The sum of responses Q0, Q1, Q2 and Q3 corresponds to the **total hydrocarbon** content, **Qt** (expressed in $g.kg^{-1}$ of soil).

In its mobile version, dedicated to the diagnosis of contaminated soils on site and in real time, the analysis method is defined by a single pyrolysis stage, which starts at 50°C in the refrigerated furnace and ends at 650°C.

The analysis provides two complementary quantitative and qualitative pieces of information in real time:

– the concentration of organic pollutants in the soil;

– the nature of the pollutant petroleum cut in terms of carbon composition, which provides information on the depollution methods that may or nor be eligible for their treatment (see Chapter 12).

2.7.3. *Calculations and parameters used by the Pollut-Eval® method*

The analysis of the thermograms produced by Pollut-Eval® during the pyrolysis stage allows the pollutants to be rapidly identified. In fact, given the relatively wide pyrolysis temperature range, oil products lead to profiles that are characteristic of their distillation cut. Figure 2.4 shows the pyrograms obtained from four distillates.

The identification of pollutants can also be deduced more precisely by determining the relative value (Rn) of each of the integration zones (Qn), relative to the sum of the integration zones (Qt), according to the formula:

$$Rn = Qn/Qt \times 100$$

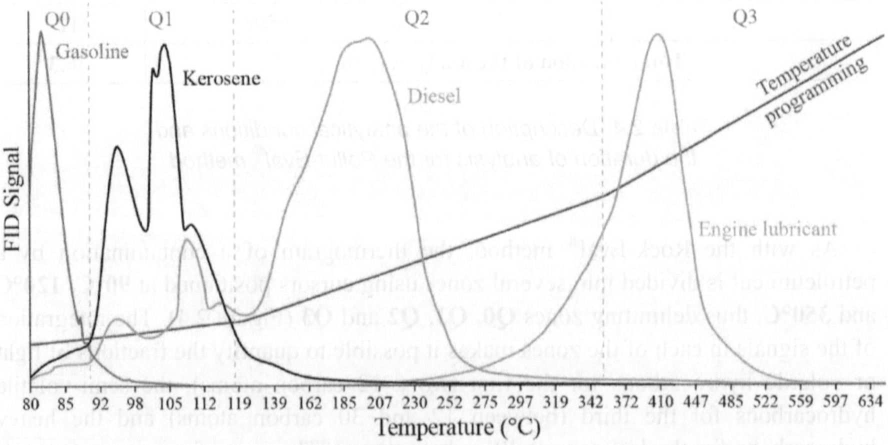

Figure 2.4. *Example of pyrograms characteristic of different petroleum cuts analyzed with a Pollut-Eval®. For a color version of this figure, see www.iste.co.uk/baudin/rock-eval.zip*

Thus, R0 will correspond to integration zone Q0, R1 to zone Q1, etc. Table 2.5 collates the values of these characteristic ratios for the four distillates.

	R0 (%)	R1 (%)	R2 (%)	R3 (%)
Gasoline	>90	<10	≈ 0	0
Kerosene	60–90	10–40	<10	0
Diesel fuel	5–15	30–40	50–60	<1
Lubricating oil	0	0	30–55	45–70

Table 2.5. *Characteristic ratio, R, used in the identification of different petroleum cuts found in polluted soils*

These ratios are characteristic of a temperature scheduling method. If the method is modified, this has an influence on the pyrolysis temperature of hydrocarbons. A comprehensive database describing the different applications of this method is presented in Chapter 12.

2.7.4. *Positioning of integration cursors*

The identification and quantification of pollutants require the positioning of the cursors to be set at 90°C, 120°C and 350°C. The surface delimited by two cursors is called the integration zone (identified by the letters Q in pyrolysis and Q' in oxidation), and the zones are numbered from 0 to 3 in pyrolysis and 0 to 2 in oxidation. We quantify: in zones Q0 and Q1, the gasoline and light jet-fuel, in zone Q2, the intermediate fraction of domestic fuel and diesel, and in zone Q3, the heavy fraction often associated with lubricants (engine oils) and heavy fuel oils. Zones Q0, Q1 and Q2 essentially correspond to the thermovaporization of hydrocarbons and zone Q3 to the cracking of heavy products. In order to validate the carbon composition of the hydrocarbons detected in each of the Qn zones, a gas chromatography method was systematically used in parallel to determine their carbon composition. This complementary analysis makes it possible to associate pyrolytic profiles and chromatographic profiles. For each of the two methods, the characteristic indices are calculated, which make it possible to identify the petroleum cut based on the boiling points of the compounds constituting it.

3

Parameters and Illustration of Results

Lauric CÉCILLON[1], Pierre BARRÉ[2], Yves BENOIT[3], Yoann COPARD[4] and Maria-Fernanda ROMERO-SARMIENTO[3]

[1] *ENS/Université de Rouen Normandie, INRAE, Rouen, France*

[2] *Géologie ENS, CNRS, Paris, France*

[3] *IFP Energies nouvelles, Rueil-Malmaison, France*

[4] *M2C, Université de Rouen Normandie, Rouen, France*

3.1. Introduction

The Rock-Eval® method gives abundant information relatively quickly. Indeed, in about an hour, the device provides five thermograms, which can be divided into several zones. Since the method's initial developments, parameters have been developed to reduce this information to a few parameters of interest, which are more easily interpretable. The methods of calculating these "conventional" indicators are detailed below and can notably be found in Espitalié et al. (1977a, 1985a, 1985b) and Lafargue et al. (1998a). Over the years, as the areas of use of Rock-Eval® have expanded, Rock-Eval® parameters have multiplied. The development of Rock-Eval® 7S, which enabled sulfur gases to be monitored, has also led to the construction of new parameters (Lamoureux-Var et al. 2019). The purpose of this chapter is to provide as complete an overview as possible of the Rock-Eval® parameters used, particularly those mentioned in this book. The areas of use of different parameters and their calculation methods are also presented in this chapter. The reader may refer to the other chapters of this book for more specific information on how the different parameters are used.

3.2. Rock-Eval® signals and basic parameters

3.2.1. *Dividing thermograms into Rock-Eval® "signals"*

Rock-Eval® thermal analysis takes place in two successive stages, with the sample being subjected to pyrolysis in the first stage and oxidation in the second stage (see Chapter 2). While the start and end temperatures of each stage, the temperature rise ramps and the duration of the isotherms that may precede the temperature rises can change depending on the sample in question or the operator, the thermograms are acquired according to the same logic for all Rock-Eval® analyses. Thus, a Rock-Eval® analysis generates five or seven thermograms – depending on whether it is the RE6 or the RE7S – corresponding to the continuous measurement of gaseous effluents during the pyrolysis and oxidation phases. The gases measured during the pyrolysis phase are CO_2 and CO, as well as the other pyrolysis effluents corresponding mainly to hydrocarbons (HC). SO_2 is added to these three gases when using an RE7S (see Chapter 8). Only CO_2 and CO are measured during the oxidation phase. Oxidation SO_2 is added to this with an RE7S (see Chapter 8). These thermograms are used to derive the Rock-Eval® parameters. These parameters can be calculated from the area under a curve or part of the curve of a thermogram (parameters corresponding to the quantity of a compound), or the shape of the thermogram (parameters providing information on the thermal stability).

Using the conventional Rock-Eval® Bulk Rock – Basic cycle method, different zones corresponding to the integration of the area under the signal of a thermogram can be distinguished between for each of the five thermograms (Figure 3.1). These zones, called "S", for surface, are expressed in mg HC/g of sample, mg CO_2/g of sample or mg CO/g of sample, depending on the thermogram considered. The S1 signal corresponds to the pyrolysis effluents, mainly hydrocarbons (hence the use of the abbreviation "HC" in the remainder of the chapter), evolved during the 300°C isotherm preceding the temperature rise of the pyrolysis phase. The S2 signal corresponds to the pyrolysis effluents (HC) measured during the pyrolysis phase between 300 and 650°C. The thermograms corresponding to CO_2 and CO measured during the pyrolysis phase are divided into the S3 and S3', and S3CO and S3'CO signals, respectively. The CO_2 measured before 400°C is incorporated into the S3 signal, and the CO_2 measured afterwards belongs to S3'. The limit between S3CO and S3'CO depends on the shape of the signal. It is located at the local minimum observed between 450 and 600°C (limit arbitrarily set at 550°C in the absence of a minimum value on the curve). The thermogram corresponding to the release of CO_2 in the oxidation phase is divided into two signals: S4CO_2 and S5. The limit between S4CO_2 and S5 is located at the level of the local minimum observed between 550 and 720°C (limit arbitrarily set at 650°C in the absence of a minimum on the curve). The CO measured during the oxidation phase corresponds to the S4CO signal.

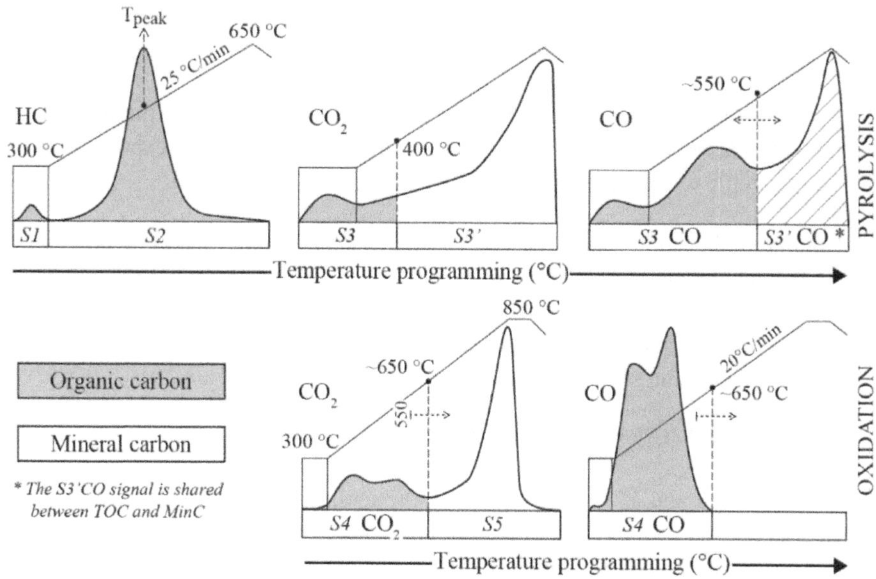

Figure 3.1. *Progression of a Rock-Eval® analysis and measurement of the five thermograms. The analysis parameters represented are those of the Bulk Rock – Basic cycle method. The bluish sections correspond to the effluents (HC, CO_2 and CO) generated by the thermal decomposition of the organic fraction. All thermograms are normalized to their maximum values, i.e. the surfaces/heights are not comparable from one thermogram to another. Modified from Behar* et al. *(2001). For a color version of this figure, see www.iste.co.uk/baudin/rock-eval.zip*

3.2.2. *Use of signals to calculate the organic and inorganic carbon contents of a sample*

The distinctions between zones S3 and S3', and S4 and S5 have been introduced to delimit the parts of the thermograms corresponding to the C coming from the pyrolysis or combustion of the organic matter in the form of CO_2 and CO from those corresponding to the C resulting from carbonate cracking. Thus, the S5 signal corresponds to the CO_2 generated by the decomposition of carbonates during the oxidation phase, while the S4CO_2 signal corresponds to the CO_2 generated by the combustion of organic matter during the oxidation phase. As the decomposition of carbonates does not produce CO, the entirety of the C in the form of CO during the oxidation phase comes from the combustion of organic matter. The S5 signal measurement enables the calculation of the quantity of C-carbonates in the sample evolved during the oxidation phase, called OxiMinC. The OxiMinC parameter is expressed as a mass percentage of C, which is calculated as follows:

1) OxiMinC = (S5 x 12/44)/10.

Factor 12/44 corresponds to the molar mass of the C divided by the molar mass of the CO_2. Factor 10 makes it possible to switch from an amount expressed in mg CO_2/g of sample to a percentage.

Similarly, the addition of the S4CO_2 and S4CO signals makes it possible to obtain the percentage of organic carbon in the sample resulting from the combustion of organic matter in the oxidation phase. This parameter is called RC, for "residual organic carbon", because this C is residual after pyrolysis. The RC is expressed as a percentage of C in the sample, which is calculated as follows:

2) RC = [(S4CO x 12/28) + (S4CO_2) x 12/44)]/10.

Factors 12/28 and 12/44 correspond to the molar mass of the C divided by the molar mass of the CO and CO_2, respectively. Factor 10 makes it possible to switch from an amount expressed in mg/g of sample to a percentage. The RC is divided into RCCO and RCCO_2, which correspond, respectively, to the residual organic carbon evolved in the form of CO and CO_2 during the oxidation phase.

The calculations of the contents (expressed as a percentage of C) of organic carbon cracked during the pyrolysis phase (PC, for "pyrolyzable organic carbon") and of carbon derived from carbonates, measured during the pyrolysis phase (PyroMinC), are somewhat less direct. Indeed, where the S3 and S3CO signals correspond to organic carbon and S3' corresponds to mineral carbon, the S3'CO signal corresponds half to organic C and half to inorganic C. Indeed, the CO_2 generated during pyrolysis can react with the C of the organic matter according to the Boudouard reaction.

$$C_{(s)} + CO_{2(g)} \rightarrow 2\ CO_{(g)}$$

This reaction results in the formation of two moles of CO from one mole of CO_2 and one mole of solid C (the organic matter here). Another complication comes from the need to convert the S1 and S2 signals expressed in mg HC/g of sample into percentage C. To do this, the C contribution of the hydrocarbons has to be determined. Analysis of many pyrolysis effluents recovered by trapping allowed Espitalié et al. (1977a) to estimate that about 83% of the mass of pyrolysis effluents measured in the S1 and S2 signals is C. The remaining 17% correspond to hydrogen and heteroatoms (O, S and N, primarily). With this in place, it is then possible to calculate PC and PyroMinC as follows:

3) PC = {(S1 + S2) x 0.83 + (S3 x 12/44) + [(S3CO) + (S3'CO/2)] x 12/28}/10;

4) PyroMinC = [(S3' x 12/44) + (S3'CO/2) x (12/28)]/10.

Once the quantities of organic and inorganic C measured during the pyrolysis and oxidation phase have been determined, it is possible to calculate the quantities of total organic carbon (TOC) and total inorganic carbon (MinC). They are obtained by adding the quantities of organic and inorganic C, respectively, measured during the pyrolysis and oxidation phases.

5) TOC = PC + RC; expressed as % by weight in the sample;

6) MinC = PyroMinC + OxiMinC; expressed as % by weight in the sample.

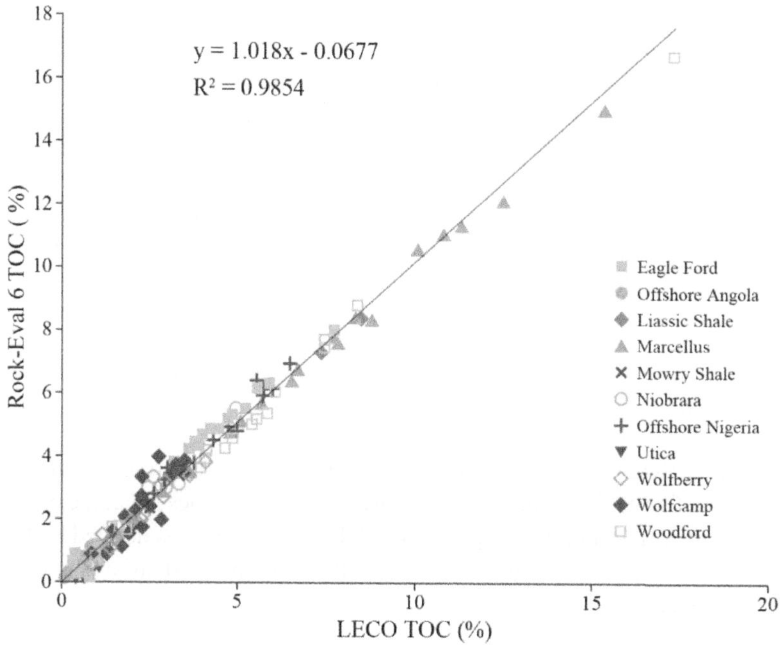

Figure 3.2. *Relationship between total organic carbon (TOC) measured by a Rock-Eval® and a LECO elemental analyzer for a series of oil source rock samples. Modified from Carvajal-Ortiz and Gentzis (2015). For a color version of this figure, see www.iste.co.uk/baudin/rock-eval.zip*

It is noteworthy that despite the approximations necessary to calculate the different parameters (limits between S3 and S3', 83% of C in S1 and S2, for example), the TOC values are very close to those of total organic C measured using reference techniques such as CHN elemental analysis (see Chapter 5). This consistency between the TOC and the CHN TOC has been observed for different types of natural samples (Behar et al. 2001; Carvajal-Ortiz and Gentzis 2015; Figure 3.2), as well as in kerogens (Behar et al. 2001) or soils (Disnar et al. 2003).

Due to the calculation method, small quantities of inorganic carbon are measured even when analyzing non-carbonated samples. This quantity of C allocated by mistake to MinC is therefore absent from the TOC. This probably explains the slight underestimation of the TOC measured by Rock-Eval®. If the operator is certain that the sample does not contain carbonates, then they can add the value of MinC (of the order of 0.1%) to the TOC (Baudin et al. 2015). In this case, the corrected TOC very slightly overestimates the TOC.

3.2.3. *Case of contamination by thermolabile carbonates*

The calculation of TOC and MinC presented above is based on the hypothesis that carbonates do not crack before about 550°C and that the carbon contained in the carbonates does not contribute to S3. While this hypothesis is reasonable for a large number of carbonates (calcite, dolomite, etc.), it is not verified for some, such as siderite. The presence of thermolabile carbonates greatly complicates the use of the S3 signals and in particular the calculation of the TOC, MinC and OI values (see Chapter 13 for further details).

3.2.4. *Parameters indicating the elemental composition of the sample: hydrogen and oxygen indices*

The Rock-Eval® method provides parameters that offer information on the H and O richness of organic matter in the analyzed sample. These parameters are the hydrogen index (HI) and the oxygen index (OI), respectively, as defined by Espitalié et al. (1977a). The HI is expressed in mg HC/g organic carbon and is calculated from the S2 signal:

7) HI = (S2x100)/TOC.

The OI is expressed in mg CO_2/g organic carbon and is calculated from the S3 signal:

8) OI = (S3x100)/TOC.

It is also possible to calculate the OICO from the S3CO signal in the same manner. The OIRE6 parameter has also been proposed (Lafargue et al. 1998a). It is expressed in mg O_2/g organic carbon and is calculated from the OI and OICO parameters:

9) OIRE6 = (16/28) x OICO + (32/44) x OI.

Although the HI and OI parameters are not direct measurements of H and O contents in organic matter, they are good indicators of the H and O content, as evidenced by their good correlations with the H/C and O/C ratios of the organic matter of the analyzed sample (Espitalié et al. 1977a; Figure 3.3.). For this reason, the graph of HI as a function of OI, which constitutes a conventional representation in many Rock-Eval® applications, is called a van Krevelen pseudo-diagram (Figure 3.4). The graph of H/C versus O/C is the van Krevelen diagram.

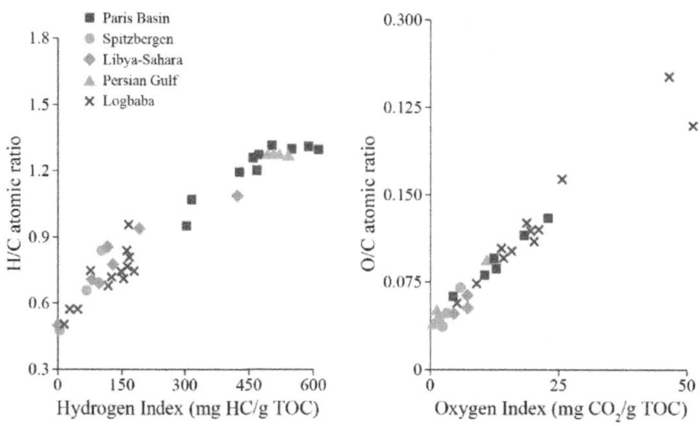

Figure 3.3. *Relationships between hydrogen and oxygen indices and H/C and O/C atomic ratios. Data from five sedimentary series show good linear relationships between the HI and H/C ratio and between the OI and O/C ratio. Modified from Espitalié et al. (1977a). For a color version of this figure, see www.iste.co.uk/baudin/rock-eval.zip*

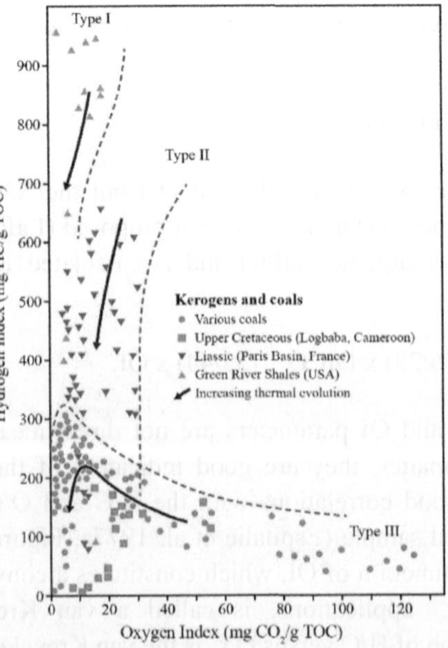

Figure 3.4. *HI–OI diagram, also known as the van Krevelen pseudo-diagram, showing the different types of kerogen (Espitalié et al. 1985b). For a color version of this figure, see www.iste.co.uk/baudin/rock-eval.zip*

3.3. Key parameters for oil exploration

3.3.1. *Rock-Eval®* *Bulk Rock – Basic cycle parameters for the characterization of source rocks*

The conventional parameters S1, S2, Tmax, HI and OI described above were initially developed to estimate the type, origin and thermal maturity of the sedimentary organic matter, as well as the oil potential of a source rock (Espitalié et al. 1986; Lafargue et al. 1998a; Behar et al. 2001; also see Chapter 6 of this book for further details).

The Tmax parameter is an excellent thermal maturity indicator mainly for Type II and III organic matter, located in the oil window. For example, the diagram of HI versus Tmax (Figure 3.5) has been widely used to identify the type and thermal maturity of organic matter from sedimentary rock samples. In general, the Tmax values are less than 430°C for immature source rocks. Organic matter in the oil-window thermal maturity is characterized by Tmax values between 430 and 460°C, while organic matter in the gas window presents Tmax values between 460 and 500°C. Beyond this, the Tmax values are no longer reliable. With respect to the hydrogen index (HI), a Type I lacustrine organic matter contains HIs greater than 700 mg HC/g TOC. An HI of about 200–700 mg HC/g TOC probably indicates Type II marine organic matter, and HIs less than 200 mg HC/g TOC suggest Type III continental organic matter. These interpretations largely depend on the geological context of the samples and integration with other geochemical parameters.

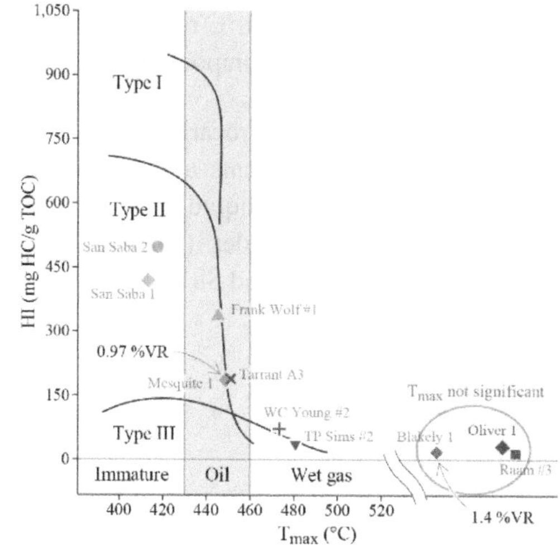

Figure 3.5. *Example of a classical HI versus Tmax diagram (according to Romero-Sarmiento et al. (2014)). For a color version of this figure, see www.iste.co.uk/baudin/rock-eval.zip*

The production index (PI) for conventional source rocks is calculated from the S1 and S2 signals and is expressed as a percentage (%):

PI = S1/(S1+S2)]x100

This PI parameter varies between 5% for immature source rocks, up to 60–100% for very mature source rocks.

3.3.2. *Rock-Eval® Shale Play™ parameters for the characterization of liquid-rich source rocks*

The parameters Sh0, Sh1 and Sh2 were developed to obtain a better estimation/evaluation of the free and/or retained hydrocarbons still present within a source rock (Romero-Sarmiento et al. 2016; Figure 3.6; also see Chapter 6 of this book for further details). The Sh0 signal corresponds to the thermovaporization effluents, mainly hydrocarbons of low to medium molecular weight released between 100 and 200°C (e.g. saturate and aromatic compounds with fewer than 20 carbon atoms). The Sh1 signal corresponds to the thermovaporization effluents, mainly the hydrocarbons of medium molecular weight detected between 200 and 350°C (e.g. saturate and aromatic compounds with fewer than 30 carbon atoms). Finally, the Sh2 signal corresponds to the pyrolysis effluents, mainly the hydrocarbons measured between 350 and 650°C resulting from the thermal cracking of organic matter (e.g. kerogen) and heavy compounds (e.g. asphaltenes and resins).

The quantity of thermovaporized hydrocarbons detected by the first two parameters, Sh0 and Sh1, therefore represents a total estimation of the free and adsorbed hydrocarbon contents available in liquid-rich source rocks. This parameter corresponds to the hydrocarbon content index (HCcont), expressed in mg HC/g initial rock, and is calculated from the Sh0 and Sh1 signals:

$$\text{HCcont} = \text{Sh0} + \text{Sh1}$$

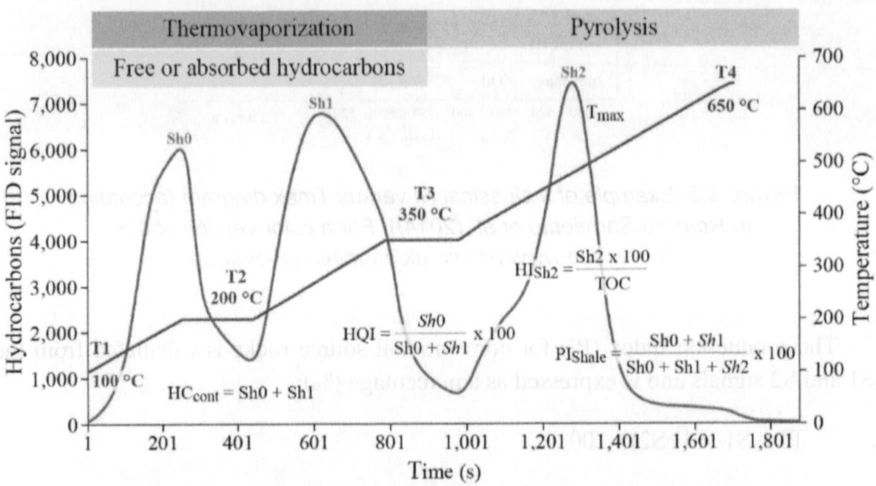

Figure 3.6. *Rock-Eval® Shale Play™ parameters developed at IFPEN (according to Romero-Sarmiento et al. (2016)). For a color version of this figure, see www.iste.co.uk/baudin/rock-eval.zip*

It is also possible to calculate the relative quantity of light hydrocarbons with low molecular weight, in order to evaluate the quality of the fluids in the reservoir intervals. The hydrocarbon quality index (HQI), expressed as a percentage (%), is calculated from the Sh0 and Sh1 signals as follows:

HQI = [Sh0/(Sh0+Sh1)] x 100

The production index for liquid-rich source rocks (PIShale) is used to define the areas of oil accumulation in a source rock–reservoir unit. This parameter is also expressed as a percentage (%) and is calculated from the three signals, Sh0, Sh1 and Sh2, as follows:

PIShale = [(Sh0+Sh1)/(Sh0+Sh1+Sh2)] x 100

3.4. Parameters developed for soil analysis

The ability to characterize a large number of samples fairly quickly has led several research groups to use Rock-Eval® for applications other than oil exploration. Thus, Rock-Eval® has been increasingly used to study soil samples for the past 20 years. While some parameters used routinely for oil exploration are also useful for soil studies such as HI and OI, other parameters have been developed to take into account the specificities of thermograms obtained during soil sample analysis. Other parameters have also been developed to respond to questions specific to soil organic matter analysis.

3.4.1. *HC thermal stability parameters*

Early studies on soil samples showed that the shape of the S2 signal was different from that conventionally observed within the context of oil exploration (Disnar et al. 2003; Figure 3.7). Indeed, the shape of the S2 signal is generally multimodal and non-Gaussian. As a result, it seems simplistic or even misleading to characterize S2 using only the Tmax or Tpeak value. Indeed, it frequently occurs that in samples taken from the same soil, one or the other of the S2 modes dominates, thus determining greatly contrasting Tpeak values. To remedy this, several approaches have been developed and fine-tuned in recent years.

In their pioneering study of around 100 soil samples, Disnar et al. (2003) explained the multimodal form of the S2 signal by the contribution of several families of molecules cracking at different temperatures. They then proposed deconvoluting the S2 signal into several Gaussians, corresponding to the signal due to the cracking of molecules of different thermal stability. Since there is no single solution for decomposing into Gaussians, Disnar et al. (2003) also proposed a more stable parameter: R-400. R-400 corresponds to the proportion of the area under the curve of the S2 signal before 400°C. The 400°C bound was set because minimally processed organic matter of vegetable origin (simple sugars, cellulose or lignin) cracks, at least when not associated with the mineral matrix, before this temperature. R-400 would therefore be an indicator of the proportion of minimally processed plant matter in relation to the total organic matter.

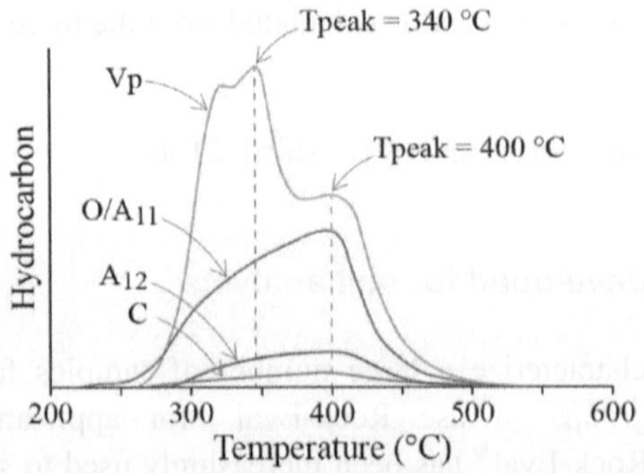

Figure 3.7. *Evolution of the shape of the S2 parameters of a grassland soil from the live plant biomass (Vp) to plant litter and soil horizons (O, A and C) (according to Sebag et al. (2006)). For a color version of this figure, see www.iste.co.uk/baudin/rock-eval.zip*

To overcome the problem of a lack of stability in the decomposition of the S2 signal into several Gaussians, Sebag et al. (2006) proposed setting the number of Gaussians and restricting their positions. On the basis of previous works, they chose to decompose the S2 signal into four Gaussians (F1, F2, F3 and F4), whose maximum values will be between 300 and 320°C, 360 and 380°C, 430 and 450°C, and above 470°C, respectively (Figure 3.8). Indeed, Sebag et al. (2006) observed that this choice allowed the S2 signals to be reproduced satisfactorily and that the addition of a fifth Gaussian did not significantly improve the fit. Changes in the relative contributions of Gaussians F1, F2, F3 and F4 in the different litter and soil horizons led Sebag et al. (2006) to propose correspondences between these Gaussians and fractions of organic matter. Thus, zone F1 would correspond to weakly resistant "biomacromolecules", zone F2 to "resistant biopolymers" such as lignin or cellulose, zone F3 to "humic substances" and zone F4 to "mature geomacromolecules". Using these distinctions, Sebag et al. (2006) proposed a biomacromolecule degradation index, the I-index. This index is calculated from the relative areas (expressed as a percentage) of the Gaussians F1, F2 and F3: $I = \log[(F1+F2)/F3]$. However, the deep questioning, in recent scientific work, of the formation of "geomacromolecules" or "humic substances" in soils and the difficulty in linking chemical characteristics of organic matter and S2 signals (see Chapter 10) suggest that significant work would be needed to robustly assign organic matter types to Gaussians F1, F2, F3 and F4.

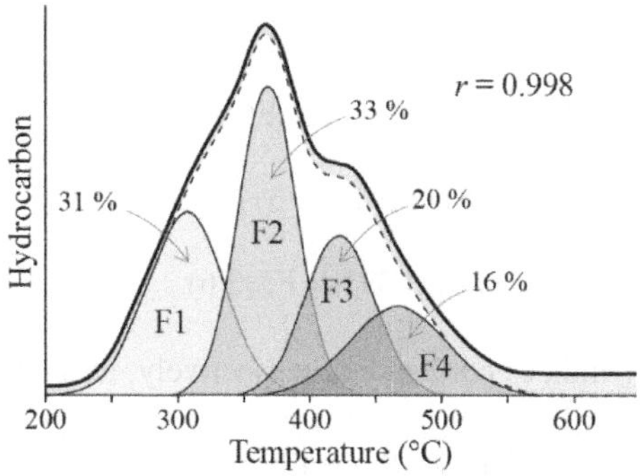

Figure 3.8. *Example of the decomposition of an S2 signal into four Gaussians (F1–F4). The gray zone corresponds to the part of the signal not taken into account in the decomposition. The decomposition into four Gaussians leaves out only a small part of the signal. Figure from Sebag et al. (2006). For a color version of this figure, see www.iste.co.uk/baudin/rock-eval.zip*

In a review article on more than 1,300 litter and soil samples, Sebag et al. (2016) slightly modified the calculation of the I-index. In this article, the authors proposed setting bounds and integrating four portions of areas under the curve of the S2 signal instead of breaking S2 down into four Gaussians. The bounds were set at 340°C, 400°C and 460°C. These bounds defined four areas: A1 between 200 and 340°C, A2 between 340 and 400°C, A3 between 400 and 460°C and A4 for the zone beyond 460°C. The authors verified that the I-index calculated from the Gaussians F1, F2 and F3 was very close to the I-index calculated using areas A1, A2 and A3. In this article, Sebag et al. (2016) also introduced the R-index. The R-index was defined as the area under the curve measured after 400°C, divided by the total area of S2. By construction, R-index = 1 – R_{400}. This is how the I- and R-indices are currently calculated.

Saenger et al. (2013, 2015) followed a similar approach to that of Sebag et al. (2006) and developed further indicators by deconvoluting the S2 signal into six Gaussian signals (F1–F6). Gaussians F1 and F2 corresponded to the thermolabile HCs cracking before 380°C; Gaussian F3 represented the HCs of intermediate thermal stability, cracking between 380 and 450°C, and finally, Gaussians F4, F5 and F6 corresponded to the thermally resistant HCs with a cracking temperature exceeding 450°C. Saenger et al. (2013, 2015) then proposed the following four Rock-Eval® parameters based on the areas of these six Gaussians:

1) C_L index = (F1+F2)/(F1+F2+F3+F4+F5+F6);

2) C_i index = F3/(F1+F2+F3+F4+F5+F6);

3) C_P index = (F4+F5+F6)/(F1+F2+F3+F4+F5+F6);

4) HS index = (F4+F5+F6)/(F3+F4+F5+F6).

The C_L, C_i and C_p indices correspond, respectively, to the relative proportions of the Gaussian areas attributed to the thermolabile HCs, the HCs of intermediate thermal stability and those that are thermally resistant. Note that the C_L index is a variant of R400 proposed by Disnar et al. (2003). It corresponds approximately to an "R380". The HS index corresponds to the fraction of highly thermally resistant HCs among the HCs that are not thermolabile. Saenger et al. (2013) hypothesized that the thermal stability of organic matter was related to its biogeochemical stability, their Rock-Eval® parameters thus becoming indicators of biogeochemical stability. This assumption was reasonably verified in the 2015 article. Indeed, in this article, they presented first a positive correlation between the stock of soil organic carbon defined as stable using the C_p index and the proportion of organic matter associated with minerals (considered more biogeochemically stable) and, second, a positive correlation between the stock of soil organic carbon defined as labile and intermediate using the C_L and C_I indices and the proportion of particulate organic matter (considered less biogeochemically stable).

3.4.2. *Fine-tuning and generalization of thermal stability parameters*

The indices presented above only consider the S2 signal. Cécillon et al. (2018) proposed new parameters that could be calculated on all the signal parts corresponding to the release of gases resulting from the evolution of organic matter in the pyrolysis and oxidation phases. These parameters, the "T-parameters", correspond to the oven temperatures for which a specific proportion of the signal has been measured (Figure 3.9). For example, the value of $T_{50\_HC\_PYR}$ is the temperature to be reached so that 50% of the hydrocarbons have been measured. In the same way, $T_{30\_HC\_PYR}$ is the temperature to be reached so that 30% of the hydrocarbons have been measured. By construction, $T_{50\_HC\_PYR} > T_{30\_HC\_PYR}$. In theory, there could be as many T-parameters as desired by the operator. In practice, only T_{10}, T_{30}, T_{50}, T_{70} and T_{90} have been used in the literature.

There are some subtleties in the calculation of T-parameters. Regarding the S2 signal, part of the pyrolysis effluents are measured while the oven is decreasing in temperature, before the sample is transferred from the pyrolysis oven to the oxidation oven. It is therefore likely that $T_{95\_HC\_PYR}$ is lower than $T_{90\_HC\_PYR}$, which would not make much sense. To avoid this, the calculation of the T-parameters for the S2 signal stops at T_{90}. It is also necessary to prevent the flow of CO and CO_2 resulting from the carbonates from polluting the measurement of the T-parameters on the signals resulting from the cracking or combustion of organic matter. For this purpose, the T-parameters are calculated on the signal cut at 560°C for the pyrolysis phase and 611°C for the oxidation phase. It should also be noted that only the parts of the signals after the 200°C isotherm are considered for the calculation of $Tx_{\_CO2\_PYR}$ and $Tx_{\_CO\_PYR}$. Examples of T-parameter calculations are shown in Figure 3.9.

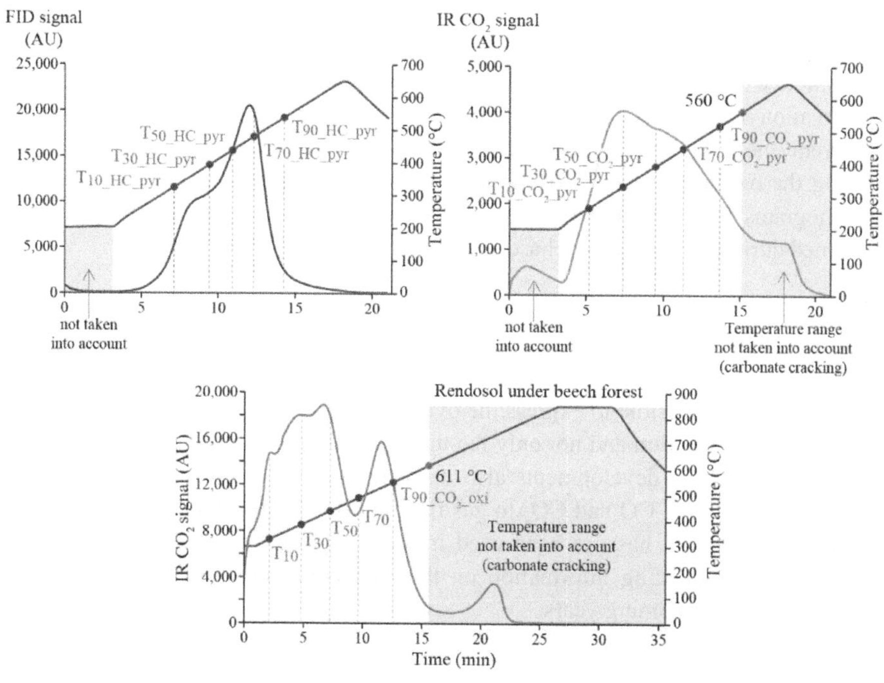

Figure 3.9. *Examples of the calculation of threshold temperatures (T-parameters) for the S2 (HC_pyr), S3 (CO2_pyr) and S4 (CO2_oxi) signals. For a color version of this figure, see www.iste.co.uk/baudin/rock-eval.zip*

3.4.3. *What should we do with the wealth of parameters on the thermal stability of soil organic matter?*

The above sections illustrated that the Rock-Eval® parameter proposals to describe the thermal stability of organic matter in soil samples have been abundant. Where oil exploration only offers Tmax, soil specialists have proposed a myriad of parameters. It is likely that in practice only the most informative parameters will continue to be used. It should also be noted that the most informative parameters will not necessarily be the same, depending on the question of interest to us researchers.

3.5. Development in progress on the mathematical processing of thermograms

Recent works on the study of soil samples have shown that the thermal stability measured with the different signals (HC, CO_2 and CO) can be informative. This observation led researchers to consider the possibility of grouping together the different thermograms obtained during the pyrolysis phase and those obtained during the oxidation phase. To this end, work has been initiated to total the three thermograms of the pyrolysis phase, on the one hand, and the two thermograms obtained during the oxidation phase, on the other hand.

The first stage in totaling the thermograms is to express the thermograms as a quantity of C and not as a quantity of HC, CO_2 or CO. Once these thermograms have been converted into the same unit, they can simply be added together. In this way, it becomes possible to express the overall thermal stability of organic matter in pyrolysis and oxidation and not only the thermal stability according to a signal (HC, CO or CO_2). These developments also make it possible to investigate the relative contributions of HC, CO and CO_2 to gas flow at different temperatures. This wealth of preliminary work has not been used in the scientific literature so far, but may provide very interesting information on the characterization of organic matter by Rock-Eval® in the coming years.

4

Guidance on Sampling, Effects of Mineral Matrix and Other Artifacts on Thermograms

François BAUDIN[1], Yves BENOIT[2] and Nicolas BOUTON[3]
[1] *ISTeP, Sorbonne Université, Paris, France*
[2] *IFP Energies nouvelles, Rueil-Malmaison, France*
[3] *Vinci Technologies, Nanterre, France*

After presenting the principles of the Rock-Eval® method and the improvements made over time with the different versions of the device in Chapter 2, and the main parameters provided by the method and their modes of representation in Chapter 3, the following pages will discuss the effects of sample collection and preparation, as well as the influence of the mineral fraction of the samples on the results obtained.

4.1. Sample collection, preparation and conservation

As organic matter is sensitive to oxidation, it is important to collect, prepare and conserve samples under the most inert conditions possible to avoid or at least minimize its weathering. The different types of samples that can be analyzed by the Rock-Eval® method and the precautions to be taken at each stage before analysis will be reviewed.

4.1.1. *Recent lacustrine or marine sediments*

In these environments where the organic matter is still relatively fresh and therefore very sensitive to oxidation, precautions need to be taken during sampling. If the sediment is in the anaerobic early diagenetic stages (e.g. sulfate-reducing zone or methanogenesis zone), samples should be collected, prepared and conserved under anoxic conditions. Using a fully filled glass vial limits evaporation losses, while rapidly placing them in a nitrogen chamber guarantees the preservation of anoxic conditions. Otherwise, it is possible to freeze them as quickly as possible after sampling, whether this sampling is carried out with an auger or by coring. Drying of the samples must be carried out without delay. It is recommended to freeze-dry recent sediment samples rather than dry them in a drying oven where the atmosphere is rarely inert. This also avoids organo-mineral transformations that can occur even at temperatures <50°C. In the case of sediments deposited in a saline context, freeze-drying or oven-drying leads to the precipitation of salts; these cause artifacts during Rock-Eval® analysis on both the organic fraction (see section 4.3) and the carbonate fraction (see section 13.3).

4.1.2. *Sedimentary rocks*

The meteoric, chemical and mechanical weathering of rocks is an active phenomenon on the surface of our planet. Its intensity depends on many factors, including climate (temperature and precipitation) and relief, and also the lithological nature of the rocks. The organic matter present in sedimentary rocks therefore undergoes oxidation that is all the more intense, the closer it is to the surface, and the higher the temperature, and the more the water and the roots penetrate easily into the rock. Before collecting samples from the outcrop, the surface should be freshened by digging to at least 30–50 cm deep in temperate climates, or more in a tropical climate where the weathered zone may be much thicker. The ideal approach is, of course, to collect samples by coring, but this process is expensive and not always easy to implement. In the absence of coring, quarries and recently cut road embankments should be favored. If the sedimentary series shows bedding, sampling of the bed/interbed contact, which often serves as a drain for meteoric waters, should be avoided. During sampling, roots and rootlets should be removed and the least-weathered facies sampled. Unrefreshed outcrops often present a very different appearance to unweathered rocks. The surface of the coals is dull and powdery, gray or whitish instead of being black and shiny, massive and consistent (also see Chapter 11). Claystones very rich in organic matter are usually fissile and bleached (presence of fine crystals of secondary gypsum), or take on a rusty to yellowish color (presence of jarosite- or copiapite-type sulfate) instead of being dark brown to black and massive. Samples collected from such outcrops would be neither physically nor chemically representative of the rocks existing at depth. Chemically, meteoric weathering results in a very significant variation in the elemental composition and thermal stability of the sedimentary organic matter. The proportion of hydrogen is found to decrease while the proportion of oxygen increases (Landais et al. 1984; Espitalié et al. 1986), which is readily apparent in the HI–OI and HI–Tmax diagrams (see Figure 4.1).

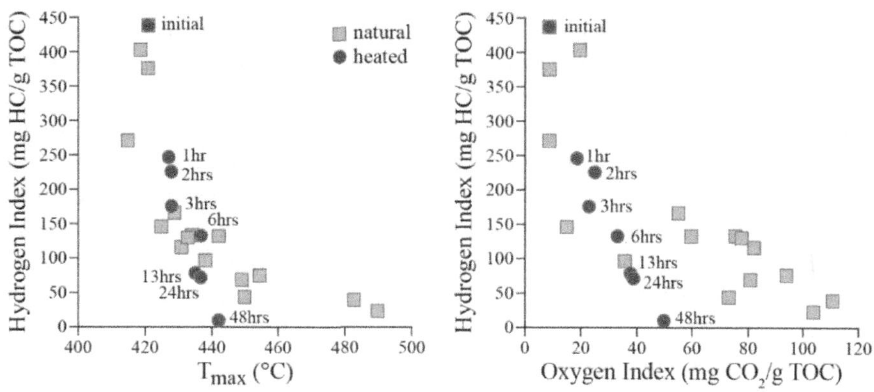

Figure 4.1. *Evolution in HI–OI and HI–Tmax diagrams of a natural series of humic coals from the Jurassic of Utah (USA) showing the effects of supergene weathering. The intensity of the weathering is marked by a decrease in HI, which correlates with an increase in OI and Tmax. Artificial oxidation of a sample of unweathered coal from this series was conducted in the laboratory in air heated to 200°C with increasing heating times. The two series, natural and artificial, show the same paths in the HI–OI and HI–Tmax diagrams. Modified from Landais et al. (1984) in Vandenbroucke and Largeau (2007). For a color version of this figure, see www.iste.co.uk/baudin/rock-eval.zip*

Under the effect of oxidation, the percentage of TOC generally tends to decrease, but paradoxically, it can increase, especially in the case of rocks rich in pyrite (Littke et al. 1991). Indeed, pyrite weathers faster than organic matter and its weathering products are leached or form less-dense minerals, which by mass balance increases the organic matter content. The presence of unweathered small pyrite grains in a rock is a good quality criterion during sampling.

If the samples collected contain water, they need to be dried in a drying oven (50°C for at least 12 hours) before being reduced to a powder whose particle size will ideally be less than 200 µm. Before grinding, it is recommended to examine the samples under a binocular microscope and remove any rootlets present using tweezers.

Reducing the samples to a powder facilitates the filling of the crucibles and ensures a larger reaction surface when the temperature rises in the ovens. Otherwise, certain parameters may be disturbed with a notable delay in the thermal cracking of the sedimentary organic matter or of the carbonates, and an increase in the corresponding Tpeak (TpeakS2 and TpeakS5). Thus, Espitalié et al. (1985a) note that Tmax increases by 3–4°C for samples whose mean grain is close to 2 mm compared to the same sample in powder form. The influence of the sample's particle size was studied in detail by Inan et al. (1998) for different lithologies with average grain sizes ranging from 0.1 to 8 mm (Figure 4.2). More recently, Hazra et al. (2017) studied the combined influence of particle size and weight of the sample on Rock-Eval® parameters in the case of coal analysis (Figure 4.3). In most lithologies (clay, marl, limestone and coal), S1 and S2 decrease as grain size increases, indicating less-efficient thermovaporization when the sample is granular. There are, however, some notable exceptions to this rule, particularly in facies that are very rich in algal organic matter (torbanite and lamosite) and certain lignites (Inan et al. 1998). Depending on the lithology, Tmax is either constant or decreases as the grain size decreases (Inan et al. 1998; Hazra et al. 2017).

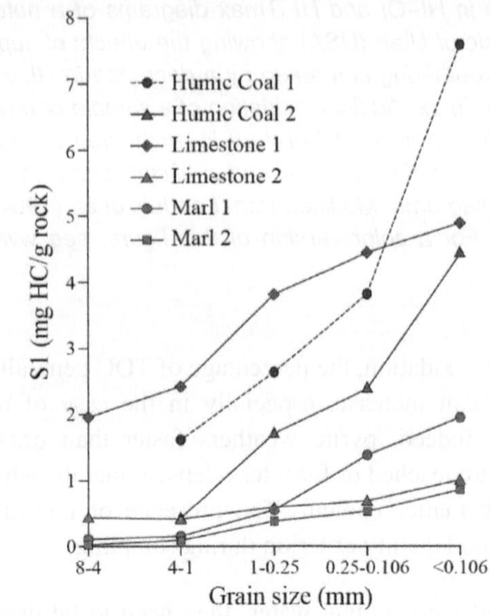

Figure 4.2. *Evolution of S1 according to the particle size of the sample for different lithologies (according to Inan et al. (1998)). For a color version of this figure, see www.iste.co.uk/baudin/rock-eval.zip*

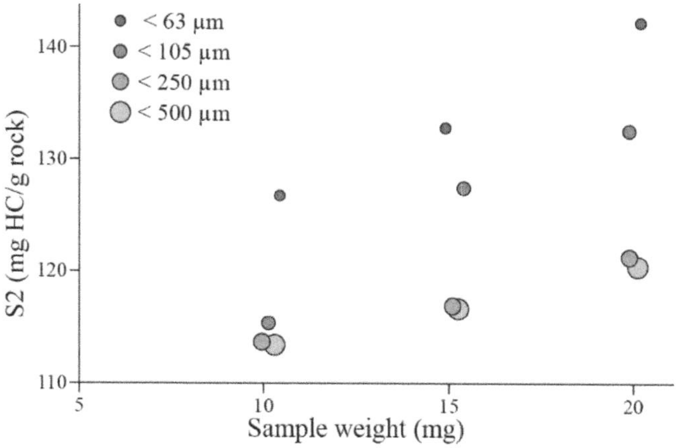

Figure 4.3. *Effect of the particle size and weight of the test portion on the S2 parameter in the analysis of a coal (according to Hazra et al. (2017)). For a color version of this figure, see www.iste.co.uk/baudin/rock-eval.zip*

After grinding, the samples need to be kept dry in airtight vials and, if possible, away from light. Even under these conditions, the weathering of the organic matter continues, which is marked by the evolution of certain Rock-Eval® parameters. In poorly dried samples, the development of fungi is reflected in an increase in the S1, S2 and S3 parameters according to Peters (1986). Similarly, S1 can halve simply through evaporation during storage of the samples, even under suitable conditions. Espitalié et al. (1985a) report a loss of up to 40% of S1 in a sample of the Liassic of the Paris Basin after 10 months of storage, Jiang et al. (2016) report a loss of 30% of S1 less than 24 hours after grinding in a sample of Ordovician claystone from Quebec (Figure 4.4) and Carvajal-Ortiz and Gentzis (2015) report a loss of 22% only two hours after grinding.

It was mentioned above that samples collected by coring should be preferred to any other type of sampling. However, for reasons of convenience and cost, many studies are conducted on samples taken from the outcrop. When the sampling precautions mentioned above are followed and the weathered zone is limited (as is the case in temperate climates), it has been shown that there is not a great difference between the results obtained on outcrop samples and those obtained from a core sampled in the same horizons (Bréhéret and Herbin 1987; Figure 4.5; Littke et al. 1991; Figure 4.6).

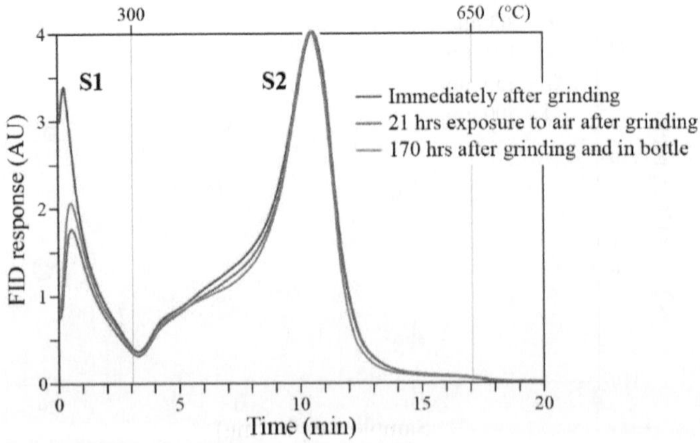

Figure 4.4. *Evolution of the S1 and S2 parameters in a claystone analyzed just after grinding, 21 hours and 170 hours after grinding. After one week (170 hours), a drop of nearly 50% of S1 can be observed, while the S2 loss is very limited (according to Jiang et al. (2016)). For a color version of this figure, see www.iste.co.uk/baudin/rock-eval.zip*

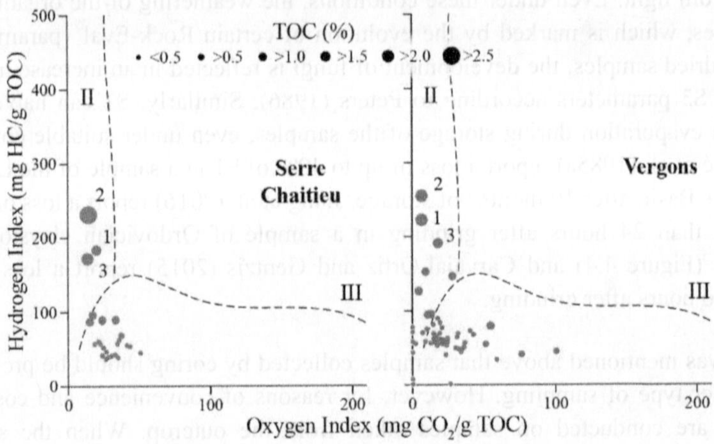

Figure 4.5. *Comparison of the distribution of samples of Aptian blue marls collected from the outcrop (Serre Chaitieu) and core (Vergons) in an HI–OI diagram. The similarity of the TOC, HI and OI values suggests that meteoric weathering is limited at Serre Chaitieu, thus validating studies on samples collected from the outcrop (according to Bréhéret and Herbin (1987)). For a color version of this figure, see www.iste.co.uk/baudin/rock-eval.zip*

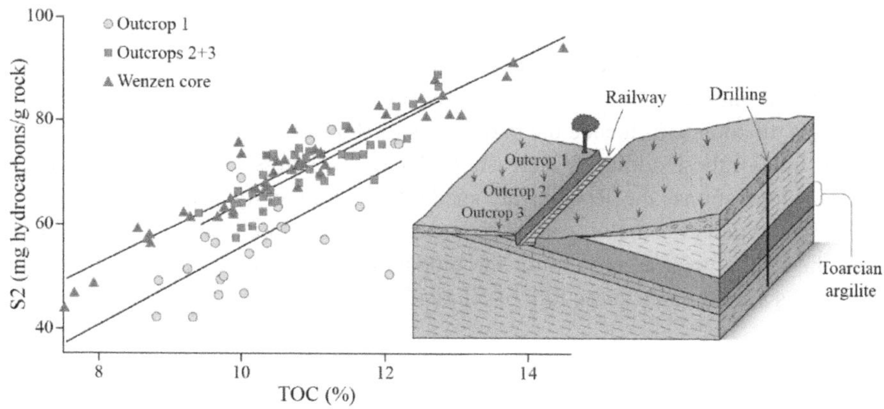

Figure 4.6. *Example of variations in Rock-Eval® parameters between a drilling and three outcrops in organic claystones of the Lower Toarcian in Germany. Outcrop 1 is a naturally eroded zone, while outcrops 2 and 3 are railway line embankments cut in 1865. These outcrops were subjected to refreshing at the time of sampling. There is clearly great consistency between the Rock-Eval® parameter values for the drilling and for outcrops 2 and 3. Outcrop 1, on the contrary, is more oxidized, demonstrated by lower TOC and S2 values (according to Littke et al. (1991)). For a color version of this figure, see www.iste.co.uk/baudin/rock-eval.zip*

4.1.3. *Cores and drill cuttings*

The cuttings should be prepared as soon as possible after they have been brought up to the drilling site (Espitalié et al. 1985a). Washing with water and sieving will remove soluble pollutants and fine particles. Grains of cuttings 1–2 mm in size will be selected from the sieve retention, avoiding allochthonous solid debris (fresh plant fibers, balls of fat, etc.). The selected grains will be air- or oven-dried at a temperature of 30°C maximum for 30 minutes. If the cuttings cannot be isolated and prepared on-site, they will be kept wet until they arrive at the laboratory where they will be subjected to the same preparation stages, as quickly as possible. Grinding should be avoided to limit losses of light hydrocarbons (Espitalié et al. 1985a).

During drilling operations, it is not uncommon for organic additives to enter the composition of the mud used to lubricate the drill bit, lift the cuttings and maintain hydrostatic pressure with the geological formations crossed by the well. These additives in water- or oil-based mud are as diverse as walnut shells, alginates, starch derivatives, cellulose or tannins, polymers (polyethylene, polyvinylalcohol, glycol, etc.), and even natural oil products (gilsonite) or refined oil products (light fuel, diesel, etc.). Some of these additives are removed by simply washing with water (starch, certain tannins, etc.), others can be removed with tweezers during examination under a binocular microscope (walnut shell fragments), but this is not the case if diesel or other hydrocarbons are added to the drilling mud. These products, of course, influence the parameters of Rock-Eval®, notably S1, S2, S3 and Tmax, and all these additives increase the HI value (Peters 1986). In the context of offshore operations, the Rock-Eval® method has frequently been implemented to identify and quantify the oil (diesel) content of drilling muds before their discharge into the sea, as well as to monitor the biodegradation of these additives within the sediments accumulated at the foot of oil platforms.

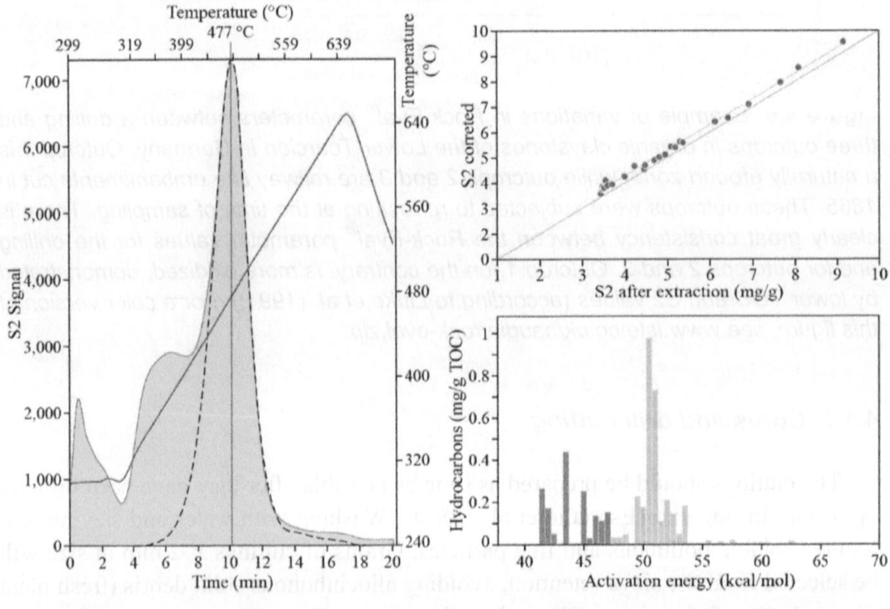

Figure 4.7. *Comparative study between the software method and the Soxhlet extraction method showing an excellent correlation between the two methods. For a color version of this figure, see www.iste.co.uk/baudin/rock-eval.zip*

Extraction of the cuttings with an organic solvent (e.g. dichloromethane or methanol-dichloromethane mixture) before their analysis is possible, but it is rarely 100% effective, and it also causes the solubilization of part of the organic matter of the samples. Another solution is to analyze the organic additive using Rock-Eval® in order to characterize it and subtract its signal at least on the S1 and S2 parameters. The Shale-Play method (Romero-Sarmiento et al. 2014) thus makes it possible to address these drilling-mud pollution problems (see Chapters 3 and 6). When analyzing soils polluted by oil distillates, an identical approach including a prior extraction with cyclohexane-acetone allows the polluting fraction to be extracted in order to identify and quantify the response of the natural, non-polluting organic matter of the analyzed matrix (humus, fulvic acid, hemicellulose and lignin). This will then be deduced from the response of the non-extracted soil. This analytical approach is lengthy and expensive, but remains essential for polluted sites presenting high natural organic matter contents (forests and peatlands) whose responses to Rock-Eval® interfere with those of distillates (see Chapter 12).

A software method, called CleanSim, can now be used to simulate the extraction of polluted samples with solvent, using Soxhlet, for example, which is often too expensive in terms of operator time. Deconvoluting the S2 parameter by kinetic equations makes it possible to manually remove the part of the S2 parameter that is due to pollutants, and recalculate the value of S2 and the Tpeak of the unpolluted sample (Figure 4.7).

4.1.4. *Specific cases of reservoir rocks and oils*

An impregnated reservoir rock needs to be frozen to ensure that the hydrocarbons that it contains are well preserved. As crude oils are a highly controlled commercial product, there is an international standard (ISO3170:2004) describing the conditions for their collection, packaging and storage. Indeed, if crude oil is sampled in open containers, this can lead to the loss of light fractions, with repercussions on S1 in particular, and also on the OI. When analyzing an oil, it is necessary to fill the crucible with ~100 mg of Fontainebleau sand[1], place the product on top of it and introduce it into the pyrolysis oven as quickly as possible after weighing (Trabelsi et al. 1994). If sufficient care is taken, it is possible to analyze even light oils and show that the S1 or S1r+S2a value can approach 1,000 mg HC/g oil.

4.1.5. *Soils and polluted soils*

With the exception of peatlands and certain soil types, pedological organic matter is generally under aerobic conditions. Its sampling therefore does not need to be carried out under an inert atmosphere. It is, however, important to dry soil samples rapidly after they have been collected; drying can be carried out in air or, preferably, overnight in a drying oven at 50°C. To collect polluted soils, a glass bottle, inert to hydrocarbons and equipped with a teflon-butyl septum, is specially

1. Before use, it is recommended to roast the Fontainebleau sand in a muffle oven up to 850°C to eliminate any organic trace.

adapted to the collection of samples polluted by petroleum cuts. However, any drying before analysis must be prohibited, since it leads to the irreparable loss of hydrocarbon fractions with the lowest vapor pressures ($\leq C_{10}$).

4.1.6. *Kerogens*

For organic geochemists, sedimentary organic matter consists of an organic fraction insoluble in organic solvents, called kerogen (Durand 1980; Vandenbroucke and Largeau 2007), and a soluble fraction, called bitumen. Kerogen is isolated from the mineral fraction of the sediment by acid attacks (HCl and HF) to destroy carbonates and silicates successively (Durand and Nicaise 1980). After isolating the kerogen, it is necessary to conserve it in fully airtight vials, under an inert atmosphere if applicable, and possibly also to freeze them. Indeed, mineral matter very often protects organic matter from oxidation. Thus, Salmon et al. (2000) showed that an organic matter fossilized for 95 million years was still reactive once the mineral phase was eliminated and the kerogen was isolated from its protective clay gangue. Kerogen stored in the dark for two years saw its OI increase from 9 to 30 mg CO_2/g TOC, while the HI and TOC were hardly affected over this period of time.

4.2. Mineral matrix effects

When Rock-Eval® analysis is performed not on pure organic matter (kerogen, oil, organic pollutant, etc.) but on soils, sediments or sedimentary rocks, the results obtained may be disturbed to a lesser or greater extent by the mineral fraction of the sample. Depending on the type and quantity of organic matter, the minerals of the sample may, in some cases, interfere to varying degrees with the pyrolysis parameters (S_2, PC, Tmax, etc.) and more rarely with the oxidation parameters (RC, etc.). These interferences, grouped under the term "mineral matrix effect", cover different phenomena. They may include:

– Temporary retention of certain pyrolysis effluents (*n*-alkanes < C_{30}, CO_2, etc.) that are released with delay without any major transformation. Tarafa et al. (1983) showed that the action of different minerals on the pyrolysis of *n*-alkanes up to *n*-C_{32} had the main effect of delaying their vaporization, yet without resulting in a very marked retention.

– Retention of heavy or aromatic hydrocarbons and heteropolar compounds at the reactive surface of minerals and their coking during pyrolysis (Figure 4.8). The neoformed organic products are then detected during the oxidation phase through the release of CO/CO_2 and SO_2/SO_3.

– The release of oxygenated compounds, in particular CO_2, by decomposition of certain carbonates (oxalate, nahcolite, siderite, etc.) at relatively low temperatures during pyrolysis (see Chapter 13). The decomposition of iron oxides also releases CO_2 at the beginning of pyrolysis, and thus greatly influences the OI.

– Interactions during heating between certain clay minerals and carbonates.

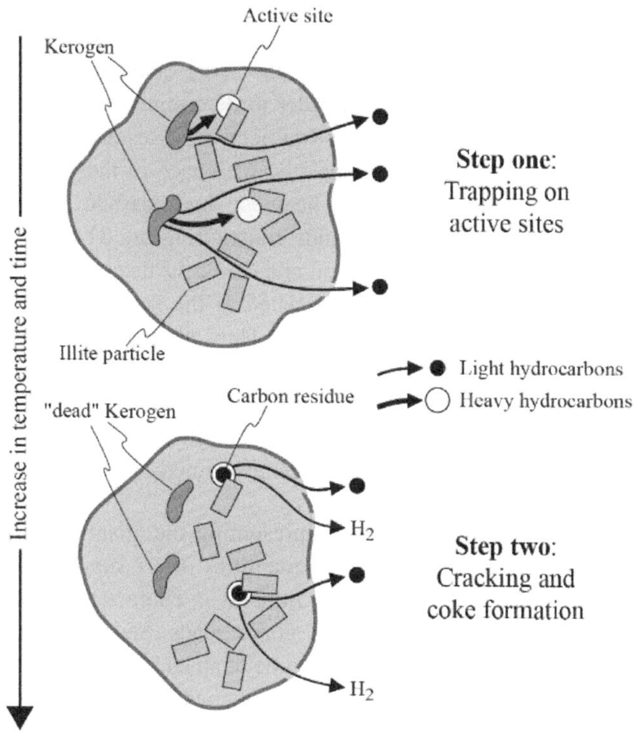

Figure 4.8. *Conceptual model of the mineral matrix effect leading to the retention of heavy hydrocarbons at the active surfaces of certain minerals (mainly clays), and their transformation into coke during pyrolysis. Modified from Espitalié et al. (1980b). For a color version of this figure, see www.iste.co.uk/baudin/rock-eval.zip*

Temporary retention has little impact on the final results of the Rock-Eval® analysis, except for a slight shift in Tmax. The effects of the early cracking of carbonates are detailed in Chapter 13. The effects of partial hydrocarbon retention by minerals and their implications on the results of Rock-Eval® analysis will be reviewed in detail in the following sections, starting by examining how these effects can be identified and quantified.

4.2.1. *Identification of mineral matrix effects on hydrocarbon compounds*

Historically, these effects were studied using Rock-Eval® pyrolysis or pyrochromatography of synthetic mixtures between a kerogen and different minerals (Espitalié et al. 1980b, 1984a; Horsfield and Douglas 1980; Larter and Douglas 1982). The basic idea was to compare the carbon balance and the relative proportions of different hydrocarbon fractions between a kerogen (or a rock very rich in organic matter) analyzed alone and the same quantity of kerogen covered with a known quantity of a mineral, in order to determine their effect on pyrolysis products. After pyrolysis, the organic and mineral residues are analyzed separately to determine the quantity of carbon associated with each of these two fractions. All of these authors observed that the heaviest hydrocarbon compounds (C_{20+} hydrocarbons and heteroatomic compounds constituting the S1 and S2 parameters) are the most easily retained on the mineral and coked surfaces during pyrolysis. The intensity of this phenomenon is directly related to the specific surface area of the mineral matter (Espitalié et al. 1980b, 1984a). Thus, attapulgite/palygorskite (which are fibrous clays), illite and smectites (such as montmorillonite) show a greater hydrocarbon retention capacity than kaolinite or calcite (Table 4.1). Certain minerals, such as iron oxides, present an even higher effect than clay minerals, whereas others, such as silica, do not generate any mineral matrix effect.

For given organic matter, the curve representing the quantities of hydrocarbon compounds retained as a function of the quantities of S2 supplied by the organic matter added in variable proportion to a mineral is characteristic of this mineral (Figure 4.9(A)). For each mineral, this curve tends toward an asymptote that corresponds to the maximum quantity that can be retained by the mineral considered. It can be seen that the proportion of hydrocarbons retained is all the greater, the lower the organic matter content of the mixture, which is clear from the steep slopes close to the origin of the diagrams in Figure 4.9. The quantity retained also depends on the type of kerogen. Thus, the hydrocarbon compounds resulting from the pyrolysis of a Type I kerogen will be retained much less than those resulting from the pyrolysis of Type II and III kerogens (Figure 4.9(B)). This is due to the fact that pyrolysis products have higher *n*- and *iso*-alkanes/alkene contents for Type I than in the other two types, which are richer in polar compounds. The degree of maturation of the organic matter does not appear to have an effect on the quantity of hydrocarbons retained when they are derived from a Type- I or -II kerogen. On the contrary, in the case of Type III kerogen, the proportion decreases by 20–40% when entering the wet-gas formation zone, i.e. Tmax values of about 470°C (Espitalié et al. 1985b). This is due to the fact that a Type III kerogen that has reached this stage of maturity gives hydrocarbon compounds of lower molecular weight, therefore less easily retained by minerals.

Organic material + added mineral matter	% Pyrolyzed Corg	% Residual Corg	% Retained Corg	% Pyrolyzed + retained Corg	% Total Corg	% Pyrolyzed Corg
Oil shale	0.99	0.89	0.00	0.99	1.88	0.99
+ Gypsum	0.94	0.88	0.10	1.04	1.92	0.94
+ Kaolinite	0.85	0.88	0.17	1.02	1.9	0.85
+ Montmorillonite	0.48	0.87	0.63	1.11	1.98	0.48
+ Montmorillonite Ca^{2+}	0.49	0.87	0.62	1.11	1.98	0.49
+ Montmorillonite Na^{+}	0.42	0.87	0.67	1.09	1.96	0.42
+ Illite	0.35	0.88	0.75	1.10	1.98	0.35
+ Attapulgite	0.23	0.88	0.86	1.09	1.97	0.23

Table 4.1. *Carbon mass balance for the analysis of a single oil shale compared to the analysis of the same amount of shale to which different minerals have been added. Data extracted from Espitalié et al. (1984a)*

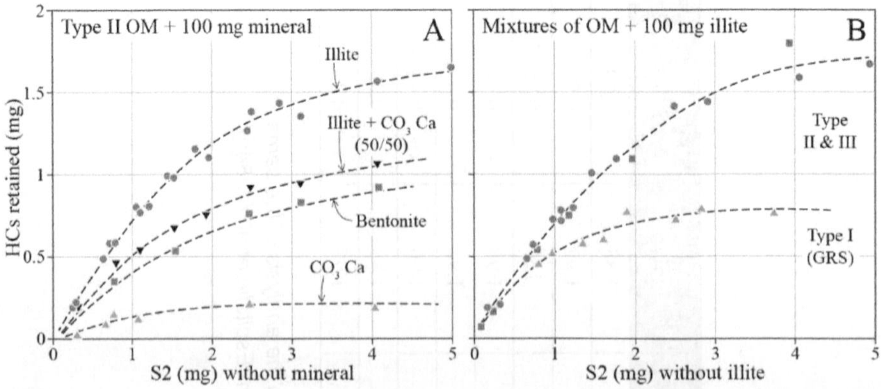

Figure 4.9. *(A) Effect of the type of mineral on the retention of the hydrocarbon compounds constituting the S2 parameter. (B) Effect of the type of organic matter (Type I vs. Types II and III) on the retention of the hydrocarbon compounds constituting the S2 parameter. Modified from Espitalié et al. (1985b). For a color version of this figure, see www.iste.co.uk/baudin/rock-eval.zip*

4.2.2. Impact of matrix effects on Rock-Eval® parameters

The matrix effect has consequences on the Rock-Eval® parameters, notably the HI, making it complicated to determine the types of organic matter from the HI–OI (Figure 4.10(A)) and HI–Tmax (Figure 4.10(B)) diagrams. The HI logically decreases with the intensity of mineral matrix activity since it is S2 hydrocarbons that are retained. The OI is not affected, unless the matrix effect concerns the early cracking of thermally fragile carbonates such as siderite, hydromagnesite, etc. (see Chapter 13). The HI–OI diagram in Figure 4.10(A) shows that the HI can drop by nearly 500, from 620 mg HC/g TOC to 130 mg HC/g TOC, simply by a mineral matrix effect. Comparison of the HIs (and OIs) measured on bulk samples with their equivalents measured on kerogens isolated from these samples is a good way to measure these matrix effects. Although the mineral matrix effect is always more pronounced in samples with low TOCs, it can even be felt in samples with TOC > 5% (Saint-Germès et al. 2002; Figure 4.11).

In the case of the matrix effect, the behavior of Tmax is variable and depends on the nature of the minerals present in the sample (Table 4.2). Silica and calcite/dolomite have little effect on Tmax, while it tends to increase with clay minerals. For a very active mineral such as illite, Tmax increases when the illite/organic matter ratio increases, while not exceeding 5–6°C for a Type II kerogen and 10–12°C for a Type III kerogen (Espitalié et al. 1985b). This is explained by the retention of the first hydrocarbons formed by the cracking of kerogen and retained by the mineral matrix modifying the shape of the S2 parameter, which causes a shift in its Tpeak toward higher temperatures. For high illite/organic matter ratio values (greater than 10), Tmax decreases again (Figure 4.10(B)). The degree of thermal maturation of the organic matter does not appear to significantly modify the mineral matrix effect. In other words, an active mineral leads to the same increase in Tmax whether the kerogen is immature or in the oil window.

Mineral/shale ratio	Mixture TOC (%)	Tmax (°C) in the mixture			
		Illite	Calcite	Sand	Ill.+Sand (50/50)
20	0.57	415	415	419	420
10	1.09	422	414	418	421
6.6	1.56	424	416	417	423
5	2	424	416	417	423
4	2.4	415	418	419	423
3.3	2.77	425	418	418	420
2.8	3.11	422	418	420	420
2.5	3.43	421			419
2	4.0	420			419
1.6	4.5	420			419
0	12		419	419	

Table 4.2. *Effects of mineral matrix/organic matter interactions on Tmax during pyrolysis of an oil shale (Type II) mixed in different proportions with different minerals (according to Espitalié et al. (1985b))*

Mineral matrix effects on the production index (PI) are more complex because they involve both S1 and S2. Espitalié et al. (1985b) noted that when the S1 values (and the S2 values) are high, S2 hydrocarbon retention is less effective because the mineral matrix is as if saturated with S1 hydrocarbons and retains S2 hydrocarbons less. In this case, a decrease in the PI is observed. In the case where S1 and S2 are low, the fraction of hydrocarbons constituting the S1 parameter that can be retained is in insufficient quantity to saturate the mineral matrix. It is then the hydrocarbon compounds of S2, rich in easily condensable compounds (alkenes, aromatics and heavy polar compounds), that will be more retained, proportionally speaking. This results in an increase in the PIs.

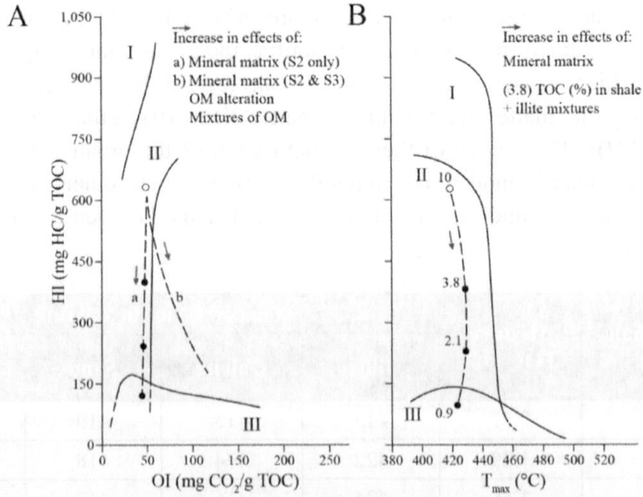

Figure 4.10. *Effect of mineral matrix activity on sample position in the HI–OI (A) and HI–Tmax (B) diagrams (according to Espitalié et al. (1985b)). For a color version of this figure, see www.iste.co.uk/baudin/rock-eval.zip*

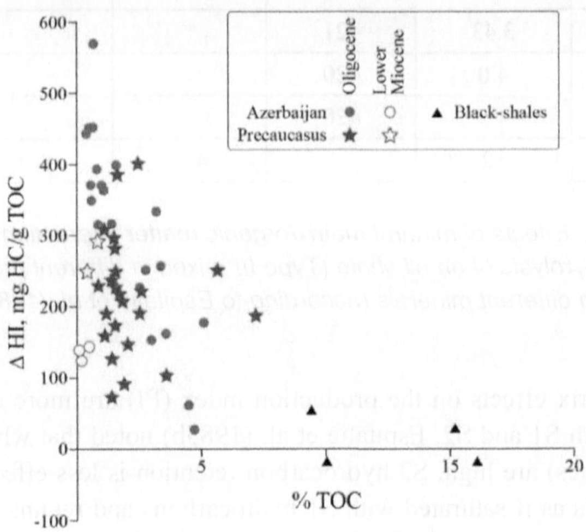

Figure 4.11. *Effect of mineral matrix as a function of TOC content in the pre-Caucasus and Azerbaijan Oligo-Miocene series. ΔHI indicates the difference between the HI measured on bulk rock and the HI measured on kerogen. It can be observed that the fall in HI can exceed 250 mg HC/g TOC, even for samples rich in organic matter (> 5% TOC) (according to Saint-Germès et al. (2002)). For a color version of this figure, see www.iste.co.uk/baudin/rock-eval.zip*

These mineral matrix effects were also observed during the analysis of polluting petroleum cuts in anthropized soils of variable mineralities. To avoid any diagnostic error and to take into account the shift of Tpeak toward higher temperatures, a database was created by artificially contaminating siliceous or clay-sandy matrices with reference oil distillates (gasoline, kerosene, diesel and heavy fuel oil) whose composition and C distribution were determined beforehand by FID/GC analysis. In order to identify the fraction and proportion of hydrocarbons presenting shifts in their vaporization temperatures, the total FID signal was split into four zones, Q0 to Q3 (Q0 $\leq$ 80°C; Q1 between 80 and 120°C; Q2 between 120 and 350°C; Q3 between 350 and 850°C). For each polluting petroleum cut and each mineral matrix, the proportion of the signal within each zone (Q0 to Q3 expressed as a % of the total Qt) made it possible to visualize the matrix effects and quantify them. For example, for a gasoline, the very high proportion of the signal before 80°C is observed in the presence of siliceous sand (93.3 %), while in clay this proportion drops to 49.9%, thus showing the high retention of clay fractions.

4.2.3. *Measurement of the activity of a mineral matrix and means of correcting its effect*

To rapidly quantify the activity of a mineral matrix on hydrocarbon compound retention, Langford and Blanc-Valleron (1990) proposed examining the results of Rock-Eval® analyses in an S2 versus TOC diagram. This representation is of interest in that it highlights a correlation line (linear regression not forced to go through the origin, i.e. $y = ax+b$) for a given data set, resulting, for example, from the analysis of a sedimentary core. It is, of course, essential for the number of data to be sufficient in order to be statistically representative and for these data to also be pedologically or geologically representative, i.e. we do not mix different types of soils or sedimentary facies that are too different. In theory, in the absence of inert carbon or black carbon (see Chapter 11) in the samples, even a very small quantity of kerogen needs to release a hydrocarbon fraction during the pyrolysis phase, leading to a value of S2 and a corresponding organic carbon fraction (fraction of PC associated with S2, i.e. $PC_{S2} = 0.83 \times S2$). In theory, the regression lines in an S2 versus TOC diagram should therefore all pass through the origin. Yet this is rarely the case and it is very frequently observed that the regression line intercepts the x-axis (Figure 4.12), as if a small quantity of carbon had to be present before the first hydrocarbons were detected. This obviously does not apply if the sample contains kerogen or thermolabile organic matter. This is just the adsorption of hydrocarbons by the mineral matrix and the intensity of this effect is evaluated by the TOC value on the x-axis. In the case of clay series, this matrix effect is often found to represent 0.3 to 0.8% TOC. By determining the equation of the line, which takes the form

$y = ax–b$, it is even possible to quantify the part of S2 (–b) that is missing (Langford and Blanc-Valleron 1990).

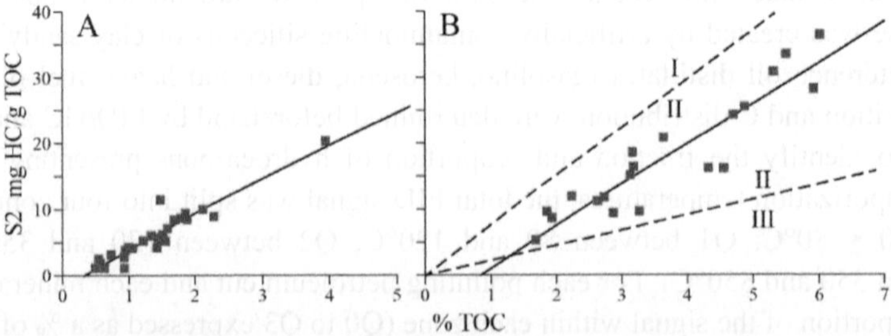

Figure 4.12. *S2 versus TOC diagram for two sedimentary series. (A) Cretaceous carbonates from the Gulf of Mexico and (B) Miocene claystones off the coast of California. It can be observed that the matrix effect is less in carbonates than in claystones since the value of the intercept of the lines on the x-axis is 0.28% TOC for A and 0.87% TOC for B, respectively. The retained fraction of S2 is equivalent to 1.4 mg HC/g of rock for A and 8 mg HC/g for B. Modified from Langford and Blanc-Valleron (1990). For a color version of this figure, see www.iste.co.uk/baudin/rock-eval.zip*

It is possible to quantify the mineral matrix effects of a rock or, in other words, its mineral activity with respect to organic matter by mixing this previously pyrolyzed rock with a kerogen whose S2 parameter characteristics are well known. Thus, by difference, it is possible to assess quantitatively (fall in the value of S2) and also qualitatively (change in shape of the S2 parameter) the matrix effect of the sample in question. This method would only be applicable to samples with TOC < 5% according to Espitalié et al. (1985b), and it should be kept in mind that prior pyrolysis of the sample does not necessarily leave the minerals unaffected. Mineralogical transformations can occur, giving the pyrolyzed rock an additional matrix effect.

4.2.4. *Conclusions on the effects of mineral matrices*

Most Rock-Eval® parameters (HI, OI, Tmax, PI, etc.) are therefore impacted by the mineral matrix effect. The intensity of this impact depends on the type of organic matter in the sample, its degree of maturation and of course the type of mineral at the origin of the effect and its proportion in the natural material or in the synthetic mixture. Most matrix effects are retardants with respect to the release of hydrocarbons, but some minerals also have catalytic effects accelerating the thermal degradation of organic matter or other minerals. Other minerals, or sometimes the same, can provide oxygen during pyrolysis or release water. In a natural material where the mineral matrix is generally made up of several minerals, these effects are combined, by either being added together or partially canceled out. It is clear that no pyrolytic method applied to a natural material can characterize only the organic matter, but that it provides characteristics of its combined organic and mineral fractions.

It should be kept in mind that vaporization and/or rapid pyrolysis in an open environment, as with the Rock-Eval® method, differs greatly from the slow evolution of organic matter in soils or the hydrocarbon generation process during the catagenesis and metagenesis stages. Organo-mineral interactions do not disappear during burial diagenesis, but they are certainly different in terms of intensity and velocity (see Chapter 7, kinetics). Moreover, the presence of water in soils and sedimentary basins certainly constitutes a fundamental difference between natural processes and those that take place during a Rock-Eval® analysis. The presence of water, of course, has repercussions on both organic matter and certain mineral phases.

The fact remains that pyrolytic techniques such as Rock-Eval® are the only ones capable of rapidly simulating hydrocarbon generation, and indeed they make it possible to study the mineral matrix effects. In the field of pedology, where it is not a matter of thermal evolution linked to burial, but of carbon stabilization on a secular or even millennial scale on the surface of minerals, the analysis of the mineral matrix effect using the Rock-Eval® method is therefore also of great interest (see Chapter 10).

4.3. Other artifacts on thermograms related to the presence of minerals

Chlorides and other salts are problematic in Rock-Eval® analysis because their volatilization during the pyrolysis phase causes parasitic ionization that disrupts the FID. It is not a matter of an interaction between organic matter and salts, but rather an analytical artifact. Thus, during the analysis of evaporatitic series or recent sediments rich in salts, a rise of the FID signal is observed after the S2 parameter generated by the cracking of kerogen, thus creating a second S2 peak (Figure 4.13). This second S2 peak is at times more intense than the first and its abnormally high Tpeak then predominates over the kerogen cracking Tpeak/Tmax. Occasionally, the rise of the FID signal occurs late during the pyrolysis phase, and it ceases as soon as the oven cools, thus creating a parameter whose Tpeak is 600°C or 650°C, depending, respectively, on whether a Rock-Eval® 2 or 6 is used.

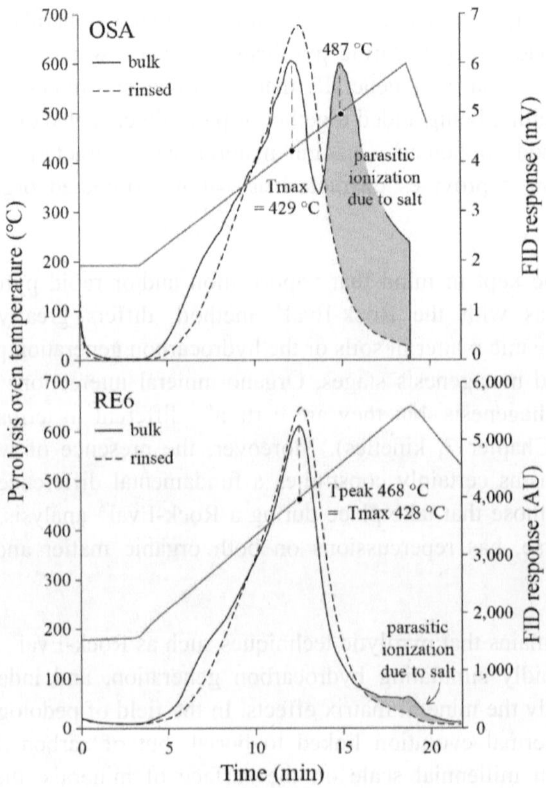

Figure 4.13. *Comparison of FID responses in the analysis of a recent salt-rich marine sediment sample with a Rock-Eval® OSA and a Rock-Eval® 6. Less parasitic ionization is observed in the case of Rock-Eval® 6, which is related to a lower temperature at the gas splitter at the outlet of the pyrolysis oven. The dotted lines show the FID signal of the same samples analyzed under the same conditions, but after being freed of their soluble salts by washing with distilled water. Modified from Baudin et al. (2015). For a color version of this figure, see www.iste.co.uk/baudin/rock-eval.zip*

It is surprising that this effect occurs at temperatures lower than those of the melting point of halite (NaCl) or sylvite (KCl), which are the main salts in marine waters. Indeed, they melt at 800°C and 790°C, respectively, whereas the phenomenon of parasitic ionization can appear from 500°C during the pyrolysis phase. The pyrolysis of 10 mg of NaCl, pure or mixed with a marine sediment previously freed of its salts, never causes parasitic ionization as strong as that demonstrated on non-rinsed samples. It is therefore likely that other salts (sulfates, borates, iodates, etc.), even present in very low concentration, play a role in this phenomenon. In this respect, it is worth noting that lithium metaborate ($LiBO_2$) is used as a fondant in materials science to prepare samples (rocks, cements, ceramics, etc.) for their chemical analysis. In any case, the temperatures reached at the end of pyrolysis do not signal the end of salt-related artifacts. These can still have an effect during the oxidation phase, but this time on the carbonate fraction (see section 13.3).

Regardless of the intensity of this parasitic ionization, this phenomenon increases the value of S2 and therefore, proportionally, the TOC and the HI (Figure 4.14); the other basic parameters of Rock-Eval® (S3, S4 and S5) are not affected. This problem was identified during the design of Rock-Eval®, which is why Espitalié et al. (1985b) recommended reducing the temperature of the splitter and washing the samples with water before their analysis in order to eliminate all or part of this artifact. For this reason, the temperature of the splitter was reduced from 550°C or 500°C in Rock-Eval® versions II and III to 400°C in Rock-Eval® 6. It can be seen from the example in Figure 4.13 that this decrease in the temperature of the splitter has largely led to the disappearance of this effect. However, this effect does not disappear entirely, and there are still cases where parasitic ionization remains significant, even when using a Rock-Eval® 6.

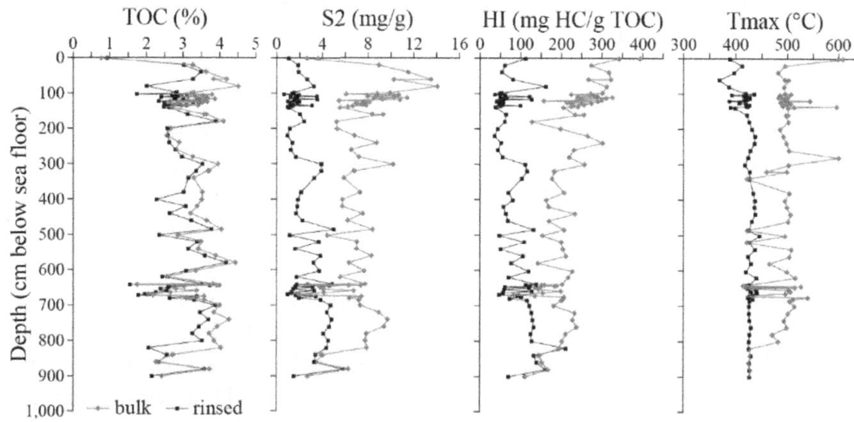

Figure 4.14. *Vertical evolution of Rock–Eval® parameters (TOC, S2, HI and Tmax) along a 9 m core in turbiditic sediments of the terminal lobe of the Congo deep-sea fan. The data were acquired on a Rock-Eval® OSA device on both the bulk sediment and on the sediment rinsed twice with distilled water to remove the salts. Modified from Baudin et al. (2015). For a color version of this figure, see www.iste.co.uk/baudin/rock-eval.zip*

These bimodal S2 parameters have been interpreted by some researchers as a mixture of two organic fractions with very different thermal stabilities, one fresh and thermally labile and the other thermally resistant and probably older, or in any case more mature. Thus, based on a false interpretation of an incorrect S2 parameter, Holtvoeth et al. (2001, 2003, 2005) defined two populations of organic matter in recent marine sediments deposited off the mouth of the Congo River. These authors considered the high Tmax fraction to be due to erosion of the deep horizons of the soils of the Congo drainage basin, referring to the work of Disnar et al. (2003), which mentions that tropical ferralsols sometimes present S2 parameters with a high Tpeak[2]. However, a key difference between the data of Disnar et al. (2003) and those of Holtvoeth et al. (2001, 2003) concerns the HI, which is very low in the deep horizons of ferralsols (between 50 and 100; Disnar et al. 2003) while it averages 200 mg HC/g TOC and even reaches 400 mg HC/g TOC in recent African margin sediments analyzed by Holtvoeth et al. Such an HI value for Tmax values of 540°C is not compatible with substantially degraded organic matter. Finally, sediments from the same zone under the influence of the Congo River see the second S2 parameter disappear after being washed (Baudin et al. 2010, 2017a, 2017b; Stetten et al. 2015; Figure 4.14). It is therefore imperative to wash recent salt-rich sediment samples to avoid these analytical problems (Espitalié et al. 1985b; Baudin et al. 2015). Another good reason for this washing operation is that chlorides produce acid vapors that, under certain conditions, attack carbonates (see section 13.3) and in all cases corrode the crucibles, ovens and tubings of Rock-Eval®.

Nevertheless, washing with distilled water is not inert with respect to organic matter. Indeed, part of the latter is still water-soluble in recent sediments; this fraction is eliminated during washing of the samples. This disappearance is perceptible in the shape of the S2 parameter where a shoulder visible in the bulk sediments (= with salt) is absent from the rinsed samples (Figure 4.15). On sediments collected off Angola and containing terrigenous organic matter, this loss represents only 3.7% of the surface of the S2 parameter (Baudin et al. 2015), which remains negligible[3]. This change in the S2 parameter shape has little effect on TOC% and the HI values, but has an impact on Tmax because the water-soluble fraction removed is thermally labile. In the study by Baudin et al. (2015), an increase of 5°C is noted for Tmax. The loss of the hydrolyzable fraction also induces a slight decrease in the OI values, since this fraction is generally rich in oxygen.

---

2. It is very likely that this high Tmax value results more from a mineral matrix effect in the presence of iron oxide than from a very advanced state of transformation of soil organic matter.

3. The water-soluble fraction is likely to be more significant for algal organic matter.

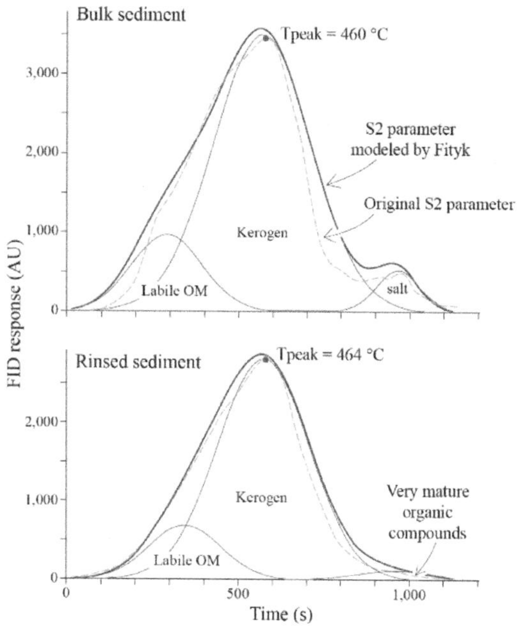

Figure 4.15. *Example of deconvolution of the S2 parameter acquired on bulk sediment and rinsed (= salt-free) sediment from the terminal lobes of the Congo deep-sea fan. The three Gaussians were obtained by deconvolution using the Fityk software (Wojdyr 2010). The first Gaussian corresponds to labile organic matter, the second to protokerogen and the third is mainly due to parasitic ionization by salts. After rinsing, this third Gaussian disappears almost entirely, while the first decreases a little due to the loss of soluble fractions of organic matter. The shoulder visible at the beginning of the S2 parameter on the bulk sediment likely corresponds to this fraction that has disappeared, since the shape of the S2 parameter is more symmetrical for the rinsed sediment (according to Baudin et al. (2015)). For a color version of this figure, see www.iste.co.uk/baudin/rock-eval.zip*

Despite the loss of the water-soluble fraction, rinsing salts increases the TOC and most other parameters by mass balance. It is likely that Rock-Eval® values measured on rinsed sediments are closer to those after lithification when porosity has become low and salt from pore water no longer has a great influence in terms of weight.

4.4. Other artifacts and abnormal signals of instrumental origin

Each sample has a particular signature before having particular parameters. It is therefore imperative to observe the curves before validating the results (Espitalié et al. 1985b; Peters 1986). A trained eye will be immediately alerted to a possible problem on the instrument or sample. Some instrumental problems, related to a maintenance defect, or wear, can indeed alter the shape of the curves produced (Figure 4.16(A)). A good laboratory practice is for the operator to conduct regular maintenance of their instrument and have a logbook for the instrument, in which the usual maintenance and complete revisions of the instrument are recorded.

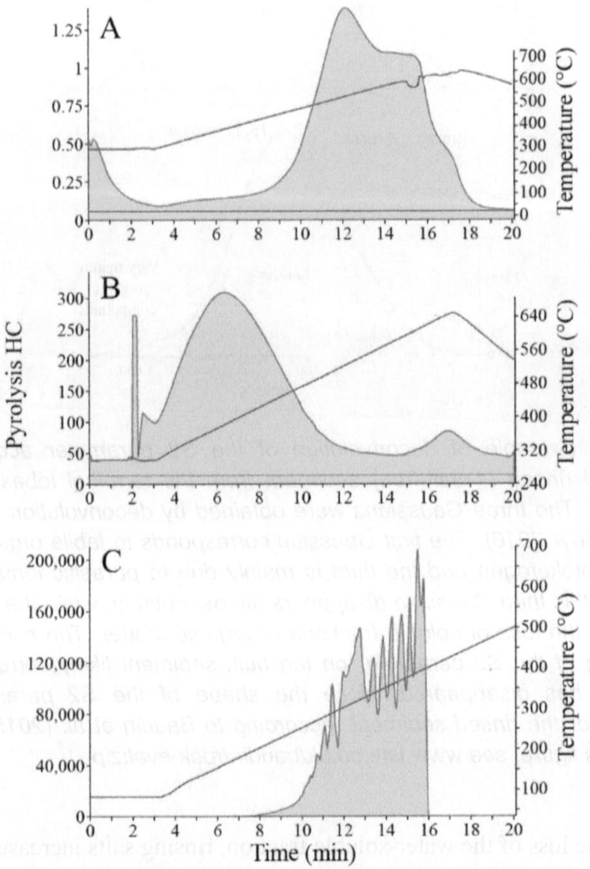

Figure 4.16. *Examples of instrumental artifacts. (A) A defective thermocouple may change the appearance of the parameter. The example here is taken from the S2 parameter during pyrolysis, but this problem can also occur on the oxidation oven. (B) Spike in the FID signal related to a particle touching the cathode of the detector. (C) Sawtooth FID signal linked to flickering of the flame, then the signal stops abruptly as the flame is blown out. For a color version of this figure, see www.iste.co.uk/baudin/rock-eval.zip*

It is also strongly recommended to regularly analyze a reference sample, every 10 analyses, for example, enabling validation of the analyses of unknown samples conducted in the meantime. This causes a little time to be wasted, yet in fact ultimately saves time if a problem is detected in an analysis of a reference sample, since it will be possible to better date the origin of the problem and therefore only re-analyze those samples for which confidence in the result is low. Similarly, it is useful to count the number of analyses carried out per crucible and to keep a new one aside for validation purposes, if a curve is abnormal or a numerical result out of specification, in the absence of any other instrumental defect.

Spikes are known in gas chromatography, and can appear when a particle escapes from the column and touches the cathode. A sudden, brief signal is recorded, which has the characteristic of starting at a right angle (Figure 4.16(B)). A spike problem on the Rock-Eval® FID can distort the area under the curve, overestimate the S2 value, and above all give a very low Tpeak value. Indeed, instead of determining the peak of a Gaussian, it is the first parameter, which is higher, that is automatically chosen to calculate Tpeak, thus giving a false Tmax value.

A sample can naturally produce a large amount of very light gases or hydrocarbons. The flow of gas arriving at the flame can then change suddenly, causing the hydrogen flame of the FID to flicker, giving a sawtooth signal (Figure 4.16(C)). When the gas flow is large and sudden, it can sometimes blow out the flame of the FID. Too much water in the sample can also extinguish the flame of the FID.

The FID signal can also be saturated without showing any particular symptoms on the parameter shape. When the signal exceeds 100,000 pA/μV, the flame is saturated with ions, and parasitic electric fields appear within the detector, which prevents proper detection of the electrons and CHO^+ ions. In order to check for possible saturation of the FID, it is possible to multiply the S2 value in mg/g by the sample weight in mg. If the value is greater than 2,000, it is recommended to repeat an analysis, reducing the weight in the crucible (Carvajal-Ortiz and Gentzis 2015).

Depending on the type of sample analyzed, and therefore its presumed richness in organic matter, it is recommended for the following weights to be respected:

– bulk rock or soil: 50–70 mg and 90 mg for Rock-Eval® 7;

– coals: 10–20 mg;

– kerogens (or TOM, or pure organic matter): 2–10 mg.

For coals and kerogens, a little Fontainebleau sand should be added (previously roasted at 850°C).

Sometimes the issue can come from the nature of the sample itself. At times during the oxidation phase, spontaneous, uncontrolled combustion occurs in the crucible. The sample ignites, resulting in unusually fine and high CO and CO_2 parameters (Figure 4.17).

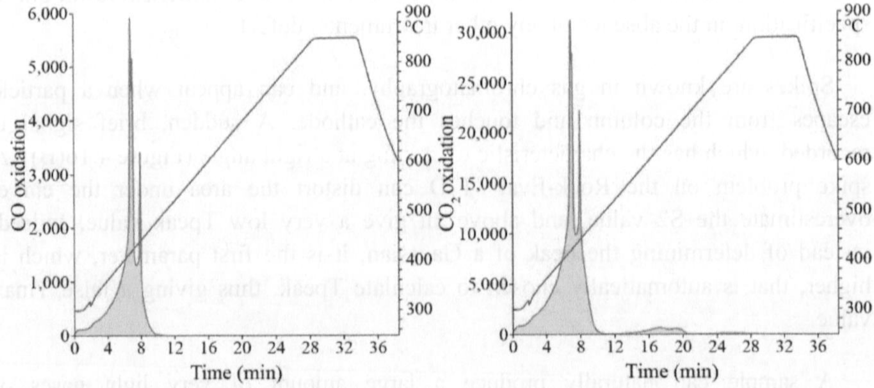

Figure 4.17. *Effect of spontaneous combustion of residual carbon during the phase of oxidation in air. CO and CO_2 parameters are released, with no effect on the RCCO and $RCCO_2$ calculations, but with errors on the Tpeak positioning. For a color version of this figure, see www.iste.co.uk/baudin/rock-eval.zip*

Finally, the signal is occasionally not correct because the choice of method is unsuitable. For example, the S1 parameter may be truncated and some of the hydrocarbons may be lost before the flame ignites (Figure 4.18). A Bulk Rock method – Basic cycle analysis starts at 300°C by default. If the sample contains a significant quantity of light oil, it escapes while the crucible is being transported by the piston into the oven, and before the flame of the FID detector ignites. It is therefore necessary to re-analyze the same sample using the "Reservoir" method, which makes it possible to start at a lower temperature (80–180°C for Rock-Eval® 6, or even 15–20°C above the ambient temperature of the laboratory with a Rock-Eval® 7) (see Chapter 3).

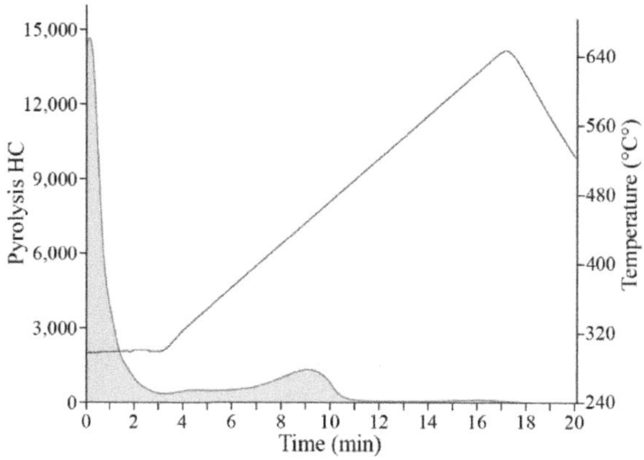

Figure 4.18. *Pyrolysis starts at excessively high temperature, resulting in a truncated FID signal. For a color version of this figure, see www.iste.co.uk/baudin/rock-eval.zip*

5

Comparison with Other Methods

Nicolas BOUTON and Jérémie AMMOUIAL
Vinci Technologies, Nanterre, France

5.1. The Rock-Eval® method at the crossroads of classical analytical methods

In the laboratory, the main instruments encountered can be categorized among the following broad families of analytical methods:

– spectroscopy, or spectrometry, was probably the first, and was used by physicists to detect atoms via their absorption and/or emission bands: when a photon (UV–visible–IR) interacts with an electron in an atom or molecule, electromagnetic interaction follows the laws of quantum physics;

– these spectroscopic analyzers are often coupled to sample vaporization systems (flame, graphite oven or plasma) to study absorption or emission lines, and trace back to the – often metallic – elements constituting the sample;

– when the wavelength of the light used in spectroscopy is within the X-ray range, the matter can be studied with X-ray fluorescence, or X-ray diffraction (XRD) techniques, for the study of minerals;

– today, spectroscopy also encompasses mass spectrometry methods, in which molecules are first ionized before they are separated by an electromagnetic field and detected based on their mass (m/z ratio, with *m* as the mass and *z* as the charge of the ion), with this technique notably enabling isotopic measurements;

– chromatography, which can be in either the gas or liquid phase, remains the most widely used separation method in laboratories, in order to study the molecules constituting a mixture; it can be coupled to mass spectrometers as detectors and to pyrolyzers as injectors;

– elemental analysis, which allows the molecular formula of a molecule to be known by giving access to the stoichiometry of the elements (carbon, hydrogen, oxygen, sulfur, nitrogen, etc.) constituting an organic molecule (e.g. butane: C_4H_{10});

– thermal analysis to study the behavior of matter as a function of temperature: thermogravimetric analysis (TGA) is a balance coupled to an oven that accurately measures the mass loss of a sample over time under the effect of temperature.

Other techniques of course exist. For example, the "loss-on-ignition" method, still used mainly in pedology, consists of measuring the weight of a sample before and after passing it through an oven, in order to quantify the carbon mass lost by combustion, therefore a priori the organic carbon mass of a soil (see Chapter 10).

The particularity of Rock-Eval® is that this instrument sits at the crossroads of thermal analysis, elemental analysis and chromatography, and, more anecdotally, uses spectrometry to detect and quantify CO and CO_2, in the infrared and SO_2 in the UV via their absorption band.

Another particularity is that this instrument is mainly found in geochemistry laboratories. Developed by geologists and geochemists for oil exploration (see Chapter 1), this analysis method is largely unknown to the other branches of chemistry in general: heavy chemicals, fine chemicals, cement works, fertilizers, petrochemicals (refining and plastics), pharmaceutical industry, cosmetics, etc. A tenuous link exists with another branch of chemistry: biochemistry, through the analysis of recent sediments and soils, and through the link that exists between oil and biomarkers.

5.2. Elemental analysis

Elemental analysis often consists of the combustion of a sample in an oven, at high temperature, with an oxidizing carrier gas (air or pure oxygen). The combustion reaction produces gases: mainly CO_2 and H_2O for all molecules containing carbon and hydrogen atoms, and also SO_2 for sulfur-containing compounds, while nitrogen forms NO_x (mainly NO and NO_2). The oxygen contained in the molecules can also be quantified, but under neutral carrier gas, by measuring the total CO of the sample (after reduction of CO_2 and CO).

The detection and quantification of CO_2, H_2O, NO_x and SO_2 makes it possible to measure the C%, H%, N% and S% content of the sample. Elemental analysis is therefore a direct measurement, and a fairly simple calibration of the instrumentation is sufficient to obtain an absolute value of the average molecular formula of the sample.

Since carbon is either of organic or inorganic origin (carbonates, see Chapter 13), it has become usual to categorize the total carbon (TC) content into TOC (Total Organic Carbon) as a weight %, on the one hand, and inorganic carbon, sometimes called TIC (Total Inorganic Carbon), as a weight %, on the other hand, with TC = TOC + TIC. For samples containing both mineral and organic carbon, it is therefore necessary to perform two successive elemental analyses:

– an elemental analysis on an aliquot of bulk sample;

– an elemental analysis on an aliquot of sample that has previously undergone an acid attack to remove the mineral fraction of the carbon.

The Rock-Eval® method is inherently different. The sample first undergoes a pyrolysis phase during which the carbon is detected in the form of hydrocarbons, on the one hand, and in the form of CO and CO_2, on the other hand (see Chapter 2). The pyrolyzable carbon (PC) is obtained by calculation in weight %. The sample remaining after pyrolysis is then burned in spontaneous and controlled combustion, which produces CO in addition to CO_2. The sum of these two signals enables the calculation of a residual organic carbon, called RC, as a weight %.

Carbonates (inorganic carbon) decompose to produce CO_2, quantified under the name MinC (Mineral Carbon).

The recombination of these different Rock-Eval® parameters makes it possible to recalculate the TC and TOC of a classical elemental analysis since TOC = PC + RC and TC = TOC + MinC. The mineral carbon is therefore quantified without the need for prior sample preparation (acid attack).

The Rock-Eval® method is therefore not, strictly speaking, an elemental analysis method, although rigorous work conducted during the development of Rock-Eval® 6 enabled the validation of the comparison with elemental analysis methods on all petroleum source rock samples (see section 3.2.2).

Metrology work carried out by IFPEN (Lamoureux-Var et al. 2019) during the development of Rock-Eval® 7 Sulfur showed that the comparison between the sulfur content measured by Rock-Eval® 7S and elemental analysis, which remains a reference method, is excellent (Figure 5.1).

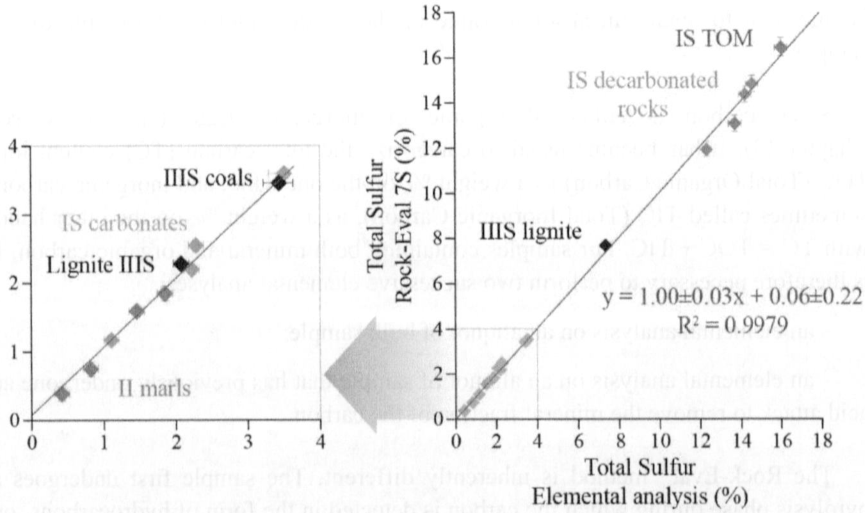

Figure 5.1. *Metrology of sulfur quantification by Rock-Eval® 7S. For a color version of this figure, see www.iste.co.uk/baudin/rock-eval.zip*

5.3. Chromatography

To a chromatography user, the Rock-Eval® instrument strongly resembles a simplified GC. The pyrolysis oven is a type of injector widely used for the analysis of solid samples, and the FID detector records Gaussian peaks: only the separation column is missing (bearing in mind that the column is at the heart of the chromatography technique). Rock-Eval® is therefore a simplified GC.

This analogy can also be made with regard to pyro-GC, which is a laboratory technique that aims to pyrolyze rocks to analyze their composition by chromatography, although this technique does not in itself allow the content of a crude oil to be simulated. Indeed, pyrolysis creates numerous unsaturated compounds (notably alkenes and alkynes) that do not exist in crude oil, as these are created in refinery units.

Moreover, the Shale Play method (Romero-Sarmiento et al. 2015) is an attempt to find a chromatography signal of an oil contained in a rock (see Chapter 6).

5.4. Thermal analysis

Thermal analysis is a laboratory analysis method that is mainly divided into two categories:

– TGA records the weight loss of a sample during a temperature ramp under either an oxidizing or neutral atmosphere: weight loss due to water loss or decomposition of matter.

– DSC (Differential Scanning Calorimetry) analysis makes it possible to record heat flows. Using thermocouples astutely placed around the crucible containing the analyzed sample, and compared to an empty crucible, the heat flows due to the reactions, which are either endothermic (water loss is revealed by an enthalpy of vaporization) or exothermic (a heat-producing combustion reaction), are recorded.

These instruments are sometimes coupled to a spectrometer, which continuously records the molecules, most often H_2O and CO_2, from the sample.

One of the most common uses of these techniques is to study chemical reactions, in order to extract a kinetic model, which makes it possible to calculate a material's lifetime, according to its storage conditions, for example.

The Rock-Eval® method enables us to study the thermal cracking reaction of kerogen during pyrolysis under laboratory conditions, in order to simulate the same reaction that takes place in nature on the geological time scale. While it does not record weight loss or heat flow, it continuously records signals (FID, CO, CO_2 and SO_2 in the case of Rock-Eval® 7S). These signals are often used by kinetic modeling software (see Chapter 7), and to study the cracking kinetics of carbonates (see Chapter 13), sulfates and sulfur-containing organic matter (see Chapter 8). Therefore, it is indeed a method of thermal analysis in this sense.

Classical thermal methods are very effective in describing the decomposition of matter, but are not very descriptive of chemical reactions. The separation methods (chromatography) coupled to pyrolyzers may, on the contrary, be too rich in chemical information and too poor in thermal information, because the separation phase is too long compared to the pyrolysis phase, and it becomes complex to reconstruct the history of the decomposition of complex matter. As Rock-Eval® lies between these two extremes, it provides chemical information that is relatively simple to interpret in real time during the pyrolysis and oxidation thermal phases, making it a success.

5.5. Comparison with other analyzers of the same type

The history of Rock-Eval® began in the 1970s (see Chapter 1), first addressing a "niche" market, that of oil exploration. The number of units sold is therefore not comparable to other instruments commonly encountered in laboratories. As Rock-Eval® was the only instrument available to solve a specific problem, and effectively so, Vinci Technologies found itself in a monopoly position, which attracted the attention of other players present in the oil geochemistry market.

Competing instruments appeared in the 1980s, such as Pyromat, followed in the 1990s–2000s by the Source-Rock Analyser (SRA), which was developed by D. Jarvie and subsequently sold by what is now Stratum Reservoir (formerly Weatherford), then the Hawk, also developed by Jarvie with Wildcat Technologies.

The major difference between Rock-Eval® and its competitors is the use of an oven dedicated to pyrolysis and another oven dedicated to oxidation. The other instruments use only one oven for obvious reasons of cost and ease of instrumental process: this first serves as a pyrolysis oven under helium or nitrogen, to obtain the parameters S1, S2 and S3, before being cooled and swept in air to start an oxidation cycle.

Some question the accuracy of certain Rock-Eval® parameters (Jarvie 2014), and while comparison tests between different analyzers have been able to be conducted, no solid publication has ever demonstrated these claims.

However, cases have been reported by clients who use both Rock-Eval® data and data from other instruments in their database in order to carry out geological studies on different sedimentary basins. It is then possible that datasets acquired with different analyzers superpose poorly in a geological study. Indeed, a small difference in S2, S3 and/or TOC can lead to significant differences in hydrogen (HI) and oxygen (OI) indices.

A real case is presented below. A total of 16 samples, which had initially been analyzed with a Hawk instrument, gave results that seemed sufficiently strange for geologists to decide to send them back for analysis with a Rock-Eval® 6, and with an elemental analyzer, for five of them, in order to obtain independent validation in case of a deviation between the results of the two pyrolyzers. Figure 5.2 shows the TOC (%) results obtained on the 16 samples: differences of up to about 3% TOC are observed on certain samples (*D* and *G*). Elemental analysis always confirms the TOC value obtained with the Rock-Eval® analyzer.

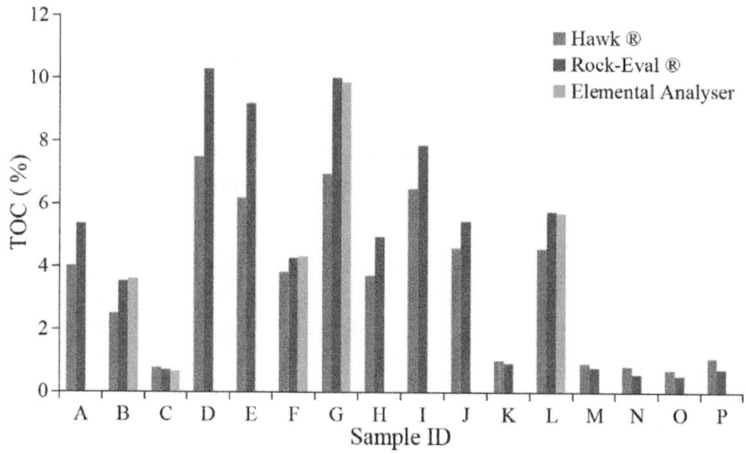

Figure 5.2. *TOC measurement of 16 oil samples analyzed on Hawk and Rock-Eval®, five of which with an elemental analysis result. For a color version of this figure, see www.iste.co.uk/baudin/rock-eval.zip*

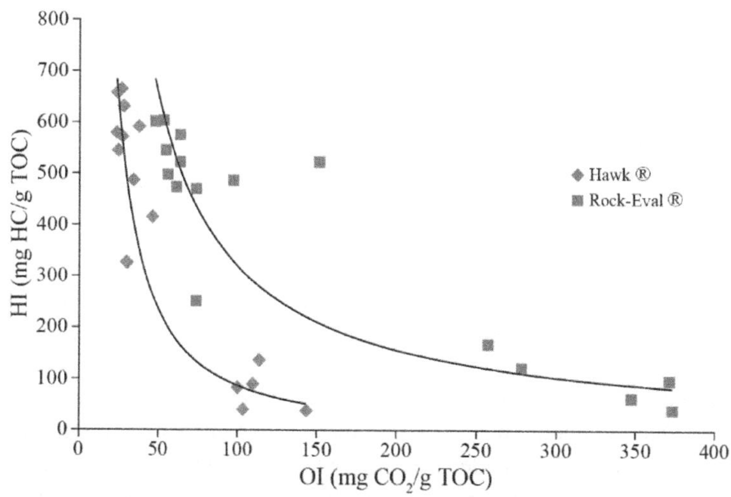

Figure 5.3. *Superposing of the HI–OI diagrams created from the analyses of the 16 samples on both instruments: small differences in the value of S2 or S3 can lead to significant differences in interpretation. For a color version of this figure, see www.iste.co.uk/baudin/rock-eval.zip*

When the HI and OI values are calculated with the TOC, very large differences are therefore observed on the van Krevelen pseudo-diagram between the two analyzers for the 16 samples (Figure 5.3).

Figure 5.4 shows that the Tmax values also differed slightly between the two instruments, enough for the points to be arranged randomly in the HI–Tmax diagram as a first approximation.

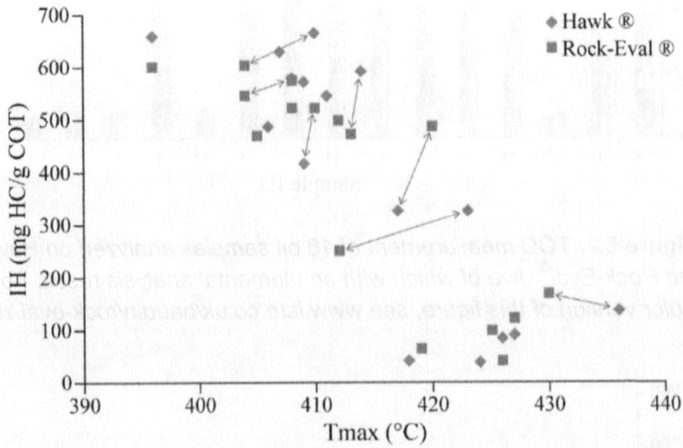

Figure 5.4. *Superposing of the HI–Tmax diagrams drawn from the analyses of the 16 samples on both instruments: small differences in the value of S2 or Tmax can lead to significant differences in interpretation. For a color version of this figure, see www.iste.co.uk/baudin/rock-eval.zip*

An internal study at Vinci Technologies focused on exploring the criticisms leveled at Rock-Eval®: the accuracy of the S1 parameter, the use of two ovens, etc.

The technical choice to use two separate ovens in Rock-Eval® was retained in order to be able to conduct pyrolysis in a more stable environment than if the oven had been used alternately under an inert and oxidizing atmosphere during each analysis. Indeed, during the pyrolysis of organic matter, some of the pyrolyzed products (mainly heavy hydrocarbons and coke) does not reach the detector and is deposited on the inner walls of the oven (for two-thirds of the deposits), as well as in the splitter (for one-third of the deposits). The quantity of deposits formed depends on the average molecular weight of the vaporized pyrolytic product and consequently on the molecular weight of the organic matter at the origin of their production during thermal cracking.

Figure 5.5 shows that coals and asphaltenes deposit more matter on the inner walls of the pyrolysis oven than rocks, with equal quantities of pyrolyzed organic matter. Figure 5.6 confirms this observation with the analysis of oils ranging from the lightest to the heaviest hydrocarbons. By extrapolation, if we hypothesize that pyrolysis would only generate methane, this would be efficiently transported to the FID and no deposit would appear inside the instrument. In contrast, if pyrolysis generates heavy, polyaromatic hydrocarbons, logically part of them would be poorly vaporized and transported by the carrier gas to the FID, and part of them would be deposited on the walls.

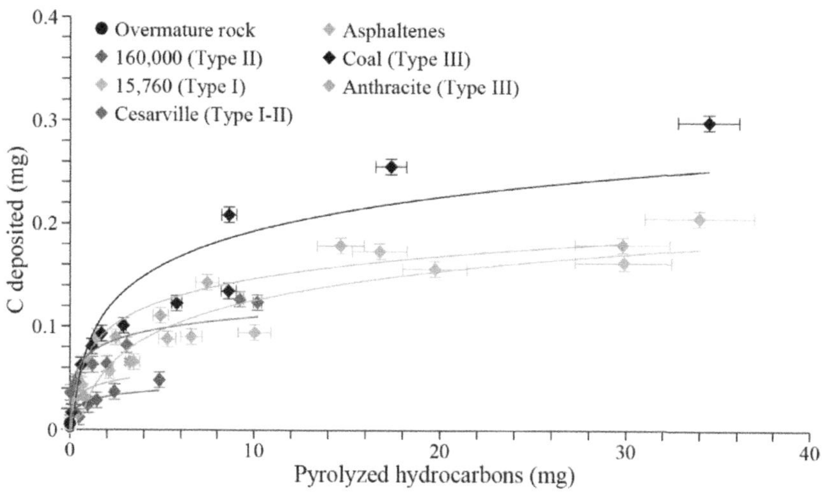

Figure 5.5. *Measurement of deposits on the inner walls of the pyrolysis oven, resulting from pyrolyses of different types of organic matter. For a color version of this figure, see www.iste.co.uk/baudin/rock-eval.zip*

If air enters the pyrolysis oven between two analyses, it will oxidize these deposits, and during the next analysis the S2 parameter will then be slightly underestimated as part of the hydrocarbons produced will be deposited again on the oven, as shown in Figure 5.7.

In conclusion, carrying out pyrolysis and oxidation successively in the same oven would not allow the preservation of the deposits naturally and inevitably produced by pyrolysis. As a result, the quantity of hydrocarbons detected by the FID will be underestimated. This phenomenon is well known to Rock-Eval® users, since if a device has not been in operation for a period of time, the first analyses of the reference sample will show a slight hydrocarbon deficit.

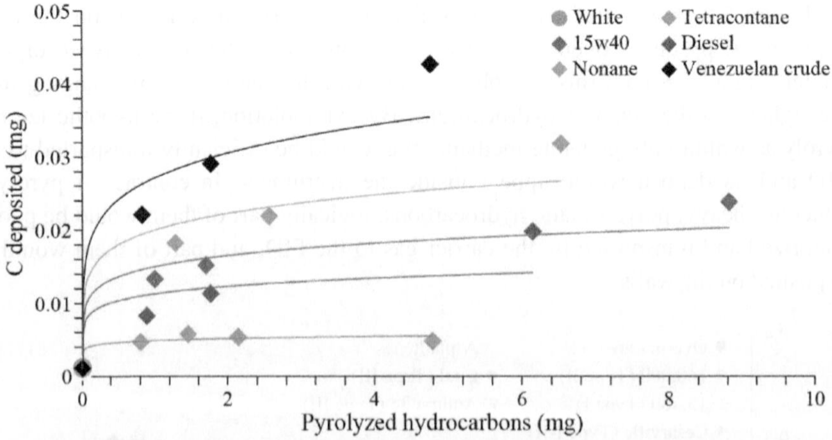

Figure 5.6. *Measurement of carbon deposits on the inner walls of the pyrolysis oven, resulting from pyrolyses of different oils (factor 10–100 with respect to Figure 5.5). For a color version of this figure, see www.iste.co.uk/baudin/rock-eval.zip*

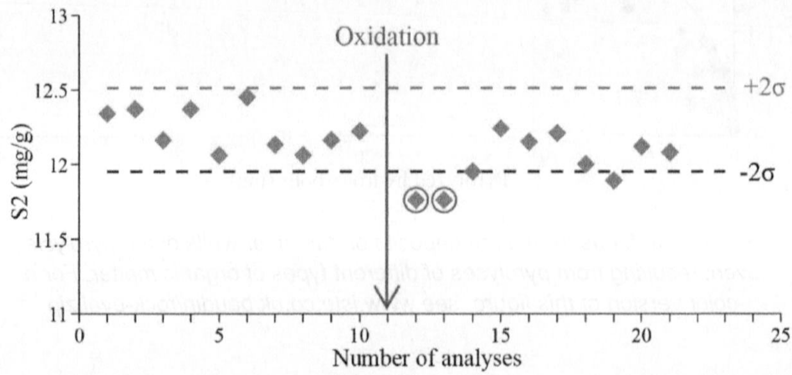

Figure 5.7. *Effect of an oxidation in a pyrolysis oven on the S2 values of subsequent pyrolyses. For a color version of this figure, see www.iste.co.uk/baudin/rock-eval.zip*

5.6. Comparison with other types of samples

The results of the Rock-Eval® method compare less well with elemental analysis for samples of recent sediments or soils. Thus, Saenger et al. (2013) showed that the Rock-Eval® TOC was about 30% too low for recent organic matter of soil origin, and while Carrie et al. (2012) showed the relevance of Rock-Eval® parameters (notably OI, HI, etc.) in distinguishing the origin of certain pure or natural organic matter (humic acids, polysaccharides, cellulose, etc.), the TOC values deviate from the actual values (see Chapter 9).

With regard to pyrolysis, the reasons for these biases stem from the non-equimolar response of the FID to compounds containing heteroatoms, in particular to oxygenated compounds, and the degradation mode of oxygenated groups that do not only produce CO and CO_2. Regarding oxidation, incomplete combustion produces unburned substances, which are therefore not detected in CO and CO_2 form.

Once the causes were identified, Vinci Technologies developed a new version of Rock-Eval® 7 that will allow the analysis of any type of material, as in elemental analysis, thus retaining the historical advantages of thermograms and their meaning (patent and publication pending).

6

Characterization of Petroleum Source Rocks

Maria-Fernanda ROMERO-SARMIENTO
IFP Energies nouvelles, Rueil-Malmaison, France

6.1. Introduction

Within the context of oil exploration and exploitation, source rocks have been extensively characterized using the Rock-Eval® technique to identify the occurrence, type and degree of thermal maturation of organic matter present in these hydrocarbon-generating rock intervals. This technique has also been used to quantify total organic carbon (TOC) and mineral carbon (MinC) content in source and reservoir rocks (Espitalié et al. 1986; Lafargue et al. 1998a; Behar et al. 2001). Applied for conventional oil exploration purposes, the S2 parameter of the classical Basic Bulk Rock and Pure Organic Matter methods (described in Chapters 2 and 3) represents the residual oil-generation potential of organic matter from a given source-rock interval, i.e. the total quantity of hydrocarbons from cracking of kerogen. When reservoir oil samples are the main objective of the study, the Rock-Eval® device also has a pyrolysis temperature program allowing better characterization of volatile components, using the "Reservoir" method. For example, it was shown that this method provided an estimate of the API oil grade (Trabelsi et al. 1994).

More recently, unconventional oil resources, in particular liquid-rich source rocks, are simultaneously considered to be exploitable source rocks and oil reservoirs. A better understanding of thermal cracking of sedimentary organic matter, and of the mechanisms of production/generation, expulsion, storage and retention of hydrocarbons in these deposits is essential for the estimation and exploration of oil and gas in place (free and/or adsorbed) present in these rocks. To this end, a new analysis and interpretation method for Rock-Eval® was developed at IFPEN: the Rock-Eval® Shale Play™ method (French patent IFPEN 14/55.009; Romero-Sarmiento et al. 2014, 2016a; Figure 6.1). This method provides relevant operational parameters dedicated to the exploration and production of these plays.

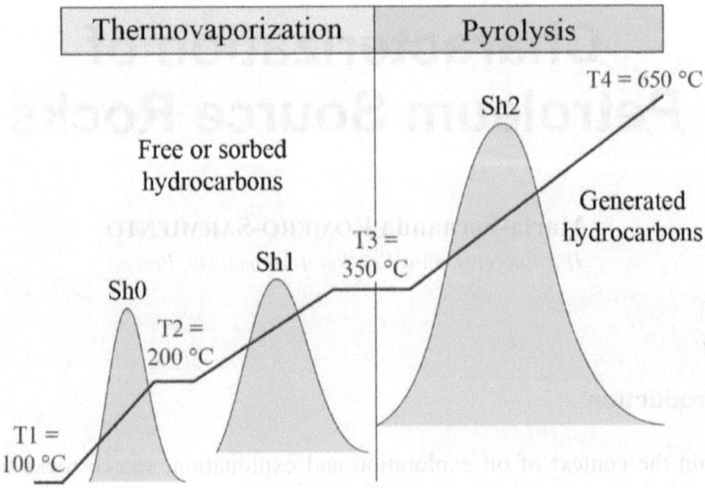

Figure 6.1. *The Rock-Eval® Shale PlayTM method developed at IFPEN (according to Romero-Sarmiento et al. (2016a)). For a color version of this figure, see www.iste.co.uk/baudin/rock-eval.zip*

The Shale Play™ method has been tested on samples of source rocks that are widely recognized as potential plays for unconventional resources, including: Silurian Shale (Algeria), Mississippian Barnett Shale (USA), Cretaceous Vaca Muerta (Argentina), Eagle Ford Formation (USA), Early Jurassic of the Paris Basin (France), Late Jurassic Bazhenov Shale (Russia) or the Eocene Green River Shale (USA), all at different stages of thermal maturity. Table 6.1 summarizes the fields of application of different Rock-Eval® methods for the characterization of various samples from petroleum systems.

Methods	Parameters	Applications
Basic Bulk Rock	S1, S2	Conventional source rocks
Pure Organic Matter	S1, S2	Isolated organic matter – kerogens
Reservoir	S1r, S2a, S2b	Oil samples
Shale Play	Sh0, Sh1, Sh2	Hydrocarbons from source rocks

Table 6.1. *Fields of application of existing Rock-Eval® methods in the oil industry*

6.2. Characterization of conventional oil source rocks and isolated organic matter (kerogens)

The Rock-Eval® S2, Tmax, HI, OI, TOC and PI parameters provided by the classical Basic Bulk Rock method have been widely used to estimate the type, origin, thermal maturity and oil potential of a conventional source rock. These parameters have also been applied for the geochemical characterization of samples of isolated organic matter concentrates and kerogens from source rocks using the "Pure Organic Matter" method (Behar et al. 2001 – see Chapter 3).

In the oil industry, the evolution of sedimentary organic matter during burial is mainly followed either by the modified van Krevelen diagram, which presents the HI versus OI parameters, or by the HI versus Tmax diagram. These two diagrams allow rapid identification of the three main types of kerogens and their corresponding thermal maturity levels. Figure 6.2 presents an example of these two classical diagrams.

In general, a Type I kerogen is very hydrogen rich. It is characterized by HI values greater than 700 mg HC/g TOC. Type II kerogen is moderately hydrogen-rich. It is therefore characterized by HI values between approximately 200 and 700 mg HC/g TOC. Finally, Type III kerogen is very hydrogen-poor with HI values below 200 mg HC/g TOC. Based on the Tmax parameter (Figure 6.2), it is globally accepted that immature source rocks are characterized by Tmax values below 430°C. Organic matter in the oil window is characterized by Tmax values between 430 and 460°C, while organic matter in the gas window presents Tmax values between 460 and 500°C. Regarding the oxygen index (OI), slightly oxidized organic matter is considered when the OI values are less than 50 mg CO_2/g TOC. In Figure 6.2, the Rock-Eval® results (HI, OI and Tmax) acquired on samples from three different wells indicate a typical evolution of Type II organic matter as a function of thermal maturity and probably depth from the oil window to the beginning of the gas window. This type of interpretation depends largely on the geological context of the samples analyzed and the integration of these Rock-Eval® data with other geochemical parameters available for a specific case study.

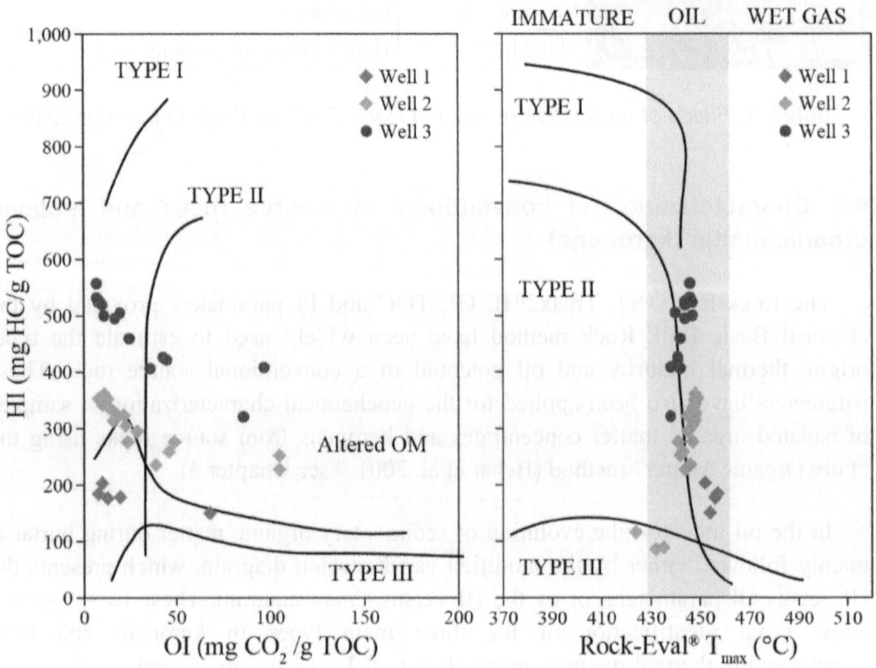

Figure 6.2. *Examples of modified van Krevelen diagrams HI versus OI and HI versus Tmax (according to Romero-Sarmiento, IFPEN internal data). For a color version of this figure, see www.iste.co.uk/baudin/rock-eval.zip*

To improve the evaluation of the qualitative and quantitative oil potential of source rocks, geochemical characterizations on kerogen are sometimes necessary. In source rocks, kerogens are closely associated with a mineral matrix. The results of Rock-Eval® analyses classically used to evaluate this oil potential can sometimes be influenced by the mineral matrix (see Chapter 4). To avoid these effects related to the mineral matrix, it should be separated from the kerogen. For this purpose, the method described by Durand and Nicaise (1980) makes it possible to destroy the mineral matrix of a source rock by HCl/HF hot acid attack while preserving the kerogen. The type and oil potential of kerogen can then be estimated using the "Pure

Organic Matter" method. Figure 6.3 shows a comparison of the OI values obtained on rocks and their corresponding kerogens before and after acid attacks in order to illustrate the effects of the mineral matrix on Rock-Eval® analyses. The OI values measured on kerogens confirm that the samples analyzed mainly contain Type I organic matter.

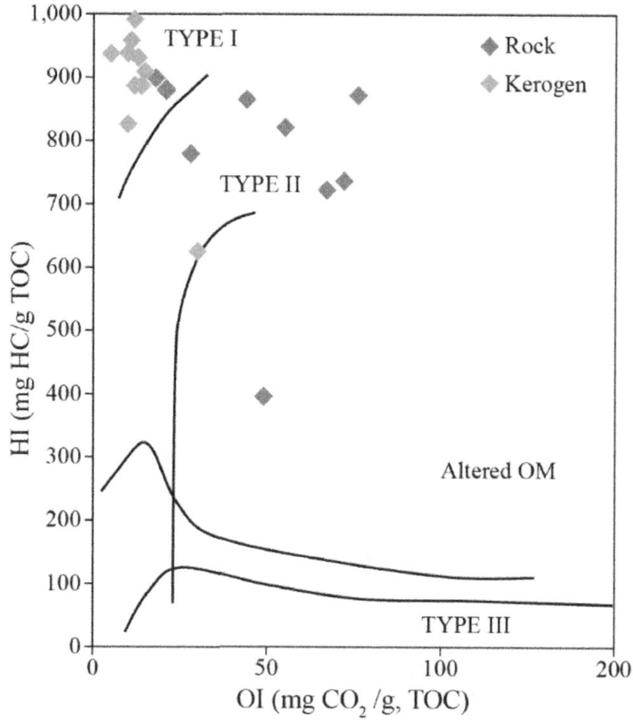

Figure 6.3. *Comparison of OIs obtained on rocks and their corresponding kerogens before and after acid attacks in order to illustrate the effects of the mineral matrix on Rock-Eval® analyses (according to Romero-Sarmiento, IFPEN internal data). For a color version of this figure, see www.iste.co.uk/baudin/rock-eval.zip*

Finally, to define the potential areas of oil interest in a source-rock interval, the Rock-Eval® S2, Tmax, HI, TOC and PI parameters are generally represented along a profile called a geochemical log. Figure 6.4 shows the evolution of sedimentary organic matter as a function of depth. A potential prospective area needs to present a set of parameters conducive to oil exploitation, i.e., for example, Tmax values in the oil window indicate that the source rock has generated hydrocarbons; a total organic carbon (TOC) content, as a minimum, of more than 2%; a production index (PI) and an oil potential (S2 parameter) greater than 25% and 30 mg/g, respectively, are desirable values in order to better target these areas (Figure 6.4).

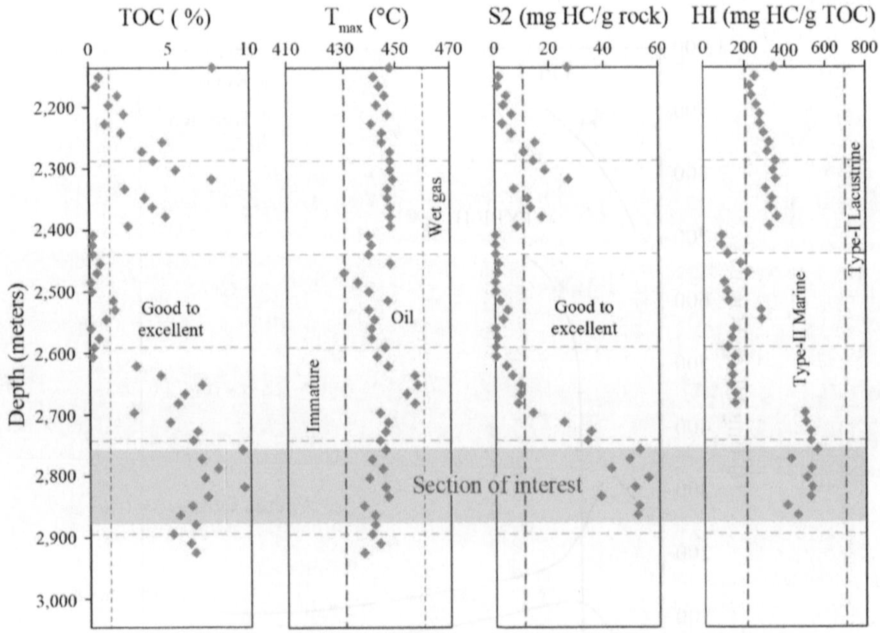

Figure 6.4. *Example of a Rock-Eval® profile as a function of depth (according to Romero-Sarmiento, IFPEN internal data). For a color version of this figure, see www.iste.co.uk/baudin/rock-eval.zip*

6.3. Characterization of oil samples

For an overall understanding of the petroleum system, in order to better characterize oil samples from reservoir rock intervals, the Rock-Eval® "Reservoir" method was also developed with the aim of specifically quantifying the oil content with a heating program that begins at a lower temperature (180°C). This method makes it possible to verify whether the oil analyzed corresponds to a conventional light or heavy oil by looking at the distribution of the Rock-Eval® S1r, S2a and S2b parameters (Figure 6.5). It has also been demonstrated that these parameters can be used to estimate API oil degrees by applying the formula proposed by Trabelsi et al. (1994).

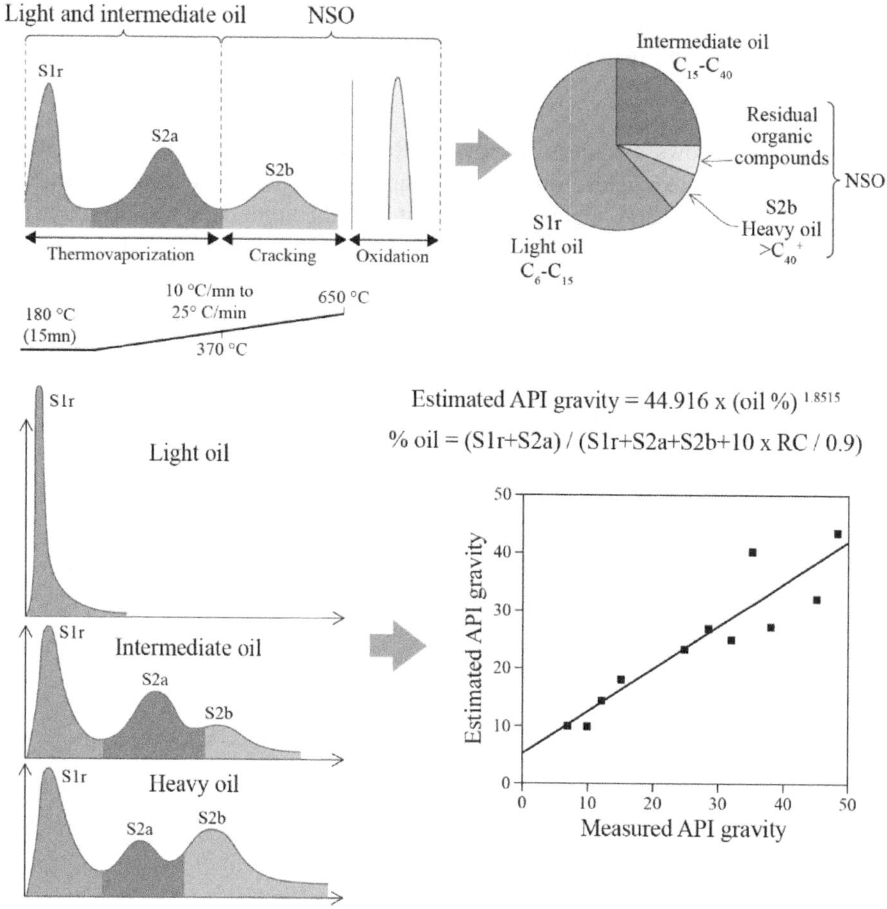

Figure 6.5. *Application of the Rock-Eval® "Reservoir" method for characterization of reservoir oil samples (according to Trabelsi et al. (1994)). For a color version of this figure, see www.iste.co.uk/baudin/rock-eval.zip*

6.4. Characterization of hydrocarbons from source rocks

It has been demonstrated that the classical Rock-Eval® Basic Bulk Rock method is not always well suited to oil-impregnated source-rock samples because, on the one hand, the deconvolution of the S2 parameter does not allow reliable determination of the Tmax parameter and, on the other hand, part of the free and/or retained light hydrocarbons (S1 parameter) is clearly lost (Figure 6.6; Romero-Sarmiento et al. 2014a, 2016a). In order to correctly determine the thermal maturity of the kerogens of these unconventional source rocks, it was therefore necessary to remove these liquid hydrocarbons in situ by extracting them with organic solvents, and then perform another Rock-Eval® analysis on the sample of extracted rock.

Figure 6.6 shows a comparison of the Rock-Eval® Basic Bulk Rock method pyrograms generated before and after organic extraction. The free or retained hydrocarbons (gas and light oil) present in the rock (S1 parameter) are underestimated because the thermovaporization stage begins at too high a temperature (300°C). With the Shale Play™ method, in contrast, the lower initial pyrolysis start temperature (T1 = 100°C) is clearly more appropriate to optimize the recovery of the majority of volatile hydrocarbons (Figure 6.6). The thermal maturity (Tmax) obtained with the Shale Play™ method on the sample of liquid-rich source rock is very consistent with the Tmax parameter obtained using the classical Basic Bulk Rock method, applied to the extracted rock. The Rock-Eval® Shale Play™ method therefore allows better determination of the quantity of oil in place (Sh0 and Sh1 parameters) and avoids organic rock extractions prior to performing a Rock-Eval® analysis (Figure 6.6).

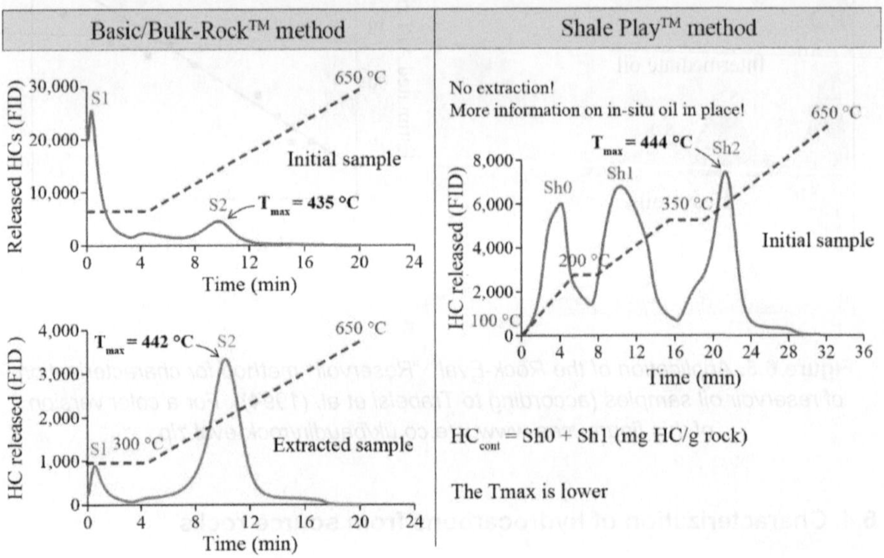

Figure 6.6. *Comparison of Rock-Eval® Basic Bulk Rock method pyrograms generated before and after organic extraction of a sample of liquid-rich source rock versus the pyrogram obtained with the Rock-Eval® Shale-Play™ method (according to Romero-Sarmiento et al. (2016a)). For a color version of this figure, see www.iste.co.uk/baudin/rock-eval.zip*

6.5. Composition of hydrocarbons detected by Rock-Eval® pyrolysis

Various analytical methodologies have been applied to determine the composition of hydrocarbons detected during the pyrolysis cycle of the Rock-Eval® technique. For example, in order to evaluate the type of compounds detected for each Rock-Eval® Shale Play™ parameter (Sh0, Sh1 and Sh2), an organic extraction procedure followed by Rock-Eval® analyses on the different fractions after extraction, was proposed by Romero-Sarmiento et al. (2016a, 2016b). The powdered rock sample is first extracted with *n*-pentane (*n*-C5) to recover the *n*-C5 extract and the residue of extracted rock. This is then extracted with the solvent dichloromethane (DCM) in order to ultimately recover the DCM extract and the corresponding sample of extracted rock, i.e. the final solid residue of two successive extractions (*n*-C5 and DCM). The *n*-C5 and DCM extracts, as well as the sample of extracted rock are then analyzed by the Rock-Eval® Shale-Play™ method. Today, it is widely accepted that the *n*-C5 extract contains saturated and aromatic organic compounds and resins that are soluble in *n*-pentane, while the DCM extract mainly contains the more-polar heavy organic compounds, which include asphaltenes and resins insoluble in *n*-pentane (Behar et al. 1989, 2008). A comparison of the Rock-Eval® Shale Play™ pyrograms obtained on the following samples: (1) initial rock, (2) pentane extract, (3) DCM extract and (4) extracted rock is shown in Figure 6.7.

In qualitative terms, the *n*-C_5 extract mainly contains the light thermovaporized hydrocarbons released between 100 and 200°C (Sh0 parameter; Figure 6.7). The *n*-C_5 extract also contains a non-negligible part of thermovaporized hydrocarbons with average molecular weight, detected between 200 and 350°C (Sh1 parameter; Figure 6.3). It can be deduced that the composition of these Sh0 and Sh1 parameters correspond mainly to saturated and aromatic compounds soluble in *n*-pentane. The compounds released by the *n*-C_5 extract at a temperature greater than 350°C likely correspond to the heavier compounds such as resins that are soluble in *n*-pentane (Sh2 parameter; Figure 6.7). The DCM extract contains, in minor proportion, the saturated and aromatic compounds soluble in DCM, and the hydrocarbon products detected by the Sh2 parameter of the DCM extract correspond mainly to heavier products such as resins and asphaltenes that are not soluble in *n*-pentane (Figure 6.7). In contrast, the sample of extracted rock contains only the insoluble organic matter. The composition of the Sh2 parameter from the sample of rock extracted after the *n*-C_5 and DCM extractions corresponds to the hydrocarbon compounds resulting from the cracking of kerogen (Figure 6.7).

In order to better characterize the composition of the hydrocarbons represented by the first two Rock-Eval® Shale-Play™ parameters (Sh0 and Sh1), additional thermal desorption (TD) analyses coupled to gas chromatography (GC) and mass spectrometry (MS) were also performed on source-rock samples (Romero-Sarmiento et al. 2016b). This system also consists of a pyrolysis oven that is interfaced with a gas chromatograph (GC) and two detectors (mass spectrometer (MS) and FID). This device can be used in such a way that volatile and semi-volatile hydrocarbon components released at any stage (temperature and range) of thermal desorption and/or pyrolysis can be selectively analyzed in-line by their molecular composition via GC–MS–FID coupling. This implies that the same rock sample can be subjected to sequential analysis with thermal desorption (TD) and/or pyrolysis performed at different (generally increasing) temperature ranges, but the sample is maintained at ambient temperature between each analysis.

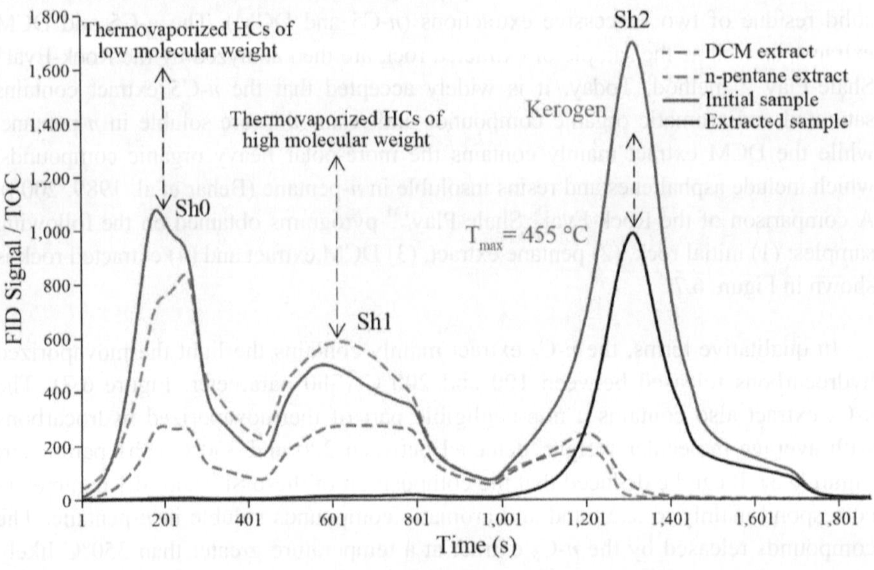

Figure 6.7. *Comparison of Rock-Eval® Shale Play™ pyrograms obtained on the following samples: (1) initial rock, (2) pentane extract, (3) DCM extract and (4) extracted rock (according to Romero-Sarmiento et al. (2016a, 2016b)). For a color version of this figure, see www.iste.co.uk/baudin/rock-eval.zip*

These analyses have already been used in previous works (e.g. Sanei et al. 2015; Jiang et al. 2016); nevertheless, the procedure described in Romero-Sarmiento et al. (2016b) proposes a modification to the thermal desorption mode in order to reproduce the temperature program of the Rock-Eval® Shale Play™ method.

As a result, TD–GC-MS–FID analyses were performed on the source-rock samples in two sequential stages, with their thermal desorption performed at different temperature ranges. The first stage of TD–GC–MS–FID was performed with a programmed thermal desorption from 100 to 200°C, with a heating rate of 25°C/min. The objective of this stage was therefore to analyze the hydrocarbon composition equivalent to the Sh0 parameters. After completion of GC analysis of the Sh0 parameters, the second stage of TD–GC–MS–FID analysis took place with a programmed thermal desorption from 200 to 350°C using a heating rate of 25°C/min in order to characterize the molecular composition of Sh1 parameters (Figure 6.8).

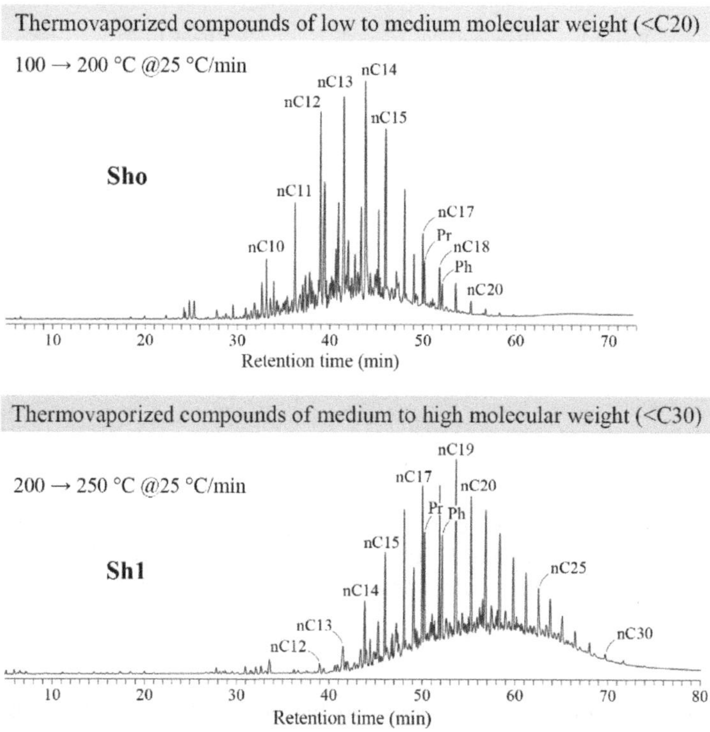

Figure 6.8. *Chromatograms showing the hydrocarbon compounds detected by the temperature intervals corresponding to the Rock-Eval® Shale Play™ Sh0 and Sh1 parameters (according to Romero-Sarmiento et al. (2016b)). For a color version of this figure, see www.iste.co.uk/baudin/rock-eval.zip*

Depending on the thermal maturity of the source-rock samples studied, it has been demonstrated that free and/or adsorbed aliphatic and aromatic hydrocarbon compounds of low to medium molecular weight (fewer than 20 carbon atoms, $<C_{20}$) are detected during the temperature interval corresponding to the Rock-Eval® Shale Play™ Sh0 parameter (Figure 6.8; Romero-Sarmiento et al. 2016b). On the contrary,

the aliphatic and aromatic hydrocarbon compounds between C_{10} and C_{30} are thermally released during the temperature sequence corresponding to the Rock-Eval® Shale-Play™ Sh1 parameter (Figure 6.8; Romero-Sarmiento et al. 2016b). More work is still needed in order to understand the proportion of light oils versus the heaviest oils detected by the two Sh0 and Sh1 parameters, and their relationship to the actual exploitable-oil production data at the basin scale.

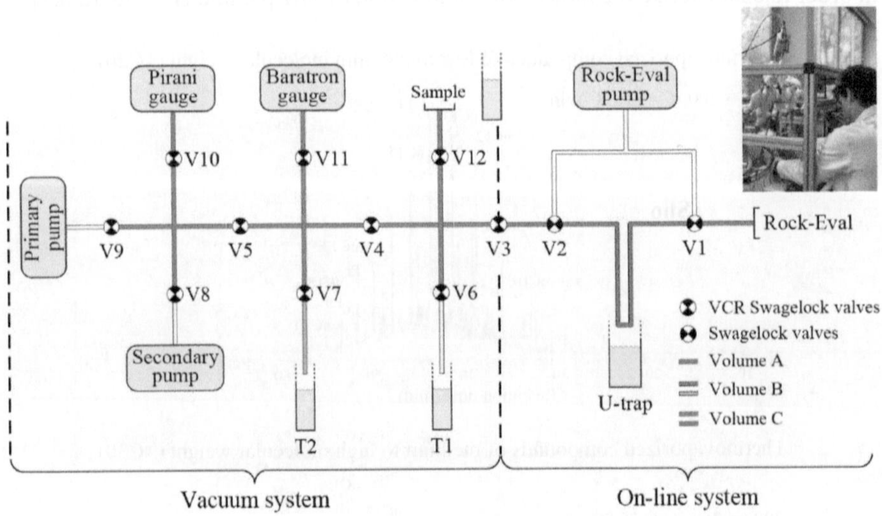

Figure 6.9. *Recovery line of the fraction of compounds obtained at the outlet of the Rock-Eval® device developed at IFPEN (according to Romero-Sarmiento et al. (2019); Cadeau et al. (2020)). For a color version of this figure, see www.iste.co.uk/baudin/rock-eval.zip*

Finally, it has thus been demonstrated that an analytical methodology is still necessary to complete the detection, recovery and quantification of chemical compounds obtained following a Rock-Eval® analysis. To this end, a line for the recovery and analysis of the fraction of compounds obtained at the outlet of the Rock-Eval® tool (during the pyrolysis and/or oxidation cycle) has recently been developed at IFPEN for a better estimation of the chemical composition within solid samples (e.g. source rocks, reservoirs, soils and recent sediments) and liquid samples (e.g. oils and lubricants), among others (Romero-Sarmiento et al. 2019; Cadeau et al. 2020). The system is composed of a first experimental in-line assembly to collect the compounds resulting from a Rock-Eval® analysis programmed according to a predefined heating sequence, and a second experimental vacuum assembly to separate and purify the liquid and/or gaseous phases contained in the recovered gas flow. The chemical composition of the compounds contained in this fraction recovered at the Rock-Eval® outlet is then determined by any type of analyzer placed downstream, such as a gas chromatograph (GC), a gas chromatograph coupled to a mass spectrometer (GC–MS) or a gas chromatograph coupled to an isotope-ratio mass spectrometer (GC–IR–MS), among others (Romero-Sarmiento et al. 2019; Cadeau et al. 2020; Figure 6.9).

6.6. Estimation of hydrocarbon reserves in place

Today, the exploration and production branches of oil companies have the mission of discovering and developing oil and gas deposits in order to meet the energy demand on a global scale. In this case, it remains necessary to make a more precise estimate of the quantity of original oil in place (OOIP) in the intervals of liquid-rich source rocks. Until now, this operational parameter (OOIP) has largely been calculated using the S1 parameter obtained by the Rock-Eval® Basic Bulk Rock method (Downey et al. 2011). Nevertheless, a chart was proposed by Romero-Sarmiento et al. (2017a) in order to easily convert the Sh0+Sh1 parameters obtained by the Rock-Eval® Shale Play™ method (mg g^{-1}) into a quantity of oil in place, expressed in barrels of oil per section (OOIP; bbl oil/acre x feet). This conversion is based on the initial formula proposed by Downey et al. (2011), but replaces the classical S1 parameter with the Shale Play™ parameters, Sh0 and Sh1. It is now possible to calculate a more accurate quantity of original oil in place (OOIP) using the diagram in Figure 6.10.

In addition, a modified oil saturation index (modified OSI = Rock-Eval® Sh0 + Sh1 x 100 / TOC) was proposed by Romero-Sarmiento et al. (2016b, 2017a) to calculate a more accurate quantity of the original oil content in geological formations of source rocks in order to determine potentially exploitable oil intervals. Figure 6.11 illustrates the free and/or adsorbed hydrocarbon content (HCcont) in relation to the total organic carbon (TOC) content of the source-rock samples studied. Based on an empirical value greater than 100 mg of oil per g of TOC (Jarvie (2012) and references therein), this diagram indicates that the areas loaded with potentially productive oils are correlated with the higher TOC values, between about 4 and 10%. This particular trend likely suggests that the oil retention capacity in the source rocks is mainly controlled by the total organic carbon content (Figure 6.11). Another very useful diagram for in-place hydrocarbon estimates is shown in Figure 6.12. The modified oil saturation index (OSI) is represented as a function of the Tmax measured by the Shale Play™ method (Figure 6.12). This chart therefore makes it possible to determine whether the source rocks studied show sufficient oil saturations in order to be a good oil prospect in a sedimentary basin.

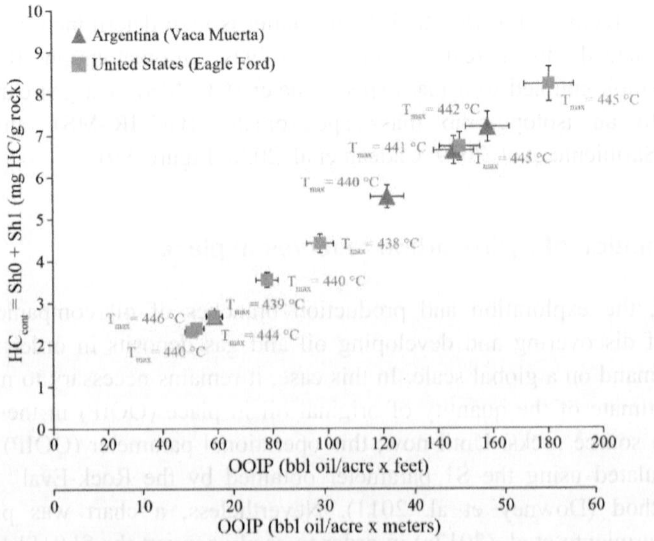

Figure 6.10. *Conversion of Sh0 and Sh1 parameters (mg g^{-1}) to the quantity of original oil in place (barrels per section) (according to Romero-Sarmiento et al. (2017a)). For a color version of this figure, see www.iste.co.uk/baudin/rock-eval.zip*

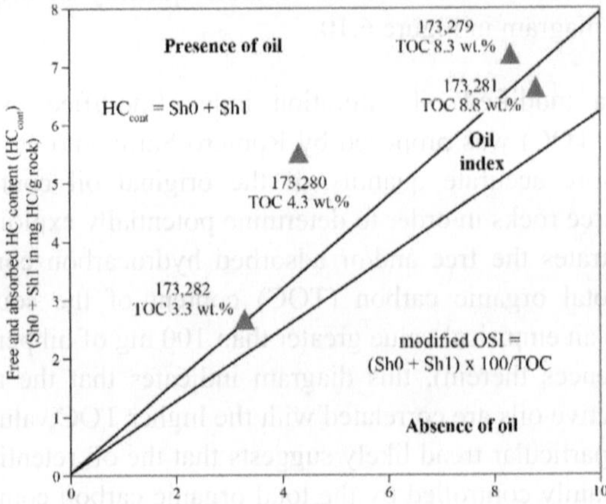

Figure 6.11. *Modified oil saturation index (OSI) showing free and adsorbed hydrocarbon content (HCcont) versus total organic carbon (TOC) content (according to Romero-Sarmiento et al. (2016b, 2017a)). For a color version of this figure, see www.iste.co.uk/baudin/rock-eval.zip*

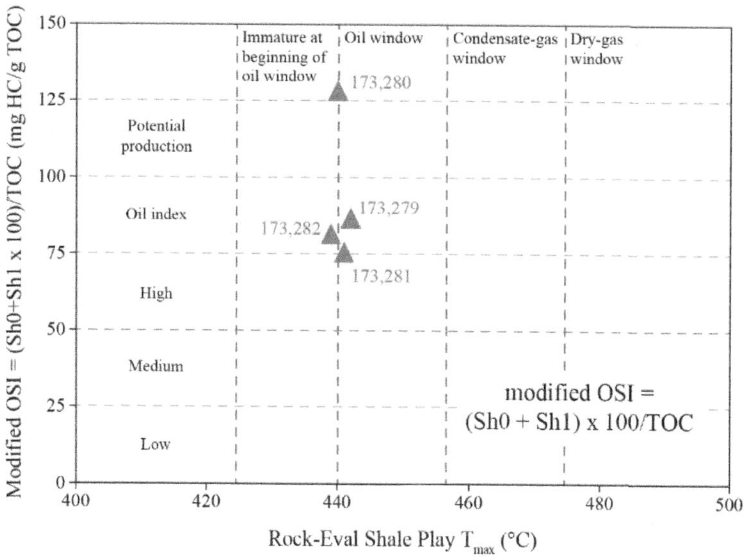

Figure 6.12. *Modified oil saturation index (OSI) versus Rock-Eval® Tmax measured by the Shale Play™ method (according to Romero-Sarmiento et al. (2017a)). For a color version of this figure, see www.iste.co.uk/baudin/rock-eval.zip*

6.7. Estimation of free hydrocarbons in porous media

The energy mix required to satisfy global needs is likely to include fossil hydrocarbons for a long time to come. Therefore, estimating their reserves, i.e. the quantity of oil and free gas present in natural porous media, remains an important issue. These free compounds present in the source-rock intervals represent the potentially recoverable oil fraction that is the easiest to extract. The adsorption phenomena within sedimentary organic matter significantly affect both this proportion and the composition and retention mechanism of oils in source rocks.

In order to better characterize this phenomenon, an analytical methodology was developed to predict the proportion of free versus adsorbed compounds. The complete analytical procedure is described in detail in the references by Romero-Sarmiento et al. (2017b) and Romero-Sarmiento (2019). It is also based on the Shale Play™ method. This involves recording the Sh0 and Sh1 parameters, i.e. the areas of the parameters measured during vaporization between 100 and 350°C. This measurement is first carried out on a sample of bulk source rock. The total sum of hydrocarbons resulting from the vaporization of the rock is considered to correspond to the free and adsorbed liquid hydrocarbons. The measurement is carried out on the organic matter isolated from the mineral matrix by means of non-oxidizing acid treatments (Durand and Nicaise 1980). Therefore, the part resulting from the vaporization of the isolated organic matter is considered to correspond solely to the adsorbed oils. The fraction of free liquid hydrocarbons is then determined from the difference between these two values (Figure 6.13).

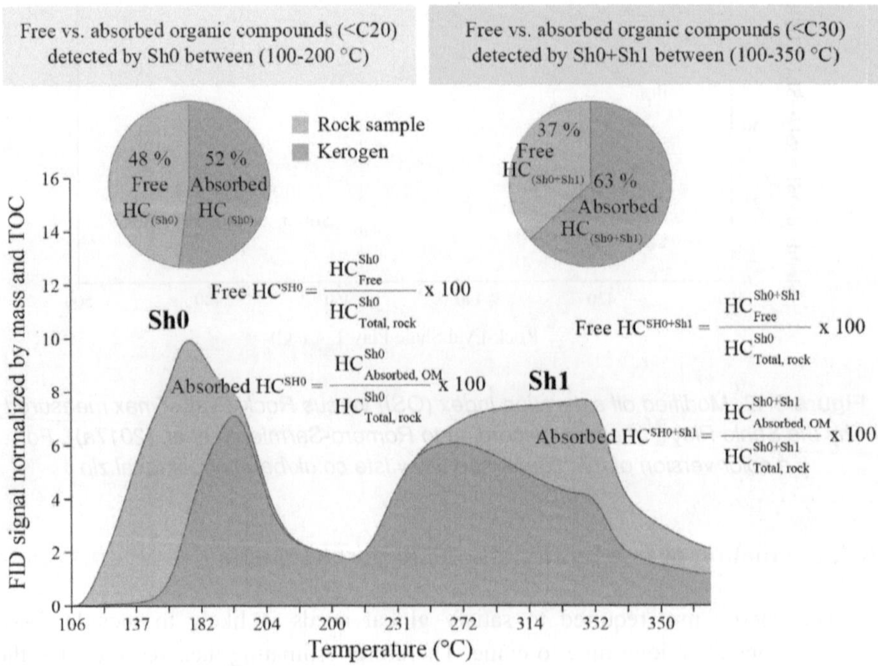

Figure 6.13. *Methodology for estimating free hydrocarbons in porous media (according to Romero-Sarmiento (2019)). For a color version of this figure, see www.iste.co.uk/baudin/rock-eval.zip*

In oil exploration, this methodology would enable the reservoir source-rock areas most suitable for oil exploitation to be identified. With adaptations to the Rock-Eval® operating conditions, such as temperature programs, analysis conditions during the pyrolysis and oxidation phases, and a possible change of detector, it will have new applications, particularly in research, for the characterization of oils and lubricants.

7

Determining the Parameters of Thermal Cracking of Fossil Organic Matter

Françoise BEHAR
TotalEnergies, Courbevoie, France

7.1. Introduction

The organic matter that is deposited in the first meters of the sedimentary column is composed partly of organic compounds that are still soluble in an aqueous medium. The humic and fulvic acids are degraded very quickly during the early stages of organic diagenesis to leave a residual organic matter insoluble in water, called protokerogen, and then kerogen at the end of diagenesis. In this chapter, the term "kerogen" refers to the organic matter that will therefore evolve during the end of diagenesis, catagenesis (or oil window) and metagenesis (or gas window).

The organic matter accumulated in the sediments undergoes an increasing burial, during which it evolves under the combined effect of temperature and time (Tissot 1969; Tissot and Welte 1984). From 70 to 80°C, this organic matter decomposes, generating essentially liquid products that can in turn undergo a thermal transformation before being expelled from the source rock. Once the majority of the products are expelled in the form of oil, the residual organic matter continues to

---

For a color version of all figures in this chapter, see www.iste.co.uk/baudin/rock-eval.zip.

The Rock-Eval Method,
coordinated by François BAUDIN.

evolve under the effect of the burial of the sediments to generate mainly gas in the form of methane. The sedimentary organic matter thus undergoes thermal cracking, the fundamental parameters of which are temperature and time. The knowledge of these two parameters defines the thermal history of the basin and makes it possible to calculate cracking rates for this sedimentary organic matter.

However, the cracking rates of kerogen cannot be determined under geological conditions, i.e. at temperatures below 200°C and time spans of the order of several million years, for the following reasons:

– Samples of source rocks covering the entire oil window are difficult or impossible to obtain. Wells are generally drilled at the top of geological structures, and therefore, the samples recovered are mostly immature or not very mature. In order to cover the oil window, samples are recovered from several geographical locations in the basin with an increased risk of mixing between different types of kerogens that do not have the same reactivity.

– The thermal history of the basin cannot be determined a priori as it results from a calibration based on assumptions and observations that are validated a posteriori. This calibration can evolve over time with the development of new geological analyses or knowledge of the basin. The uncertainties are therefore high and the results of the basin models cannot be taken as definitive to determine absolute cracking rates.

For both of these reasons, it is impossible to rely on the maturity measurements of natural samples to determine absolute thermal cracking rates for the three major sources of organic matter: lacustrine, marine and continental. Thus, the only possibility for determining the cracking rates of kerogen under geological conditions comprises the following stages:

– conduct thermal maturations in the laboratory for each of the reference kerogens, namely, of lacustrine, marine and continental origin;

– based on these experiments, determine the cracking rates under laboratory conditions;

– determine from these rates, the so-called “kinetic” cracking equations with determination of the kinetic parameters, E, activation energy in kcal/mol, and A, the frequency factor in s^{-1};

– based on these kinetic equations, calculate the cracking rates of kerogen at low temperature.

7.2. Basic equations for thermal cracking

In a cracking reaction, the compound that decomposes thermally is called the "reagent", and the compounds generated during this cracking reaction are called the "products". Cracking reactions are dependent on time and temperature. Thus, the cracking rate is expressed as follows for a given temperature:

$$-dC/dt = kC^n \qquad [7.1]$$

where C is the reagent concentration, t is the time and n is the order of the reaction.

The order, n, of the reaction is determined experimentally. The most common values are 0, 0.5, 1 and 2. Data from the literature obtained on the cracking of model compounds such as saturated and aromatic hydrocarbons have shown that for cracking reactions, the value of n is 1 (Dominé 1989; Behar and Vandenbroucke 1996; Behar et al. 1999, 2002; Lorant et al. 2000; Dartiguelongue et al. 2006; Leininger et al. 2006; Fusetti et al. 2010; Rakotoalimanana et al. 2016). Thus, the overall rate of thermal cracking can be expressed as:

$$k = \text{Log}\ (1\text{-conv})/t \qquad [7.2]$$

k = Log(1-conv)/t				
Temps h	s	conv.	1-conv	Log(1-conv)
0	0	0		0
1	3,600	0.144	0.856	-0.1555
3	10,800	0.4	0.6	-0.5108
6	21,600	0.58	0.42	-0.8675
9	32,400	0.78	0.22	-1.5141
15	54,000	0.904	0.096	-2.3434
18	64,800	0.95	0.05	-2.9957
24	86,400	0.977	0.023	-3.7723

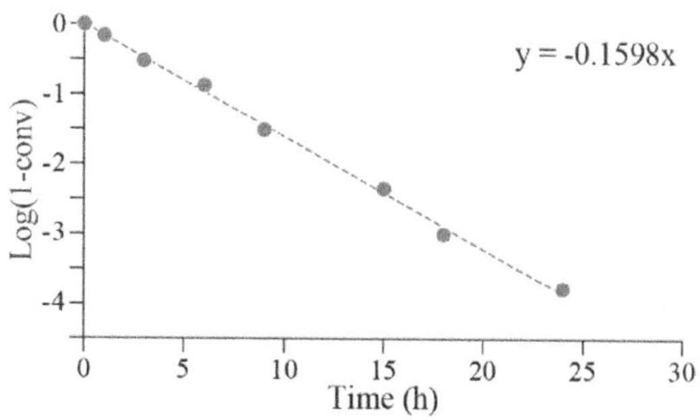

k = slope = 0.156 h^{-1} ou 4.33 × 10^{-5} s^{-1}

Figure 7.1. *Example of an order-1 diagram to determine the cracking rates of the reagent for a given temperature (400°C in the example)*

In practice, pyrolysis experiments are carried out under isothermal conditions for variable durations in order to determine at least four to five values of the conversion of the reagent for a given temperature. The experimental design is applied for the temperature interval covering the thermal decomposition of the reagent in order to verify the value of the reaction order as a function of temperature. For kerogens, the temperature interval under isothermal conditions is between 250 and 350°C, and for hydrocarbons (saturated and aromatic), it is between 300 and 450°C. The conversion values thus obtained for each temperature are plotted on a graph in which the values of Log (1-conversion) are indicated as a function of time in seconds. The slope of the line thus obtained gives the value of the overall rate of conversion of the reagent. An example is given in Figure 7.1.

Once the overall velocities have been determined, it is necessary to determine the relationship between these velocities and the temperature in order to calculate the velocities outside the experimental field here at low temperature between 80 and 200°C. In general, the relationship between cracking rates and temperature is expressed by the following equation:

$$k = A\, e^{-E/RT} \quad [7.3]$$

or

$$\text{Log } K = \text{Log } A - E/RT \quad [7.4]$$

where A is the frequency factor or Arrhenius constant (s^{-1}), E is the activation energy (Joule), R is the perfect gas constant equal to 8.314 $J.mol^{-1}.K^{-1}$ and T is the temperature (Kelvin).

Equations [7.3] and [7.4] are two expressions of the Arrhenius equation. E, the activation energy, corresponds to the quantity of energy needed to break a covalent chemical bond. Thus, the higher the value, the more difficult it is to break the bond. A, the frequency factor, corresponds to the number of collisions (per second) between molecules: the higher the value of this factor, the easier it will be to break a chemical bond. For the cracking of kerogens and of hydrocarbons, the intervals of E are between 40 and 80 kcal/mol, and for A, they are between $10^{10}\ s^{-1}$ and $10^{18}\ s^{-1}$.

To construct an Arrhenius diagram, the logarithm of the overall velocities needs to be plotted as a function of 1/RT. The slope of the line thus determined corresponds to the value of the activation energy and the y-intercept corresponds to the value of A. An example is given in Figure 7.2.

In this example, the slope equal to 26,927 corresponds to the E/R ratio.

R being equal to 8.314 J $mol^{-1}K^{-1}$, E = 26,927 x R = 223,871 J/mol.

1J being equal to 0.239006 cal, E = 223,871 x 0.239006 = 53,507 cal/mol = 53,507/1,000 kcal/mol = 53.507 kcal/mol.

Experimental results in the laboratory				
T °C	T OK	1/T	k s^{-1}	Lnk
350	623	0.0016	2.64E-06	-12.845
375	648	0.0015	1.39E-05	-11.184
400	673	0.0015	6.51E-05	-9.639
425	698	0.0014	2.74E-04	-8.201

1J	0.239006	cal
R	8.314	J mol^{-1} K^{-1}
E/R	26,927	
E	223,871	J/mol
E	53,507	cal/mol
E	53.51	kcal/mol
Log A	30.374	
A	1.55E+13	s^{-1}

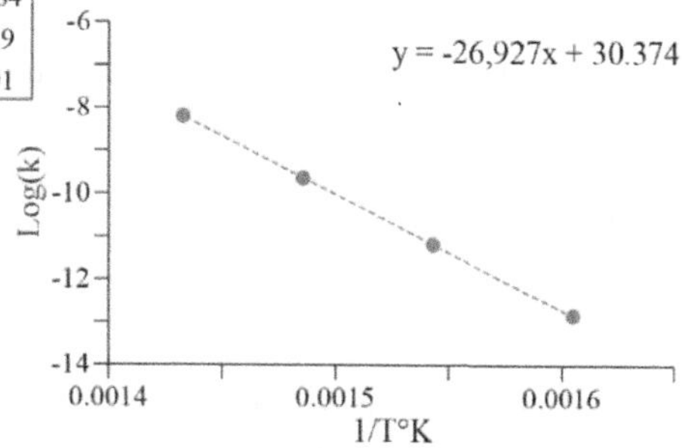

Figure 7.2. *Determination of parameters A and E from an Arrhenius diagram*

$k = A\,e^{-E/RT}$ | $Log(1\text{-}conv.) = -kt$

T °C	RT	E/RT	k s^{-1}	Ma	Year	Day	hour	second	-kt	(1-conv.) exp(-kt)	conv. %
100	741.2	72.20	6.62E-19	1	1.00E+06	3.65E+08	8.76E+09	3.15E+13	0.0000	0.999979	0.00
110	761.0	70.31	4.36E-18	1	1.00E+06	3.65E+08	8.76E+09	3.15E+13	-0.0001	0.999863	0.01
120	780.9	68.52	2.61E-17	1	1.00E+06	3.65E+08	8.76E+09	3.15E+13	-0.0008	0.999178	0.08
130	800.8	66.82	1.43E-16	1	1.00E+06	3.65E+08	8.76E+09	3.15E+13	-0.0045	0.995506	0.45
140	820.6	65.21	7.20E-16	1	1.00E+06	3.65E+08	8.76E+09	3.15E+13	-0.0227	0.977540	2.25
150	840.5	63.66	3.37E-15	1	1.00E+06	3.65E+08	8.76E+09	3.15E+13	-0.1061	0.899317	10.07
160	860.4	62.19	1.46E-14	1	1.00E+06	3.65E+08	8.76E+09	3.15E+13	-0.4617	0.630222	36.98
170	880.2	60.79	5.96E-14	1	1.00E+06	3.65E+08	8.76E+09	3.15E+13	-1.8796	0.152653	84.73
180	900.1	59.45	2.28E-13	1	1.00E+06	3.65E+08	8.76E+09	3.15E+13	-7.1922	0.000752	99.92
190	920.0	58.16	8.24E-13	1	1.00E+06	3.65E+08	8.76E+09	3.15E+13	-25.9709	0.000000	100.00
200	939.9	56.93	2.82E-12	1	1.00E+06	3.65E+08	8.76E+09	3.15E+13	-88.8248	0.000000	100.00

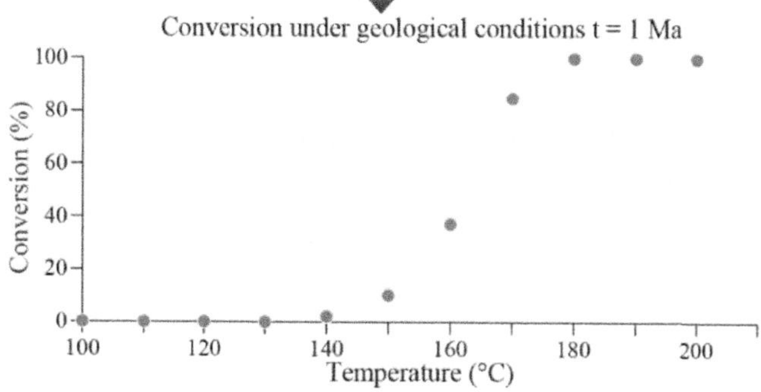

Figure 7.3. *Calculation of the conversion to geological conditions for a residence time of 1 Ma*

The y-intercept being equal to 30.374 corresponds to Log A $\rightarrow$ A = exp(30.374) = 1.55 x 10^{13} s^{-1}.

E = 53.51 kcal/mol and A = 1.55 × 10^{13} s^{-1}.

Once kinetic parameters E and A have been determined (Figure 7.2), it is then possible to calculate the cracking rates under geological conditions, i.e. for temperatures between 80 and 200°C, as illustrated in Figure 7.3 for a time of 1 million years:

$$\text{Log(1-conversion)= - k (t – to)} \qquad [7.5]$$

In summary, the kinetic parameters for calculating the cracking rates of kerogens cannot be calculated from the data determined on natural samples. They will be determined from data obtained by artificial maturation under laboratory conditions performed at a high temperature to compensate for geological times of several million years.

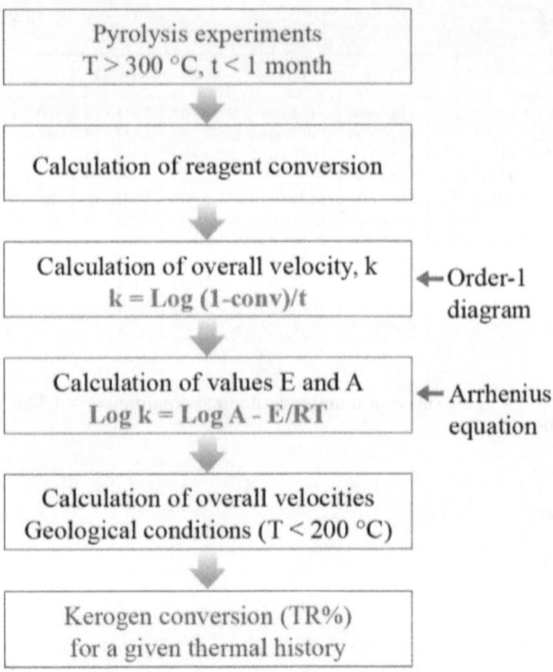

Figure 7.4. *Description of the different stages to determine the conversion of kerogens under geological conditions*

The thermal cracking rates of kerogens under laboratory conditions are determined from the order-1 diagram. Once these velocities have been determined, the kinetic parameters *E*, activation energy, and *A*, frequency factor, are determined from the Arrhenius equation. These kinetic parameters are used to calculate the cracking rates under geological conditions, i.e. for temperatures generally below 200°C. Once these velocities have been determined, it is possible to calculate the conversion of kerogens for different geological times by again using the order-1 diagrams. The complete process is summarized in Figure 7.4.

7.3. Experimental determination of kinetic parameters of kerogen

The kinetic parameters for the cracking of kerogens can be determined by carrying out pyrolysis experiments in an open and closed system, as illustrated in Figure 7.5 (Fitzgerald and van Krevelen 1959; Tissot 1969; Tissot and Espitalié 1975; Espitalié et al. 1985a; Ungerer and Pelet 1987; Burnham and Braun 1990; Ungerer 1990; Braun and Burnham 1991; Lewan 1993, 1997; Behar et al. 1997, 2001, 2003, 2008, 2010; Lafargue et al. 1998a; Lewan and Ruble 2002). Indeed, kerogens are solid products and are therefore not vaporized by the carrier gas before pyrolysis in open systems.

In an **open system**, the carrier gas evacuates the products generated to a specific detector in order to be quantified. In this system, the secondary-cracking reactions of the products generated are limited since they are evacuated from the pyrolysis chamber. This is the case for all saturated and aromatic hydrocarbons, as their boiling point allows near-complete vaporization, and therefore, a very efficient transfer to the detector. However, kerogens also generate organic structures of molecular weight high enough to be vaporized little or not at all. These products, generally containing heteroatoms such as nitrogen (N), oxygen (O) or sulfur (S) and therefore called NSO compounds, remain partly or totally in the pyrolysis chamber and therefore undergo significant secondary cracking. Among these NSO compounds, a distinction is made between resins, which may be partially vaporizable, and asphaltenes, which are solid, non-vaporizable compounds.

The general equation for cracking of kerogens in an open system is as follows:

$$\text{Kerogen} \rightarrow CO_2 + H_2S + H_2O + \text{S2 parameter} + \text{insoluble residue} \qquad [7.6]$$

In a **closed system**, the products generated remain in the pyrolysis chamber. If their thermal stability is low, they are then degraded at the same time as the reagent.

The advantage of closed-system pyrolysis experiments is that all pyrolysis products can be recovered and mass balances established. These mass balances make it possible to follow the evolution of each chemical class and establish, for each of them, stoichiometric equations that will be incorporated into an overall kinetic scheme.

A typical mass balance comprises the following families:

– in the gas fraction: all hydrocarbon (C_1 to C_4) and non-hydrocarbon (CO_2, H_2S) gases;

– in the liquid fraction: saturated and aromatic hydrocarbons, resins and asphaltenes;

– in the solid fraction: residual kerogen.

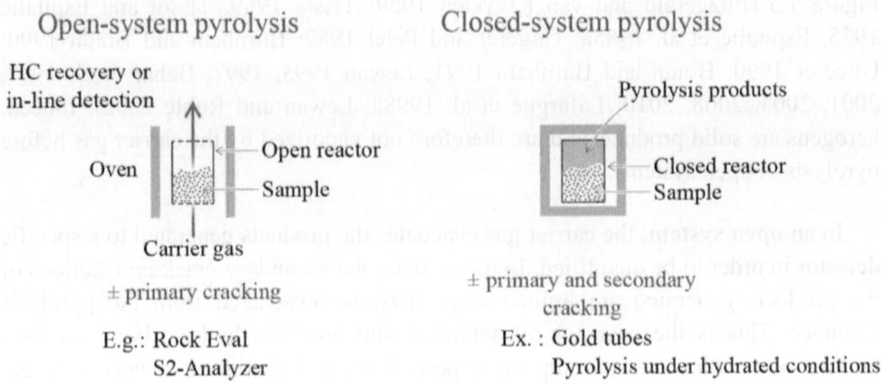

Figure 7.5. *Description of the two pyrolysis systems for simulating the cracking of kerogen under laboratory conditions*

From these mass balances, it is shown that asphaltenes are the first products generated from kerogen, associated with low production of hydrocarbons (less than 1–3%). Asphaltenes are converted both to resins and to a significant quantity of hydrocarbons (HC1) and a polyaromatic solid residue. The resins are then degraded, in turn, to generate a second quantity of hydrocarbons (HC2) and also a polyaromatic solid residue (Behar et al. 2008, 2010), as illustrated in Figure 7.6.

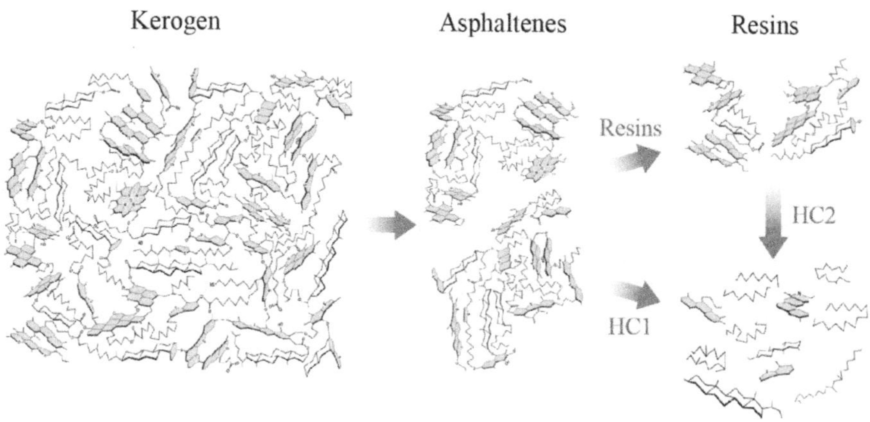

Figure 7.6. *Illustration of kerogen cracking reactions in a closed system*

The compositional scheme for the genesis of hydrocarbons is therefore as follows:

Kerogen → CO_2 + H_2S + H_2O + asphaltenes + solid residue

Asphaltenes → CO_2 + H_2S + resins + HC1 + solid residue

Resins → CO_2 + H_2S + HC2 + solid residue

with HC1 and HC2 = C_1-C_4 + C_5-C_{14} + saturated C_{14+} + aromatic C_{14+}.

Since asphaltenes and part of the resins are not vaporized out of the pyrolysis chamber in the case of the open system, these three equations can therefore be grouped into one as follows:

$$(\text{Kerogen} + \text{asphaltenes} + x\ \text{resins}) \rightarrow CO_2 + H_2S + H_2O + HC1 + yHC2 + (1-x)\ \text{resins} + \text{insoluble residue} \quad [7.7]$$

This results in the following equation:

$$\text{S2 parameter} = HC1 + yHC2 + (1-x)\ \text{resins} \quad [7.8]$$

The writing of this equation has been confirmed experimentally by recovering pyrolysis effluents swept by argon in a trap filled with liquid nitrogen (Behar et al. 1989). Thus, the argon is liquefied with the liquid effluents, the gaseous fraction being lost. In order to specifically recover the gas fraction, the same pyrolysis system is used, but connected to a vacuum extraction line, which allows the gas fraction to be quantitatively recovered, and its chemical composition to then be known through gas chromatography analysis.

The mass balances of pyrolysis effluents obtained from the combination of these two techniques for immature kerogens of lacustrine, marine and terrestrial origin are given in Figure 7.7. They show a very large proportion of vaporized NSO products with contents that can reach more than 50%. This clearly demonstrates that the Rock-Eval® S2 parameter does not contain 100% hydrocarbons and that the NSO content can reach 35–55%.

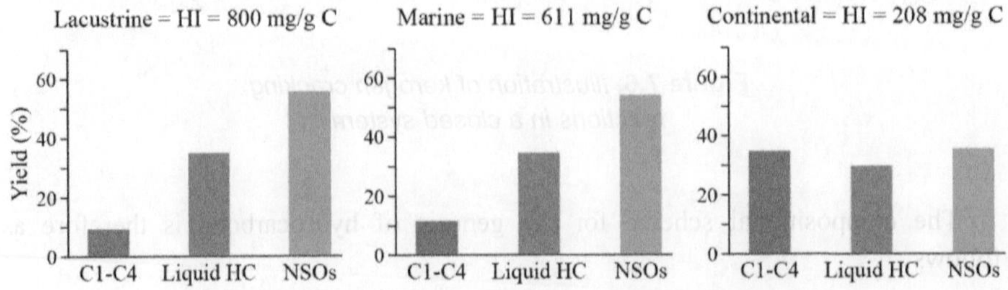

Figure 7.7. *Chemical composition of pyrolysis effluents generated from kerogens during open-system pyrolysis*

Moreover, during open- or closed-system pyrolysis, the cracking of kerogen leads to the formation of an insoluble residue very low in hydrogen in order to maintain the hydrogen balance of the overall pyrolysis equation. It has also been observed that the asphaltenes and resins that constitute the NSO products also generate a solid residue that is only slightly hydrogenated. These three sources of solid residue are therefore mixed with the initial kerogen (Behar et al. 2008; 2010). Thus, the change in the quantity of this solid residue is first reflected by a decrease due to the conversion of the initial immature kerogen, and then the addition of the new solid residue leads to a rise in the total quantity of solid residue, as shown in Figure 7.8.

For the sample of lacustrine origin, a strong decrease is observed for low maturities, which corresponds to the decrease in the initial kerogen and the low production of solid residue by kerogen and by secondary cracking of the asphaltenes and resins generated. Beyond 40% conversion, the slope of the curve increases because the secondary cracking reactions become the majority with respect to the decrease in initial kerogen. For the marine and terrestrial samples, the decrease in the reagent is much less marked due to competition of the primary cracking of kerogen and the production of solid residue either as a product generated directly from the initial kerogen or by secondary cracking of the NSO products generated.

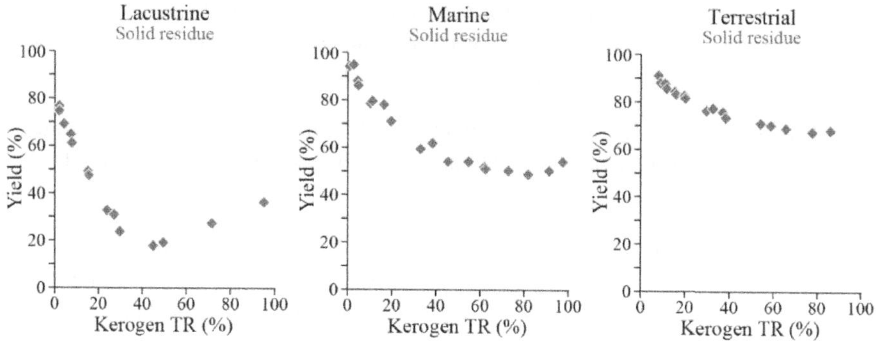

Figure 7.8. *Evolution of the curve of solid residue as a function of the conversion of initial kerogen (Behar et al. 2008, 2010)*

Thus, these results clearly show that it is not possible to define the conversion of the initial kerogen by the residual mass of solid residue. The only way to define the conversion of the initial kerogen is to follow the curves of the products generated, provided that their secondary cracking is limited. For oil exploration, one of the objectives of which is to know the mass and rate of the hydrocarbons generated, the conversion of kerogen is calculated on the generation of the Rock-Eval® S2 parameter, with the S2 measured then being the total quantity of hydrocarbons generated for 100% conversion of the initial kerogen.

7.4. Calculation of the overall kinetic parameters of kerogen in an open system

The overall rates of thermal cracking of kerogen in an open system are obtained using the Rock-Eval® technique under non-isothermal conditions, i.e. with variable heating rates: 2, 5, 10, 15 and 25°C/min between 200°C and 700°C. The experiments are carried out with a nitrogen flow rate of 100 ml/min used as carrier gas.

Before conducting these experiments, it is very important to observe the following recommendations:

– As explained in section 4.1, the mineral matrix effect of clays can very significantly affect the absolute value of the S2 parameter when TOC values are less than 4% (Espitalié et al. 1980b, 1984a; Behar et al. 2020). Thus, it is necessary either to destroy the mineral matrix (Durand and Nicaise 1980) and calculate the kinetic parameters on kerogen or to use samples whose TOC is greater than 4–5%.

– It is important to eliminate the quantity of free hydrocarbons in the kerogen in order to have a clear return to the baseline before the start of the generation of the S2 parameter.

– The initial sample quantity loaded into the Rock-Eval® crucible needs to be adjusted according to the heating velocities in order to ensure a quantitative and reproducible measurement of the S2 parameter, regardless of the heating rate. In particular, when the heating rates are low, it is necessary to increase the initial mass, as illustrated in Figure 7.9.

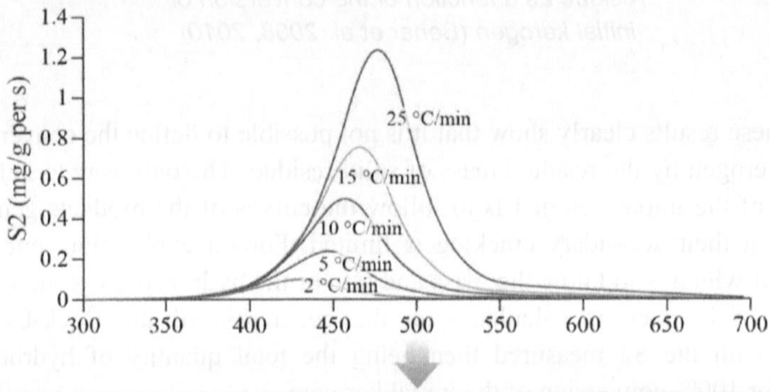

Heating ramp	25 °C/min	15 °C/min	10 °C/min	5 °C/min	2 °C/min
Sample weight	5 mg	8 mg	12 mg	25 mg	62 mg

Figure 7.9. *Influence of the heating rate on the height of the Rock-Eval® S2 parameter*

This condition was a limitation of Rock-Eval® 2. In this device, the carrier gas was not nitrogen but helium. Using this carrier gas, a systematic decrease in the value of the S2 parameter was observed for low heating rates, as shown by the values given in Table 7.1 (Behar et al. 2001).

°C/min	S2 (He)	S2 (N_2)
25	47.0	42.7
15	44.2	43.0
10	43.6	42.1
5	41.0	43.0
2	38.4	42.3
1	35.3	42.1

Table 7.1. *Estimation of the S2 parameter of a standard sample measured with Rock-Eval® 6 under helium and nitrogen for heating rates between 1 and 25°C/min (Behar et al. 2001)*

7.5. Acquisition and optimization of kinetic parameters E and A

The overall kinetic scheme for the kerogen pyrolyzed in an open system is summarized in Figure 7.10, with the equation enabling the kinetic parameters, *E* and *A*, to be derived, and therefore the kerogen transformation rates or TR (%) to be calculated. The values of the cracking rates, as well as the values of *E* and *A* are calculated according to the experimental protocol described in Figure 7.10.

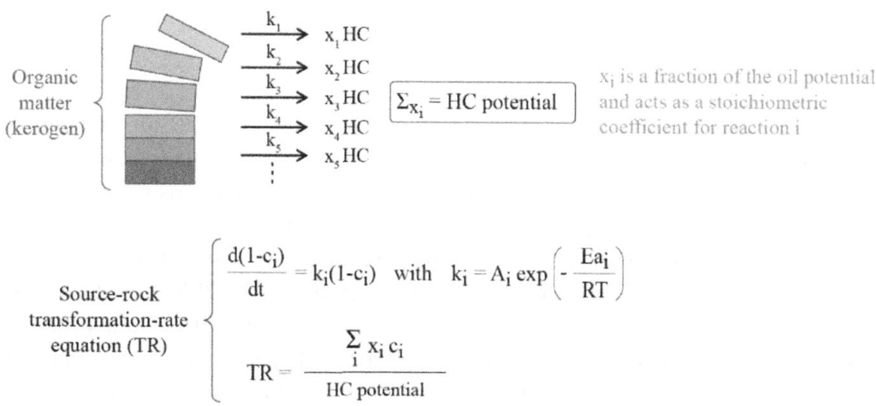

Figure 7.10. *Overall kinetic scheme for kerogen cracking and equation for calculating the kinetic parameters and transformation rates (TR%)*

Two types of calibration can be performed to calculate the values of the activation distribution, E, and the frequency factor, A, as illustrated in Figure 7.11: a Gaussian curve for the distribution of E or a discrete distribution for the distribution of E with, in both cases, a set value of A.

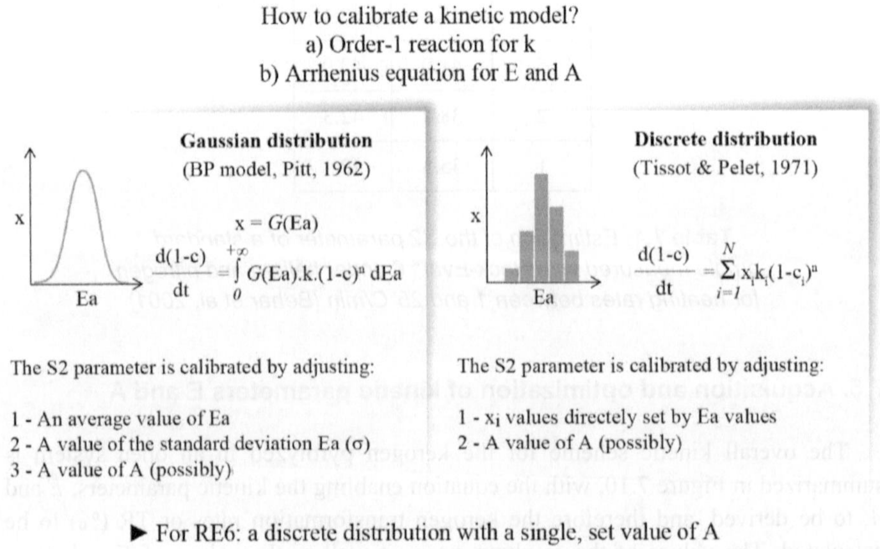

Figure 7.11. *Two methods of optimizing activation energy distributions with the value of A set for all partial reactions*

For the Rock-Eval® 6 pyrolysis technique, the values of E are determined using calibration with discrete distributions. Thus, the kinetic parameters are determined based on the hypothesis of a kinetic scheme comprising a series of parallel reactions with values of E between 40 and 80 kcal/mol, each spaced 2 kcal/mol apart. All these partial reactions have the same value of A, obtained by optimization using the Geokin Classic software developed by IFPEN or Geoworks by Vinci Technologies. The final distributions for the values of E are adjusted by comparing the S2 parameter values observed and calculated for the five heating rates selected. The protocol for optimizing these kinetic parameters is summarized in Figure 7.12.

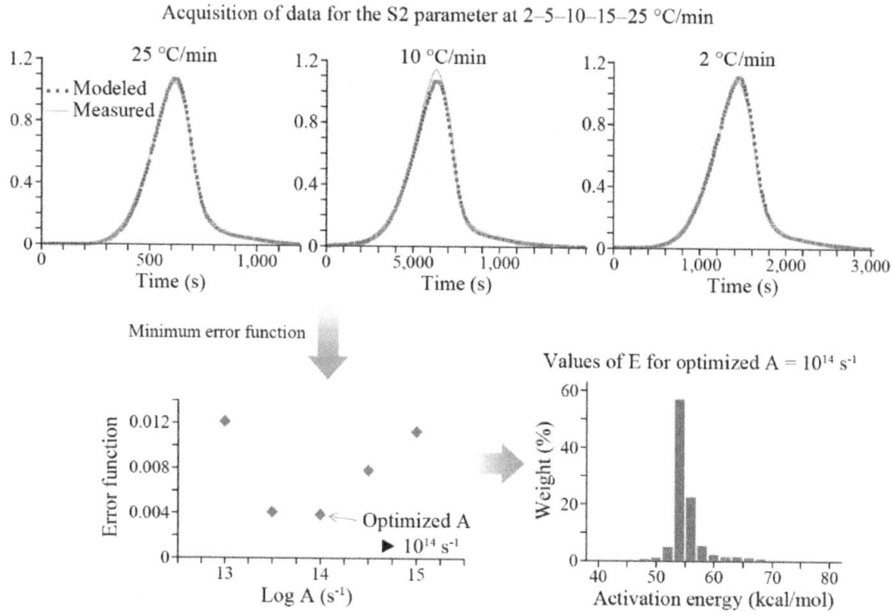

Figure 7.12. *Description of the protocol for optimizing kinetic parameters E and A using the Rock-Eval® 6 pyrolysis technique*

7.6. How are the absolute values of *E* and *A*, obtained in the laboratory, used to calculate the values of TR under geological conditions?

7.6.1. *Effect of compensation between* E *and* A *under laboratory conditions*

The values of kinetic parameters *E* and *A* determined under laboratory conditions cannot be directly used to calculate the conversion of kerogen under geological conditions, as parameters *E* and *A* have antagonistic behaviors on the calculation of the conversion. Indeed, *E* is the energy needed to break a covalent bond in a given molecule, therefore the greater the value of *E*, the more difficult it will be to break the covalent bond. In contrast, factor *A* is the number of intermolecular collisions. Thus, a high value of *E* can be "compensated" by a high value of *A*, and similarly a low value of *E* can be compensated by a low value of *A*. In other words, sets of kinetic parameters with high values of *E* and *A* may correspond to the same cracking-rate value as if obtained from a set of low E and A values. Consequently, for the same temperature, several sets of E and A will lead to the same value of k, as illustrated in Table 7.2.

E (kcal/mol)	A (s^{-1})	E/A set	T (°C)	k (s^{-1})
52	$1.95\ 10^{+12}$	1	400	$2.52\ 10^{-5}$
53	$4.11\ 10^{+12}$	2	400	$2.52\ 10^{-5}$
54	$8.70\ 10^{+12}$	3	400	$2.52\ 10^{-5}$
55	$1.84\ 10^{+13}$	4	400	$2.52\ 10^{-5}$
56	$3.87\ 10^{+13}$	5	400	$2.52\ 10^{-5}$

Table 7.2. *At the same temperature, several combinations of E and A lead to the same value of the overall cracking rate, k*

When the different sets of E and A in Table 7.2 are used to calculate the values of k under laboratory conditions, the different values of k as a function of temperature are given in Table 7.3, together with the corresponding conversion values.

In this table, the values of E and A determined at 400°C correspond to the same value of the cracking rate of $2.52 \times 10^{-5}\ s^{-1}$, as given in Table 7.2.

When these E and A datasets are used to calculate cracking rates either at lower temperatures (350 and 375°C) or higher temperatures (425 and 450°C), the results show a systematic shift in conversion values: between 0.6 and 0.8% at 375°C and between 93 and 96% at 450°C, as shown in Table 7.3. However, the observed shift remains within the margin of experimental uncertainty for the calculation of E and A. This means that under laboratory conditions, the compensation effect between E and A is high enough to predict accurate conversion values.

When this same set of kinetic parameters is used under geological conditions with temperatures below 200°C, there is no longer a compensation effect, as illustrated by the calculation of k obtained at 150°C in Table 7.4: the conversion values vary between 13 and 57%.

T	E	A	k	time lab. cond.			(1-conv.)	conv.
°C	cal/mol	(s^{-1})	(s^{-1})	h	s	-kt	exp(-kt)	%
350	52000	1.95E+12	1.11E-06	2	7200	-0.0080	0.992013	**0.8**
350	53000	4.11E+12	1.05E-06	2	7200	-0.0075	0.992493	**0.8**
350	54000	8.70E+12	9.88E-07	2	7200	-0.0071	0.992914	**0.7**
350	55000	1.84E+13	9.29E-07	2	7200	-0.0067	0.993335	**0.7**
350	56000	3.87E+13	8.73E-07	2	7200	-0.0063	0.993732	**0.6**

T	E	A	k	time lab. cond.			(1-conv.)	conv.
°C	cal/mol	(s^{-1})	(s^{-1})	h	s	-kt	exp(-kt)	%
375	52000	1.95E+12	5.63E-06	2	7200	-0.0405	0.960263	**4.0**
375	53000	4.11E+12	5.46E-06	2	7200	-0.0393	0.961455	**3.9**
375	54000	8.70E+12	5.32E-06	2	7200	-0.0383	0.962453	**3.8**
375	55000	1.84E+13	5.16E-06	2	7200	-0.0371	0.963555	**3.6**
375	56000	3.87E+13	5.00E-06	2	7200	-0.0360	0.964628	**3.5**

T	E	A	k	time lab. cond.			(1-conv.)	conv.
°C	cal/mol	(s^{-1})	(s^{-1})	h	s	-kt	exp(-kt)	%
400	52000	1.95E+12	2.52E-05	2	7200	-0.1818	0.833800	**16.6**
400	53000	4.11E+12	2.52E-05	2	7200	-0.1818	0.833800	**16.6**
400	54000	8.70E+12	2.52E-05	2	7200	-0.1818	0.833800	**16.6**
400	55000	1.84E+13	2.52E-05	2	7200	-0.1818	0.833800	**16.6**
400	56000	3.87E+13	2.52E-05	2	7200	-0.1818	0.833800	**16.6**

T	E	A	k	time lab. cond.			(1-conv.)	conv.
°C	cal/mol	(s^{-1})	(s^{-1})	h	s	-kt	exp(-kt)	%
425	52000	1.95E+12	1.02E-04	2	7200	-0.7318	0.481061	**51.9**
425	53000	4.11E+12	1.04E-04	2	7200	-0.7500	0.472382	**52.8**
425	54000	8.70E+12	1.07E-04	2	7200	-0.7719	0.462115	**53.8**
425	55000	1.84E+13	1.10E-04	2	7200	-0.7917	0.453069	**54.7**
425	56000	3.87E+13	1.13E-04	2	7200	-0.8119	0.444010	**55.6**

T	E	A	k	time lab. cond.			(1-conv.)	conv.
°C	cal/mol	(s^{-1})	(s^{-1})	h	s	-kt	exp(-kt)	%
450	52000	1.95E+12	3.72E-04	2	7200	-2.6755	0.068872	**93.1**
450	53000	4.11E+12	3.90E-04	2	7200	-2.8113	0.060127	**94.0**
450	54000	8.70E+12	4.12E-04	2	7200	-2.9667	0.051472	**94.9**
450	55000	1.84E+13	4.33E-04	2	7200	-3.1195	0.044179	**95.6**
450	56000	3.87E+13	4.56E-04	2	7200	-3.2798	0.037634	**96.2**

Table 7.3. *Compensation effect between the values of E and A under laboratory conditions*

T	E	A	k	time geol. Cond.			(1-conv.)	conv.
°C	cal/mol		(s^{-1})	Ma	s (3600)	-kt	exp(-kt)	%
150	52000	1.95E+12	2.64E-15	**10**	3.15E+14	-0.8317	0.4353	**56.47**
150	53000	4.11E+12	1.69E-15	**10**	3.15E+14	-0.5334	0.5866	**41.34**
150	54000	8.70E+12	1.09E-15	**10**	3.15E+14	-0.3436	0.7092	**29.08**
150	55000	1.84E+13	6.99E-16	**10**	3.15E+14	-0.2205	0.8021	**19.79**
150	56000	3.87E+13	4.49E-16	**10**	3.15E+14	-0.1415	0.8680	**13.20**

Table 7.4. *Derivation of the calculation of conversion between laboratory and geological conditions*

7.6.2. *Shape of the curve of the error function optimized to calculate the value of A*

Another experimental problem is the shape of the error function obtained during optimization of the kinetic parameters, as illustrated in Figure 7.13.

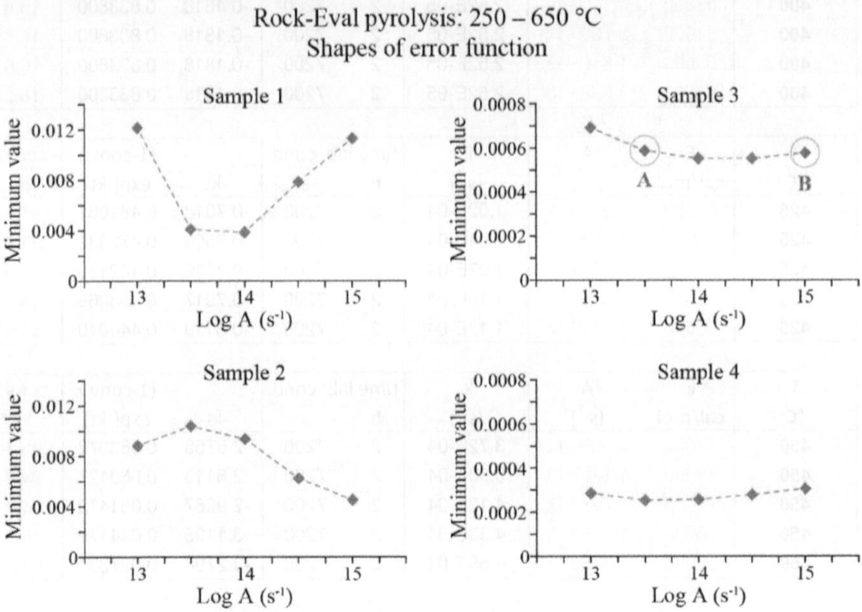

Figure 7.13. *Examples of shapes of the error function for determining the value of A*

The results show that the curves of the error function can be variable from one sample to another. For the first cases, the curve clearly shows a minimum for values of A between 13.5 and 14 for the first case, and 15 for the second case. Conversely, in cases 3 and 4, for sample no.3, the error function curve is flat, and therefore, it is impossible to favor a value of A within the range 13.5–15. The situation is even worse for case no. 4 since the minimum of the function is the same between 13 and 15. When these different sets of parameters are used to simulate the cracking of kerogen under geological conditions, the results show cracking curves that are identical under laboratory conditions and very different in geological cases, which results in an uncertainty of 13°C in temperature in the oil window (Figure 7.14). When the geothermal gradients are cold with values of 20°C/km, the error in the positioning of the beginning of the oil window will be 650 m; for a gradient of 30°C/km, the error will still be 400 m.

In general, when extrapolating cracking experiments from the laboratory to the geological case, sets of kinetic parameters with a low frequency factor and a low activation energy lead to higher cracking rates than for a set with a high frequency factor and a high activation energy.

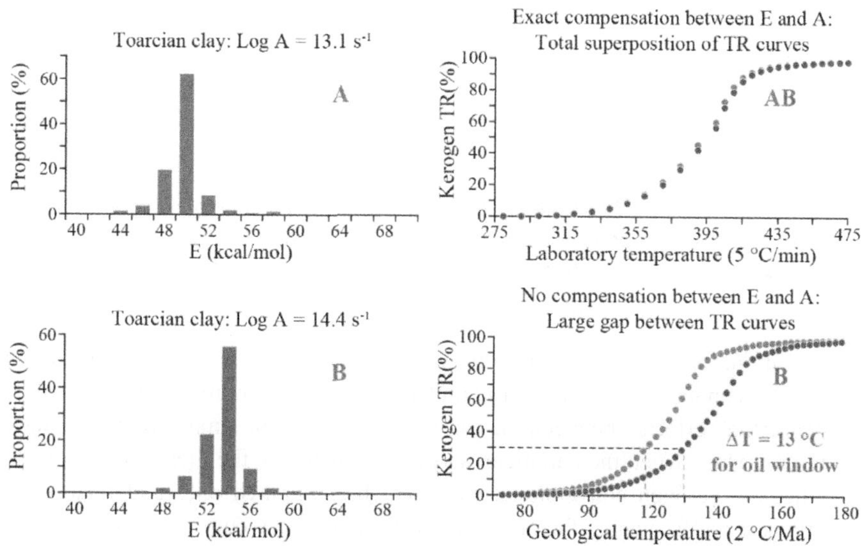

Figure 7.14. *Transformation ratio (TR) curves of a kerogen, obtained in the laboratory and under geological conditions, for which two sets of kinetic parameters were determined with the same minimum value of the error function*

The only solution to remove this uncertainty on the absolute values of the *E* and *A* pairings determined under laboratory conditions is to compare the kerogen conversion curves with other independent maturity parameters such as Tmax profiles or vitrinite reflectance values measured directly on natural samples. In the first example (Figure 7.15a), the only set that simulates the observed Tmax variations has a value of Log A at 12.9 s^{-1}, and in example 2 (Figure 7.15b), that simulating the values of the vitrinite reflectance has a Log A between 13 and 13.5 s^{-1}.

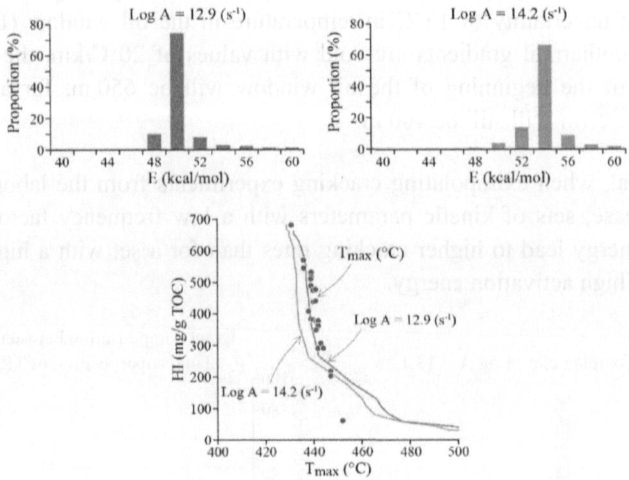

Figure 7.15a. *Example of selection of absolute kinetic parameters under geological conditions: profile of the evolution of HIs as a function of Tmax values for two sets of kinetic parameters determined in the laboratory*

In summary, kinetic parameters *E* and *A*, determined under laboratory conditions, cannot be used directly under geological conditions because of the compensation effects between *E* and *A* under high-temperature laboratory conditions. Thus, when the parameter sets are published in the literature for kerogen cracking, the kerogen conversion curves will be compared only under laboratory conditions if the frequency factors, *A*, are different. These curves can in no case be compared at low temperature, i.e. under geological conditions. The only valid comparisons for the geological case will be made with frequency factors, *A*, that are identical or very similar in the sets of kinetic parameters available.

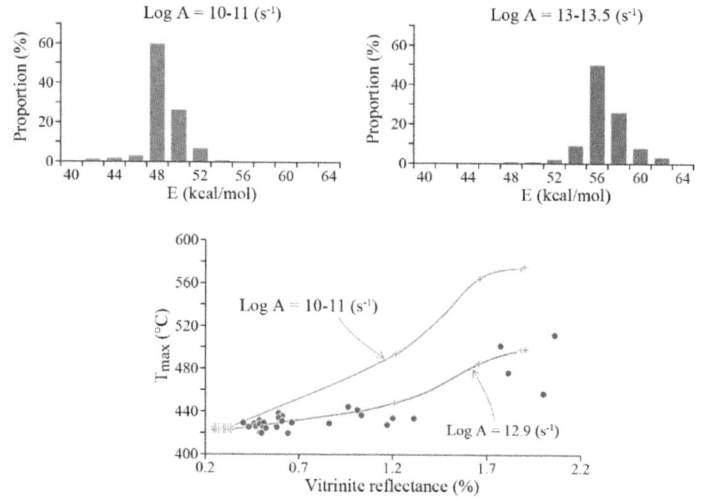

Figure 7.15b. *Example of selection of absolute kinetic parameters under geological conditions: profile of the evolution of the Tmax values as a function of the evolution of the vitrinite reflectance for two sets of kinetic parameters determined in the laboratory*

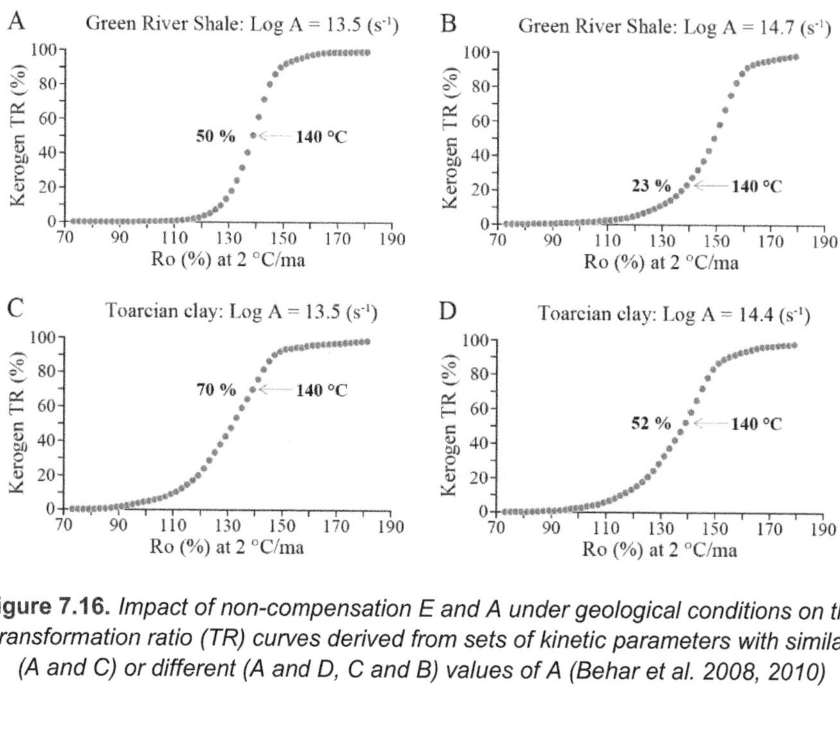

Figure 7.16. *Impact of non-compensation E and A under geological conditions on the transformation ratio (TR) curves derived from sets of kinetic parameters with similar (A and C) or different (A and D, C and B) values of A (Behar et al. 2008, 2010)*

This is illustrated in Figure 7.16. Two sets of kinetic parameters were determined on two samples, one of lacustrine origin (Green River Shale (Behar et al. 2010)) and the other of marine origin (Toarcian Shale (Behar et al. 2008)). The four transformation ratio (TR) curves are plotted in Figure 7.16. If comparisons are made with sets of kinetic parameters with the same *A* value (Figure 7.16(A) and (C)), it is clear that the kerogen of marine origin is much more unstable than that of lacustrine origin. If sets A and D are compared, the results show that the two kerogens have an equivalent thermal reactivity. Finally, if sets A and B are compared, the deviations exaggerate the differences in reactivity between the two samples.

In order not to increase or decrease the difference in thermal reactivity between two kerogens in geological cases, the transformation ratio curves will be derived from sets of kinetic parameters with the same or very close value of *A*.

7.7. Summary

The kinetic parameters for calculating the cracking rates of kerogens cannot be calculated from the data determined on natural samples. They need to be determined from the data obtained by artificial maturation under laboratory conditions.

The thermal cracking rates of kerogens under laboratory conditions are determined from the order-1 diagram. Once these velocities have been determined, the kinetic parameters, *E*, the activation energy, and *A*, frequency factor, are determined from the Arrhenius equation. These kinetic parameters are used to calculate the cracking rates under geological conditions, i.e. for temperatures between 70 and 200°C. Once these velocities have been determined, it is possible to calculate the conversion of kerogens for different geological times by again using the order-1 diagrams.

When the cracking rates of kerogen are calculated from laboratory experiments in an open system such as Rock-Eval® pyrolysis, experiments are carried out between 300 and 700°C, with heating rates between 2 and 25°C. Preliminary precautions need to be taken to ensure the reliability of the kinetic parameters determined. Rock samples will have a TOC greater than 4–5% to avoid any mineral matrix effect if clays are present and experiments are conducted with nitrogen as the carrier gas and not helium. This means that the kinetic parameters acquired with the Rock-Eval® 2 version are not valid.

The overall equation is as follows:

(kerogen + asphaltenes + x resins) → CO_2 + H_2S + H_2O + HC1 + yHC2 + (1-x) resins + insoluble residue

With S2 parameter = (HC1 + yHC2 + (1-x) resins.

The conversion of kerogen cannot be determined by its mass loss, as the insoluble residue recovered during kerogen maturation is a mixture of initial kerogen that is not yet decomposed, a solid residue generated by the kerogen itself, and a solid residue generated by secondary cracking of the asphaltenes and resins generated.

The kinetic parameters E and A, determined under high-temperature laboratory conditions, 300–700°C, cannot be used directly under low-temperature geological conditions, 100–200°C, because of the compensation effects between E and A under high-temperature conditions. Thus, when the parameter sets are published in the literature for kerogen cracking, the kerogen conversion curves will be compared only under laboratory conditions if the frequency factors, A, are different. These curves can in no case be compared at low temperature, i.e. under geological conditions. The only valid comparisons for the geological case will be made with frequency factors, A, that are identical or very similar in the sets of kinetic parameters available.

8

Characterization of Sulfur Compounds

Jean ESPITALIÉ[1], Violaine LAMOUREUX-VAR[2] and Nicolas BOUTON[1]

[1] *Vinci Technologies, Nanterre, France*

[2] *IFP Energies nouvelles, Rueil-Malmaison, France*

8.1. Introduction

Similarly to the work carried out by van Krevelen (1981) on the characteristics and thermal evolution of coal, it was possible as early as 1970–1971 to characterize the different types of sedimentary organic matter and their path of evolution during their burial based on a diagram of the H/C versus O/C ratios constructed from the analysis of the atomic elements C, H and O of kerogens (Durand and Espitalié 1973). With the Rock-Eval® methodology developed by Espitalié et al. (1977a), rapid pyrolysis of small sediment samples provided a hydrogen index (HI) and an oxygen index (OI), which could be correlated with these H/C and O/C ratios. It was therefore possible to easily characterize sedimentary organic matter (type, degree of evolution and oil potential), without even having to extract it from the sediment. However, sulfur, an important element of sedimentary organic matter, was not accessible using the Rock-Eval® methodology, whereas it was determined by elemental analysis of kerogens.

For a color version of all figures in this chapter, see www.iste.co.uk/baudin/rock-eval.zip.

The Rock-Eval Method,
coordinated by François BAUDIN.

Sulfur is a marker of sedimentation conditions (anoxic and euxinic environments) found in the sedimentary organic matter into which it is incorporated at the beginning of diagenesis (Berner 1984). This natural sulfurization of organic matter effectively preserves it from biodegradation at geological time scales (Boussafir et al. 1995; van Kaam-Peters et al. 1998) and leads to sulfurized kerogens at the end of diagenesis. Then, during catagenesis, this organic sulfur reduces the temperature of the thermal cracking of kerogen (Orr 1986; Tomic et al. 1995; Lewan 1998). Oils from these sulfur-rich kerogens are themselves high in sulfur content, especially oils generated at the beginning of catagenesis (Orr 1986). Yet, the presence of sulfur in oils impacts the entire oil industry, inducing additional exploitation costs: production and transport require corrosion-resistant installations and specific H_2S gas treatment units, and refining has to include specific desulfurization processes.

As of 1976, Madec and Espitalié analyzed the sulfur from the pyrolysis of source rocks by replacing the flame ionization detector of a Rock-Eval® 1 with a sulfur detector (Madec et al. 1978; Patent FR-2376414). They were able to differentiate the sulfur bound to organic matter from that resulting from the thermal decomposition of pyrite. In addition, it was possible to calculate, following the example of the hydrogen index, a sulfur index, an additional indicator of the state of evolution of types II and III organic matter (Madec and Espitalié 1985). Another instrument, the Geoelf Sulphur, was then designed and used to analyze organic and inorganic sulfur from sediment pyrolysis, along with the classical parameters provided by Rock-Eval® (S1, S2 and Tmax) (Bjorøy et al. 1993; Jensen et al. 1998). These two instruments were subsequently abandoned, but in the 2000s, when the production of heavy oils and oil sands resumed, the exploitation difficulties linked to their richness in sulfur revived interest in the study of sulfur in source rocks and reservoir rocks. A new version of Rock-Eval®, called 7S or 7 Sulfur, was developed with the aim of determining, jointly with the parameters already provided by the Rock-Eval® method, the entire organic and mineral sulfur in the samples. For this purpose, the different forms of sulfur, organic and mineral, are analyzed, not only during the pyrolysis phase, but also during the oxidation phase (Espitalié et al. 2010; Patents FR 2010/2937737, US 2011/0263034).

This chapter provides an overview of the Rock-Eval® 7S device, the information that this methodology provides on sulfur in kerogens, oil rocks and oils, and various applications in oil exploration and production. Finally, prospects for application in various fields are discussed.

8.2. Experimental device

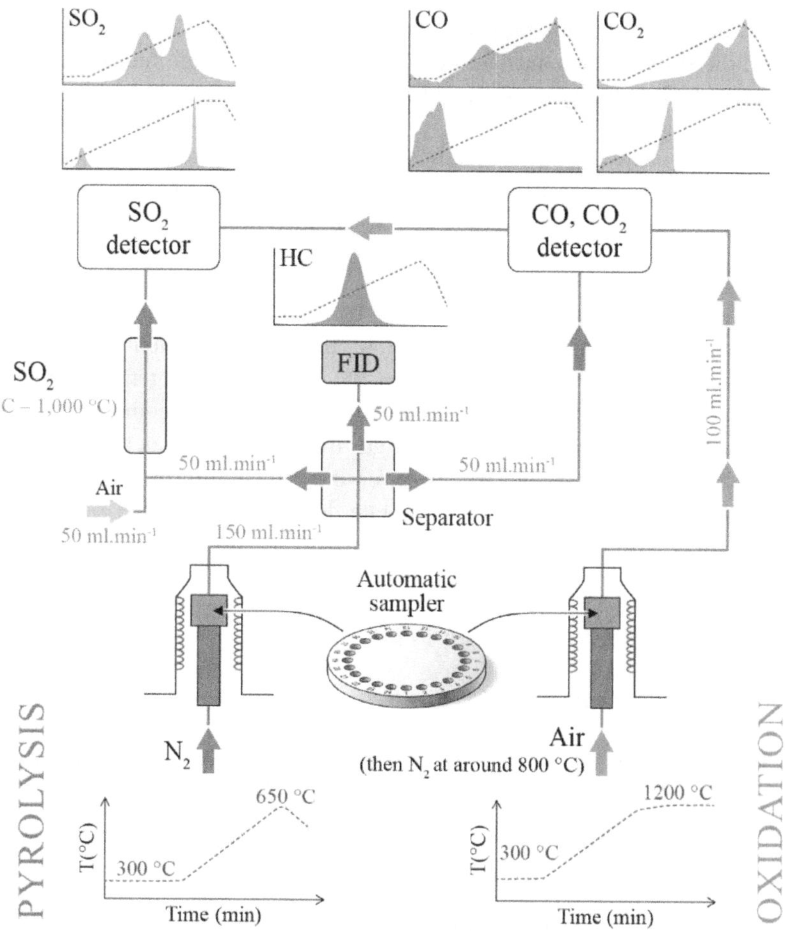

Figure 8.1. *Diagram of the Rock-Eval® 7S device in operation according to the bulk rock method*

Derived from Rock-Eval® 6, Rock-Eval® 7S not only enables the quantification of the classical parameters (S1, S2, S3, HI, OI, Tmax, TOC and MinC), but also the quantification of different sulfur contents: labile organic sulfur and pyritic sulfur released in pyrolysis; residual refractory organic sulfur and pyritic sulfur, which are then released by oxidation; finally, sulfur from the decomposition of natural sulfates or sulfur compounds neoformed during analysis. For this purpose, a new splitter line was created at the outlet of the pyrolysis oven, equipped with an extra oxidation oven, also called an SO_2 oven, in which the sulfur effluents emitted during pyrolysis are continuously oxidized in the presence of air at 850°C. They are thus transformed into SO_2, which is analyzed by means of a specific UV detector. Then, during the oxidation phase, the sulfur compounds are burned into SO_2 directly in the oxidation oven and immediately detected by the same specific UV detector (Figure 8.1).

Like carbonates, sulfates decompose under the effect of heat. Thus, the thermal decomposition of calcium sulfate follows the chemical reaction that produces sulfur dioxide, calcium oxide and oxygen as a by-product:

$$CaSO_4 \rightarrow CaO + SO_2 + \frac{1}{2} O_2$$

Thus, for this reaction to be complete, the oxidation oven must no longer be swept with air, but with nitrogen above 700–730°C. At this temperature, since the organic matter is already oxidized, and the carbonates decompose identically in nitrogen and in air, changing the carrier gas from air to nitrogen allows the complete reduction of sulfates without any effect on the detection of organic compounds nor carbonates. Since the carrier gas flow rate does not change, the calibration of the SO_2 signal by UV analysis remains identical.

8.3. Characterization of sulfur in kerogens

8.3.1. *Thermograms*

Rock-Eval® 7S analysis of type II immature kerogens (of marine origin) provides thermograms of sulfur of the type shown in Figures 8.2 and 8.3. The thermograms in Figure 8.2 are associated with a kerogen containing pyrite, while those in Figure 8.3 are associated with a kerogen containing little or none.

During pyrolysis of the pyrite-rich sample (Figure 8.2), a first sulfur peak, named S1S, is released at 300°C, simultaneously with the free-hydrocarbon peak S1. Then, a second sulfur peak, called S2S, is released simultaneously with the S2 peak of the cracking of organic matter between 300 and 500°C. These two peaks therefore represent the pyrolysis organic sulfur. The FeS peak, which appears above 500°C, originates from the sulfur released by the partial decomposition of free pyrite or included in the organic matter (framboidal pyrite). Then, during oxidation of the pyrolysis residue, a peak called Residual S is observed, which comes from the oxidation of the entirety of organic sulfur and pyritic sulfur remaining after the previous pyrolysis phase.

The thermograms of the pyrite-poor sample show an almost non-existent FeS peak in pyrolysis (Figure 8.3). In this case, the Residual S oxidation peak almost exclusively represents the organic sulfur remaining after pyrolysis.

Type I and III kerogens present SO_2 thermograms that are similar to those of type II. However, they differ in the shape of their S2S peak, which is narrower for type I and wider for type III.

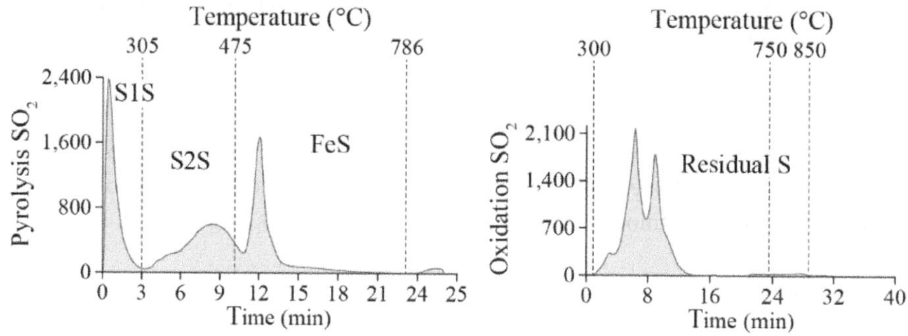

Figure 8.2. *SO_2 thermograms of pyrite-rich kerogen during pyrolysis (left) and then during oxidation (right)*

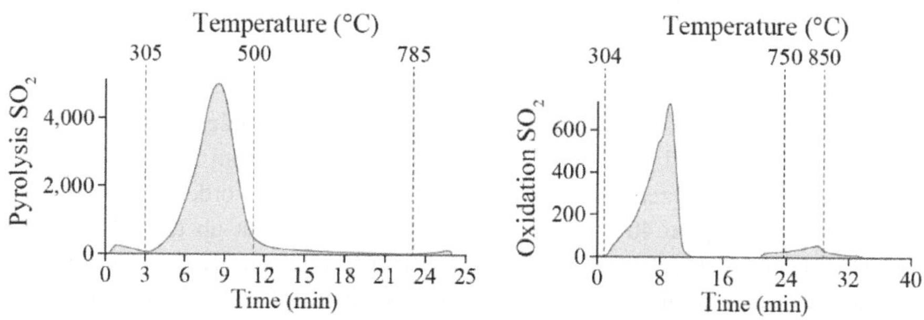

Figure 8.3. *SO_2 thermograms of a pyrite-poor kerogen during pyrolysis (left) and then during oxidation (right)*

8.3.2. *Origin of the analyzed sulfur compounds*

S1S parameter

Taking as an example the Rock-Eval® 7S analysis of an immature kerogen, the sulfur associated with the parameter S1S is quantified at 15 mg S/g TOC, while the free hydrocarbons associated with the parameter S1 are quantified at only 0.10 mg HC/g TOC. With the HC/S ratio being so disadvantageous to hydrocarbons, it is obvious that the sulfur of the parameter S1S does not originate solely from the sulfur bound to free hydrocarbons S1. It likely also comes from the early cracking, at 300°C, of sulfur-containing organic matter (Rosenberg et al. 2017). It is therefore sulfur-derived from highly labile organic compounds. As catagenesis progresses, the S1 parameter will increasingly consist of sulfur-containing hydrocarbons from the thermal cracking of kerogen, and it is this sulfur that will constitute the main part of the S1S parameter.

S2S parameter

This parameter represents not only the sulfur embedded into aromatic hydrocarbons (thiophenes, benzothiophenes, etc.), but also the sulfur of aliphatic hydrocarbons (thiols, thioethers, etc.) from the cracking of kerogen.

FeS parameter

This parameter results from the quantity of SO_2 produced in the SO_2 oven from the S_2 gas released by the thermal decomposition of pyrite to pyrrhotite $Fe_{(1-x)}S$, with $0 < x < 0.20$. A comparative study between Rock-Eval® 7S and in situ XRD showed that pyrrhotite from Rock-Eval® pyrolysis had a stoichiometry of Fe_7S_8, implying the following chemical reaction (Aboussou 2018):

$$FeS_{2\,(s)} \rightarrow {}^{1}/_{7}\, Fe_7S_{8\,(s)} + {}^{3}/_{7}\, S_{2\,(g)}$$

Pyrolysis followed by oxidation carried out on pyrite samples showed that when pyrolysis was carried out under standard conditions (from 300 to 650°C at 25°C/min), the pyrolyzed sulfur/total sulfur ratio was of the order of 28–30%, and that it increased up to 40% when pyrolysis was carried out up to 800°C. A new coefficient, called Coef p, is used to estimate the total quantities of pyrite from pyritic sulfur measured in pyrolysis. The advantage of this coefficient is that it does not vary, whatever the type of organic matter and its degree of maturity.

8.3.3. *Calculation of "sulfur" parameters*

The calibration of the device is carried out using a reference gas cylinder containing 2,000 ppm of SO_2. It is therefore easy to calculate the different percentages of sulfur relative to the weights of the samples analyzed. This allows calculation, in pyrolysis, of the S1OrgS percentage of the sulfur corresponding to the S1S parameter, the S2OrgS percentage from the S2S parameter and the PyroFeS percentage from the FeS parameter. In oxidation, the percentage of sulfur remaining after pyrolysis (organic sulfur + pyritic sulfur) is called Residual S.

Moreover, using the Coef p described above, the following calculations are performed automatically:

– the percentage of total pyritic sulfur (TotFeS), estimated from the relationship:

TotFeS = PyroFeS/Coef p;

– the percentage of residual pyritic sulfur:

OxyFeS = PyroFeS x (1-Coef p)/Coef p.

The percentages of the following parameters are derived from the previous ones:

– the pyrolysis organic sulfur (labile organic sulfur):

PyroOrgS = S1OrgS + S2OrgS;

– the residual organic sulfur (refractory organic sulfur):

OxyOrgS = Residual S - OxyFeS;

– the total organic and mineral sulfur:

Total S = PyroOrg S + PyroFeS + Residual S;

– the total organic sulfur:

TotOrgS = Total S - TotFeS;

– the ratio of labile organic sulfur to total organic sulfur, called Coef c:

Coef c = PyroOrgS/TotOrgS.

This set of parameters is shown in Figure 8.4. The Rock-Eval® 7S method thus makes it possible to rapidly estimate the percentage of labile organic sulfur in kerogens (S1OrgS + S2OrgS), which is progressively found as catagenesis proceeds in the organic sulfur compounds in oils and also in the H_2S gas encountered during drilling. The determination of organic sulfur by Rock-Eval® 7S compares very favorably to that made by elemental analysis (Figure 8.5).

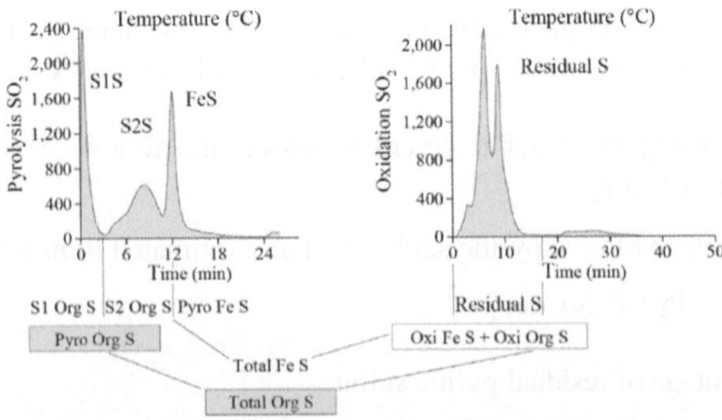

Figure 8.4. *Organic and pyritic sulfur types deduced from the SO_2 thermograms obtained during pyrolysis and then oxidation of a kerogen*

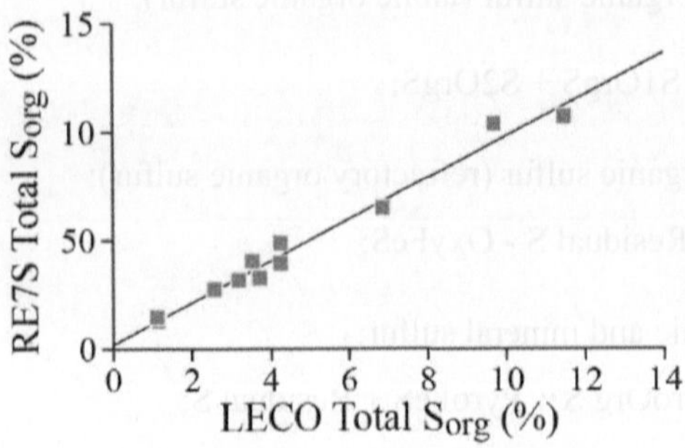

Figure 8.5. *Total organic sulfur contents of various kerogens. Comparison between elemental (LECO) and Rock-Eval® 7S analyses*

As with the hydrogen index (HI) obtained from the S2 parameter of hydrocarbons, it has now become possible to calculate an organic sulfur index, **SI,** in mg S/g TOC as being equal to S2OrgS/TOC x 100. This index allows the levels of source rocks richest in organic sulfur to be recognized, and is also a criterion of maturation because it decreases with the evolution of organic matter during the burial of the sediments (Madec and Espitalié 1985; Table 8.1). It also becomes possible to calculate the **TOS/TOC** parameter as being equal to TotOrgS/TOC. This parameter is equivalent to the S/C proposed by Orr (1986), intended to classify sulfur-rich type II kerogens in a separate class, class II-S, when S/C > 0.04. This rule is widely used, although the S/C is distorted when the sulfur is measured by elemental analysis and the kerogen contains pyrite; this S/C is then overestimated and the kerogen can be incorrectly declared to be type II-S. Conversely, the TOS/TOC index of Rock-Eval® 7S corresponds perfectly to Orr's nomenclature, since it only targets organic sulfur.

	Elemental analysis						Pyro. RE7S		Oxyd. RE7S	S calculated (%)				
Sample	Type	Tmax	TOC	OrgS	PyrS	TotS	Pyro. OrgS	Pyro. FeS	Resid. S	Tot. S	Tot. FeS	Tot. OrgS	C coeff.	SI
169,421	III	400	65	1.13	1.07	2.2	1.13	0.24	0.91	2.28	0.86	1.42	0.70	17
169,420	III	406	63.4	6.67	0.43	7.3	4.08	0.2	2.9	7.18	0.71	6.47	0.66	63
J6	III	407	65			?	3.30	0.39	2.9	6.59	1.39	1.00	0.63	50
54,016	III	430	65			8.3	1.15	1.51	5.64	8.3	5.39	2.91	0.39	17
163,000	III	422	?			0.5	0.12	0.03	0.35	0.38	0.10	0.28	0.44	
167,950			50	32	14.8	18	2.79	4.06	11.16	18.01	14.50	3.51	0.79	55
167,949	II		63	11.52	0.15	11.97	10.10	0.40	1.6	12.1	1.43	10.67	0.94	160
166,790	I	439	71	3.76	2.17	5.93	0.66	0.48	3.8	4.94	1.71	3.23	0.20	9
166,792	I	441	66.7	3.55	3.4	6.95	0.86	0.76	5.17	6.79	2.71	4.08	0.21	12
166,795	I	437	65	1.21	0.93	2.14	0.37	0.26	1.48	2.11	0.93	1.18	0.31	5
166,796	I	439	53	9.71	9.87	19.58	1.37	2.84	16.27	20.48	10.14	10.34	0.13	25
167,951	II	409	62			?	5.2	0.66	3.35	9.21	2.36	6.85	0.75	83

	Elemental analysis						Pyro. RE7S		Oxyd. RE7S	S calculated (%)				
Sample	Type	Tmax	TOC	OrgS	PyrS	TotS	Pyro. OrgS	Pyro. FeS	Resid. S	Tot. S	Tot. FeS	Tot. OrgS	C coeff.	SI
160,000	I	416	59			14.1	3.12	2.73	8.49	14.34	9.75	4.59	0.67	52
156,394	I	430	66.7			14	4.33	1.74	7.7	13.77	6.21	7.56	0.57	64
156,301	I	440	49			24.35	3.41	5	17.98	26.39	17.86	8.53	0.39	69
156,306	I	455	29.7			37.3	1.04	4.45	25.6	31.09	15.89	15.20	0.06	35
99,296	II	419	60	3.22	9.48	12.7	3.2	2.65	6.75	12.6	9.46	3.14	1.02	53
15,050	II	418	66	4.28	5.2	9.48	3.56	1.43	4.95	9.94	5.11	4.83	0.80	53
167,951	I	420	67			0	5.27	0.58	3.35	9.2	2.07	7.13	0.73	78
156,107	II	413	63.4	2.59	7.07	9.66	2.49	1.5	4.05	8.04	5.36	2.68	0.92	39
156,112	II		42	8.4	19	27.4	2.9	3.1	10.49	16.49	11.07	5.42	0.53	69
129,214	II	402	68.8	16.95	0.39	17.34	10.1	0.33	2.05	12.48	1.18	11.30	0.89	146
55,000	II	412	60			12.9	4.86	1.08	3.72	9.66	3.86	5.80	0.83	81

Table 8.1. *Sulfur types of various kerogens characterized by Rock-Eval® 7S*

In addition, it presents the advantage of avoiding the long acid attack operations of pyrite that are sometimes carried out to quantify the organic sulfur content and that may alter a part of the organic matter, inducing a bias on TOC measurements (Canfield et al. 1986; Aboussou 2018; Cohen-Sadon et al. 2021).

The benefit of analyzing sulfur contained in kerogens using Rock-Eval® 7S lies not only in the rapidity of the analyses or the determination of their total organic or mineral sulfur such as pyrite, but above all in the determination of their labile organic sulfur (PyroOrgS). Figure 8.6 shows the evolution of the PyroOrgS/TotalOrgS ratio as a function of kerogen maturation, estimated by the Tmax indicator. Labile organic sulfur is indeed difficult to measure using the classical methods of organic geochemistry. Yet, as already mentioned above, it is this sulfur that is found partly engaged in the hydrocarbons and H_2S generated by kerogens during their evolution with depth in sedimentary basins (Carvajal-Ortiz et al. 2021).

This PyroOrgS/TotalOrgS ratio, called Coef c (Table 8.1), is used to make corrections to the organic and pyritic sulfur contents obtained with Rock-Eval® 7S during pyrolysis and oxidation of bulk rocks. In fact, these contents are affected by reactions between the sulfur and the CO_2 released with certain minerals present in the rock samples, for example carbonates.

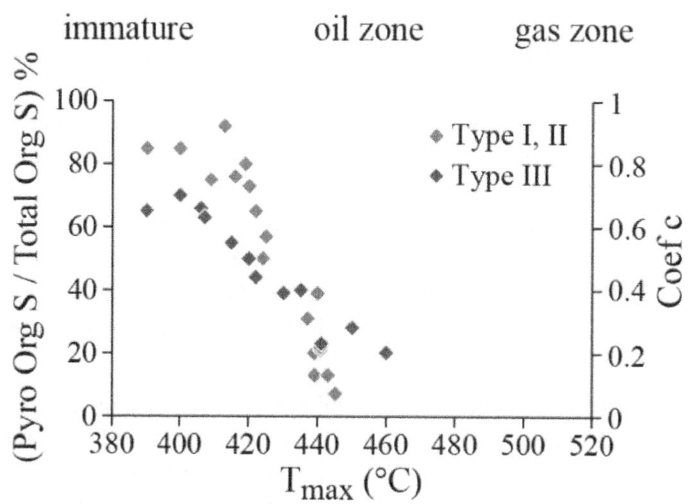

Figure 8.6. *Evolution of the PyroOrgS/TotalOrgS ratio as a function of the maturity of kerogens (Tmax) for the different kerogen types*

Immature type I and II kerogens (Tmax < 410°C) present higher pyrolyzable organic sulfur to total organic sulfur ratios (> 80%) than those of type III, which are between 60 and 70%, and these ratios decrease faster as a function of their maturation than those of type III (Figure 8.6). It is thus possible to calculate, for these three major types of organic matter, the functions that relate their Coef c values to their Tmax values (Coef c = f(Tmax)) so as to be able to calculate for rock samples, whose organic matter type and Tmax values, and pyrolyzable organic sulfur to total organic sulfur ratios are known. Thus, for each sample analyzed, the Coef c is calculated and then, having determined the PyroOrgS, the TotalOrgS content is deduced by the relation PyroOrgS/Coef c. Likewise, knowing the Coef p of pyrite (0.28), which is constant regardless of the degree of evolution, the total pyritic sulfur content is determined by the relationship:

Total FeS = PyroFeS/Coef p

These two relationships are subsequently used to better estimate the quantities of organic and pyritic sulfur in source rocks, as the effects of the mineral matrix on the results of these samples are significant.

8.4. Characterization of sulfur in oil source rocks

The thermograms of sulfur from the analysis of rocks are much more difficult to interpret than those obtained from the analysis of the corresponding kerogens, due to the presence of minerals. The main differences relate not only to the sulfur parameters recorded, as can be seen in Figures 8.7 and 8.8, but also and above all to the quantities of sulfur obtained, which are much too low compared with the values expected from the pyrolysis/oxidation of the corresponding extracted kerogens. First of all, while in pyrolysis the same parameters are found on rocks as those observed on kerogens, the same can no longer be said for the sulfur oxidation results (Figures 8.7–8.9). Indeed, new sulfur parameters are observed above 750°C, which do not exist in the thermograms of the corresponding kerogens. These new parameters are as follows:

– Two sulfur parameters between 900 and 1,000°C: these parameters of neoformed sulfur compounds result from reactions between organic and/or mineral sulfur, released during pyrolysis and oxidation, with rock minerals such as carbonates, for example; these sulfur parameters are called "*Retained S*". For some samples, only one neoformed sulfur parameter is observed in this temperature interval.

– A parameter above 1,000/1,100°C, which corresponds to the sulfur released by the thermal decomposition of natural sulfates such as gypsum ($CaSO_4.2H_2O$). Like carbonates, sulfates decompose under the effect of heat. This is a carbonate–carbon reduction and sulfate–sulfur reduction reaction. Thus, calcium carbonate and calcium sulfate observe the following chemical reactions:

$$CaCO_3 \rightarrow CaO + CO_2$$

$$CaSO_4 \rightarrow CaO + SO_2 + ½\ O_2$$

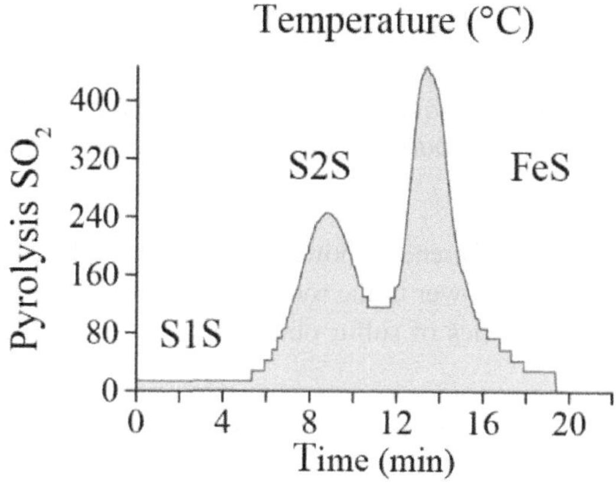

Figure 8.7. *Thermogram of pyrolysis sulfur for the IFPEN-160,000 bulk rock sample*

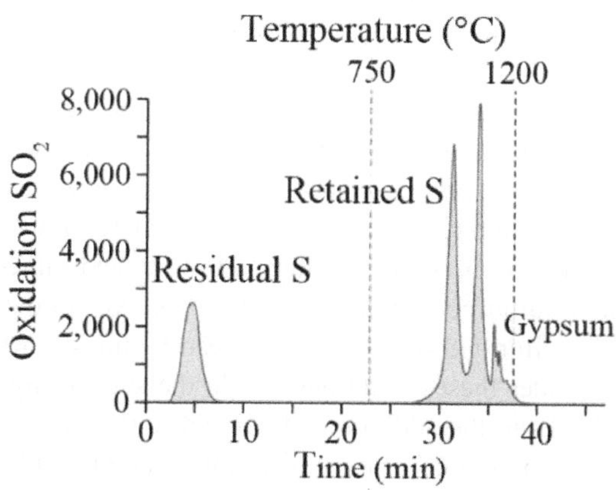

Figure 8.8. *Thermogram of oxidation sulfur after pyrolysis of the IFPEN-160,000 bulk rock*

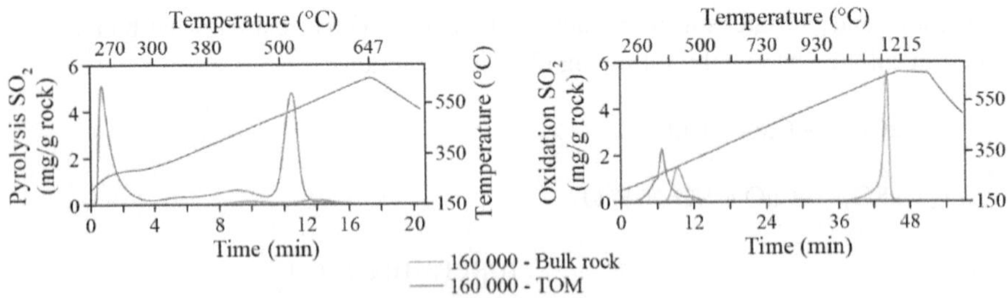

Figure 8.9. *Comparative analysis of bulk rock and extracted kerogen (IFPEN-160,000); bulk rock is in red and kerogen is in green*

The most significant differences, however, lie above all in the analyzed sulfur quantities, which are much lower in the rocks than those obtained on their kerogens. Table 8.2 shows the quantities of sulfur obtained for different kerogens and for the corresponding rocks:

– Pyrolysis sulfur and residual sulfur from bulk rocks have reciprocally lost, on average, up to 80 and 60% of their observed contents on their corresponding kerogens.

– In the temperature interval of between 750 and 1,100°C, one or more sulfur parameters (*Retained S*) generally appear, the quantity or quantities of which represent(s) the entirety of organic and pyritic sulfur that has disappeared during pyrolysis and at the beginning of oxidation (*Residual S*).

The “Retained S” parameter is the signature of the decomposition of sulfur compounds in the crucible at high temperature. These sulfur compounds are said to be neoformed as they were produced in the crucible by chemical reactions during pyrolysis and at the beginning of oxidation. Chemical reactions take place in the crucible between sulfur from organic matter and pyrite and minerals present in the rock, to produce new sulfur compounds, which do not exist in the original rock. Up to 80% of pyrolysis sulfur and oxidation sulfur can be transformed into neoformed sulfur compounds. Under these conditions, while the sum of organic S + pyritic S remains accurate thanks to the addition of Retained S, this is no longer the case for the Pyro Org S and Pyro Fe S contents, nor for the Total OrgS (TOS) and Total FeS contents, which are very often much too low. The Retained S therefore forms part of the total S of the sample and in particular of the organopyritic S. Yet, it is not possible to know what the share of organic sulfur is, compared to that of pyritic sulfur. Moreover, the quantity of Retained S formed depends on the content of organic matter and its degree of maturity, as well as those of pyrite and certain minerals.

In the Geoworks interpretation software, the method used to estimate the quantity of total organic sulfur, a parameter called Cor Total Org S%, is based on a preliminary estimate of the ratio of total organic sulfur to total organopyritic sulfur. To achieve this, Geoworks uses the new parameter, called Coef c (see Figure 8.6 and Table 8.1), which depends on the type of organic matter (I, II or III), and the organic matter maturity.

From the observed Pyro OrgS and Pyro FeS contents, and knowing the Coef c of organic matter and the Coef p of pyrite, for each sample, we can estimate the proportion of organic sulfur in the total sulfur, called the Ratio in Geoworks:

Ratio = (Pyro OrgS/Coef c)/((Pyro OrgS/Coef c) + (Pyro FeS/Coef p)).

Based on this ratio and knowing the true organopyritic sulfur content (Total S), the Cor Total Org S is easy to estimate:

Cor Total Org S = Ratio x Total S (Figure 8.10).

Total S does not include the sulfur from sulfate decomposition in these calculations.

Based on the Coef c, the pyrolysis labile sulfur can also be calculated:

Cor Pyro OrgS = Cor Total OrgS x Coef c,

as can the refractory sulfur remaining after pyrolysis:

Cor Oxi OrgS = Cor Total Org S Cor – Pyro Org S.

It is also possible to calculate the sulfur index (SI) either from the total organic sulfur of the sample (SI = Cor Total OrgS/TOC), which is the classical method, or from the pyrolysis sulfur (SI′ = Cor Pyro OrgS/TOC). The latter, which could not be carried out routinely before, has the advantage of corresponding to the labile sulfur intervening in chemical reactions. In any case, these indices are all the easier to interpret, the less evolved the organic matter.

Kerogen (mg/g TOC)						
Sample	**TOC**	**OrgS pyro**	**FeS pyro**	**Residual S**	**/**	**Total S**
166,790	71	9.29	6.76	53.5		69.55
166,792	66.7	14.4	9	77		100.4
166,795	65	5.69	4	22		31.69
166,796	53	26	54	306		386
167,951	68	76	10	49		184
160,000	59.5	52	34	142		228
156,304	65.4	75	18	118		210
Bulk rock (mg/g TOC)						
Sample	**TOC**	**OrgS pyro**	**FeS pyro**	**Residual S**	**Retained S**	**Total S**
166,790	4.48	1.78	2.45	6.02	87	97
166,792	2.06	1.45	2.91	0	97	101.36
166,795	2.03	2.18	0.49	0	41	43.67
166,796	2	3.5	23	72	217	315.5
167,951	22	68	11	39	20	178
160,000	3.3	8	12	69	169	260
156,304	1.21	58	8	0	190	256
Sulfur retained in bulk rock (mg/g TOC and %)						
Sample		**OrgS pyro**	**FeS pyro**	**Residual S**	**Retained S**	**/**
166,790		7.5/80	4.3/64	47.5/88	59.3/85	
166,792		12.9/89	6.1/67	77/100	96/95	
166,795		4.22/74	3.51/87	22/100	29/91	
166,796		22.5/86	31/57	234/100	287.5/74	
167,951		8.1/10	0	10	18.1/10	
160,000		44/85	34/74	73/51	151/63	
156,304		17/23	10.1/55	119/100	145/69	

Table 8.2. *Quantities of sulfur, by type, measured by Rock-Eval® 7S on source rocks and their extracted kerogen. This table shows, based on the analysis of the organic and pyritic sulfur of kerogens (a) and their corresponding bulk rocks (b), the quantities (in mg/g TOC) and the percentages of sulfur that reacted with the mineral matrix during the pyrolysis and then oxidation phases (c). It can be observed that the decrease in the organic and pyritic sulfur contents (Pyro Org S, Pyro Fe S and Residual S) obtained from the analysis of bulk rocks is generally very significant, reaching up to 90% of the contents on kerogens*

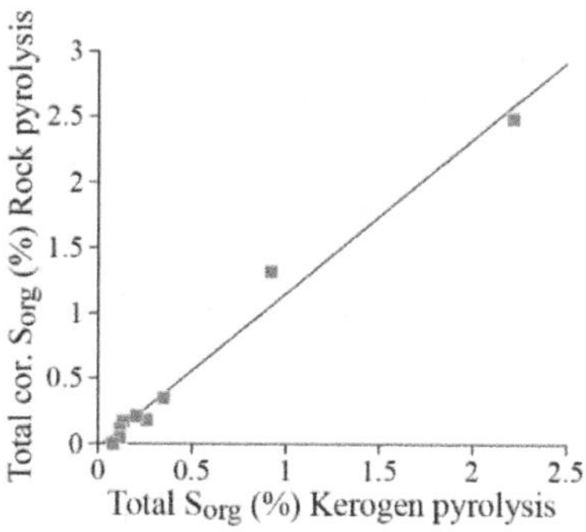

Figure 8.10. *Correlation between the total organic sulfur contents obtained from pyrolysis of kerogens and the corrected total organic sulfur contents obtained from pyrolysis of the corresponding bulk rocks. Since organic sulfur from kerogens is not affected by the presence of minerals, it is considered as a reference: the calculated parameter Cor Total Org S correctly estimates the organic sulfur content. Without these corrections, the total organic sulfur calculation would be erroneous because of the high sulfur retentions during pyrolysis of the rocks, as can be seen in Table 8.2*

An alternative approach consists of isolating organic matter from the bulk rocks, after acid attacks, in order to remove the mineral matrix in a first step, and the pyrite in a second step. The quantity of organic sulfur can then be determined in absolute terms. The values thus obtained were compared with the Rock-Eval® 7S parameters. A graph similar to Figure 8.6 was obtained (Cohen-Sadon et al. 2021). Moreover, not only is the Rock-Eval® 7S method faster than the classical method but, again according to Cohen-Sadon et al. (2021), it is more accurate, since it contains fewer individual steps, and therefore fewer potential error propagations.

Research is still needed to identify the precise origin and chemical reactions leading to parameters of neoformed sulfur compounds and to quantify the different organic and inorganic sulfur types in rocks. The work carried out to this effect by Aboussou (2018) made it possible to highlight the chemical reaction of pyrite. She demonstrated that during pyrolysis, pyrite decomposes into pyrrhotite, a thermally stable form of iron sulfide, which then decomposes during the oxidation phase, and proposed the following chemical equations:

$$\text{Pyrolysis} \quad FeS_2\,(s) \rightarrow {}^{1}/_{7}\, Fe_7S_8\,(s) + {}^{3}/_{7}\, S_2\,(g)$$

$$\text{Oxidation} \quad {}^{1}/_{7}\, Fe_7S_8\,(s) + {}^{53}/_{14}\, O_2\,(g) \rightarrow {}^{1}/_{2}\, Fe_2O_3\,(s) + {}^{8}/_{7}\, SO_2\,(g)$$

Aboussou (2018) highlighted that in the presence of carbonates, smectite and type III organic matter, this chemical balance was no longer observed, reducing the quantity of S_2 released during pyrolysis (Figures 8.11 and 8.12). It has been suggested that S_2 gas forms organosulfur compounds in reaction with type III organic matter, which are then decomposed in the oxidation phase, releasing Residual S before 750°C. It has also been suggested that in the presence of carbonates, S_2 gas forms metal sulfides, which are themselves oxidized to sulfates, which are then decomposed during the oxidation phase, releasing the Retained S at temperatures above 750°C.

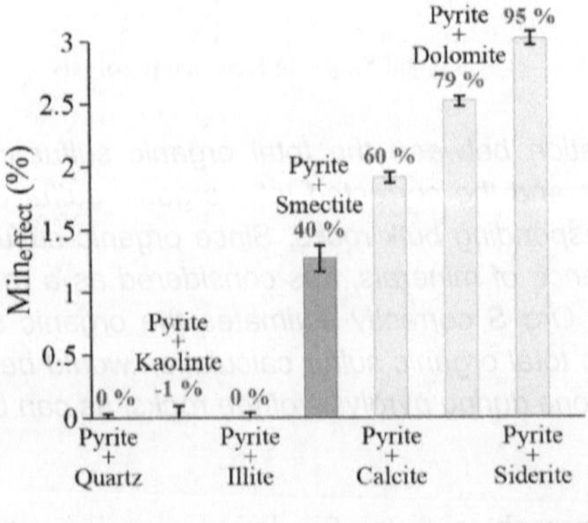

Figure 8.11. *Effect of clays and carbonates on the measurement of sulfur released by pyrite during Rock-Eval® pyrolysis (Aboussou 2018)*

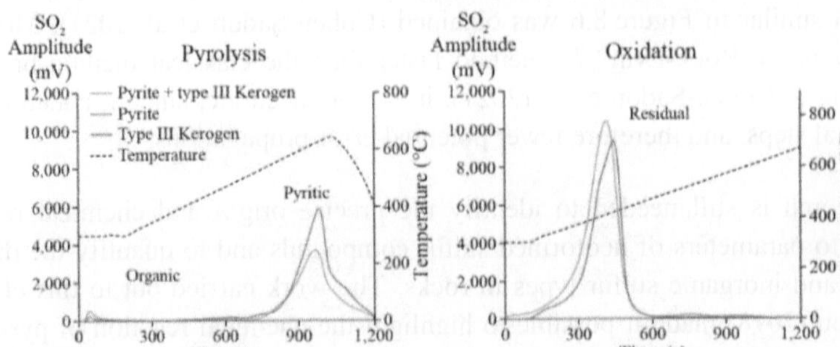

Figure 8.12. *Effect of type III organic matter on the measurement of sulfur released by pyrite during Rock-Eval® pyrolysis (Aboussou 2018)*

8.5. Kinetics of cracking of the organosulfur compounds of organic matter

When organic matter contains sulfur, its thermal cracking is favored (Orr 1986), as the covalent bond energies involving the sulfur atom are lower than those between carbon atoms: S–S (264 kJ/mole); C–S (272 kJ/mole); C–C (348 kJ/mole). The C–S bonds that break first create sulfur free radicals, which then attack the hydrocarbon chains of kerogen, and preferentially form H_2S and hydrocarbons (Lewan 1998). This reaction follows a first-order kinetic law. The observations made with Rock-Eval® 7S pyrolysis are in accordance with this theory, showing that the cracking of the organosulfur compounds of organic matter often starts at a lower temperature than that of the hydrocarbons of this same organic matter, as evidenced by the pyrograms of SO_2 and hydrocarbons from the sulfur-containing source rock shown in Figure 8.13.

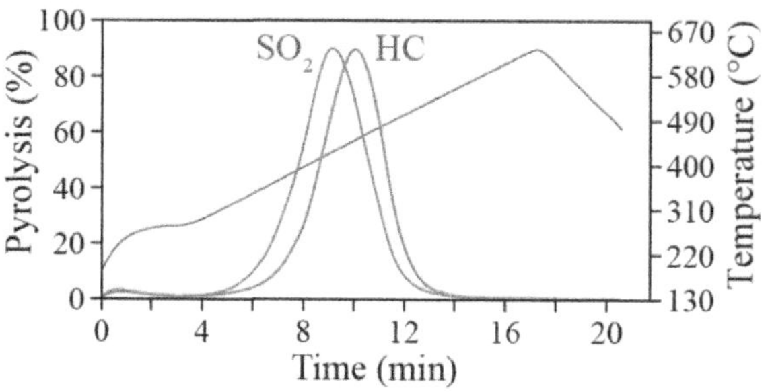

Figure 8.13. *Hydrocarbons (in blue) and organosulfur compounds (in green) emitted by the cracking of an immature source rock under Rock-Eval® 7S pyrolysis. Organosulfur compounds (in green) are emitted at a lower temperature than hydrocarbon compounds (in blue), indicating that cracking of organosulfur compounds is facilitated compared to that of hydrocarbons. The IFPEN-300,000 reference rock has a pyrolyzable sulfur content of 1.5% by weight and S2 = 70.6 mg/g*

With Rock-Eval® 7S, it is possible to estimate the cracking kinetics of organic matter based on the S2 parameter of the hydrocarbon pyrogram and the S2S parameter of the sulfur pyrogram. As with hydrocarbons, the calculation of the cracking kinetics of sulfur compounds is based on two basic hypotheses: the kinetics are first-order with respect to the sulfur content, and the cracking rate, k, follows the Arrhenius equation (see Chapter 7):

$$k = A \exp(-Ea/RT)$$

where A, the pre-exponential factor, will be set to a realistic value, T, the sample temperature, is approximated by measuring the thermocouple placed under the crucible, and R is the ideal gas constant. The activation energies, Ea, are then numerically calibrated to best reproduce the measured sulfur pyrogram. The values of the kinetic parameters (*A* and *Ea*) can then be transferred as input data for simulations of cracking of sulfur-containing source rocks in sedimentary basins.

As shown in the example in Figure 8.14, the activation energies thus obtained are generally lower for the sulfur signal than for the hydrocarbon signal. Extrapolated to catagenesis in sedimentary basins, this kinetic approach thus recognizes that the presence of organic sulfur is a factor in the early cracking of organic matter and that type IIS kerogens reach maturity at a lower temperature than type II kerogens.

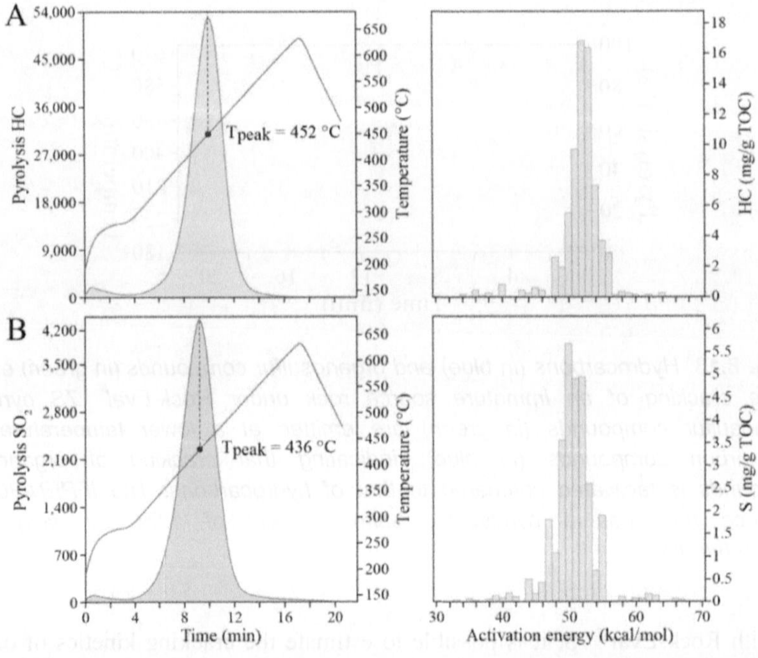

Figure 8.14. *Activation energies of cracking of hydrocarbons (A) and organosulfur compounds (B) of the same IFPEN-300,000 reference rock as in Figure 8.13. The calculations were conducted with a pre-exponential factor,* $A=10^{14}s^{-1}$*. (A) The activation energies of hydrocarbon cracking (pyrogram of hydrocarbons detected by FID) have a Gaussian distribution centered on 52–53 kcal/mole. (B) The activation energies of organosulfur compound cracking (SO_2 pyrogram) have a Gaussian distribution centered on 50–52 kcal/mole*

The application of such sulfur kinetics to basin modeling has not yet been performed, since before Rock-Eval® 7S, the tool to measure organic sulfur pyrolysis as a function of temperature was less accessible, but the mathematical tools are available.

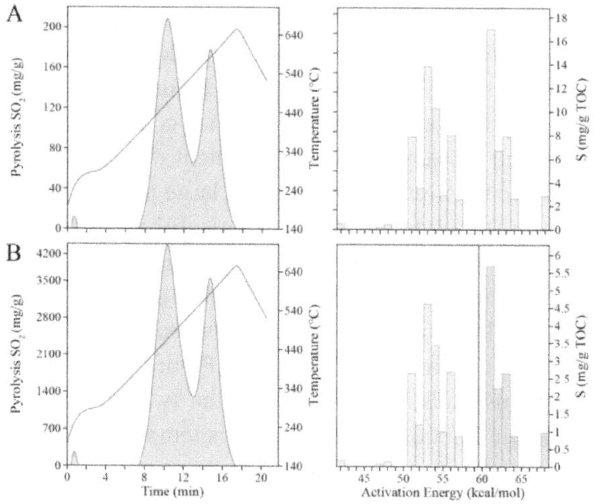

Figure 8.15. *Pyrolysis SO_2 parameters used for the calculation of kinetic parameters and associated histograms of the deduced activation energies. (A) Calculation of the decomposition kinetics of organic and pyritic sulfurs; (B) calculation of the decomposition kinetics of organic sulfur only*

As seen in the previous section, Rock-Eval® 7S pyrolysis of the IFPEN-160,000 rock produces a sulfur pyrogram presenting two parameters (Figure 8.7). They are associated with pyrolyzable organic matter (S2S) and pyrolysis pyritic sulfur (FeS), respectively. From this pyrogram, it is possible to estimate the thermal decomposition kinetics of labile organosulfur compounds of organic matter. To achieve this, the double parameter will first be deconvoluted into two distinct parameters: S2S and FeS. The Geoworks software allows this operation to be performed, and, in a second step, the activation energies of the S2S parameter to be isolated and validated, using the CleanSim module (Figure 8.15). Once the kinetic parameters have been cleaned up, further calculations applied to a basin can theoretically be conducted. This method allows the kinetics of appearance of H_2S in the basin to be estimated.

8.6. Characterization of sulfur in oils

The carbon and sulfur contents of oils and oil products such as maltenes, asphaltenes, distillation residues, etc. are quantifiable by means of a Rock-Eval® 7S analysis. This instrument, by its thermal principle, also makes it possible to distinguish classes of thermal reactivity among oil sulfur compounds, and to highlight the distribution of sulfur between these classes. This is particularly interesting when distinguishing conventional oils from heavy oils, the latter being exploited with more difficulty and at higher cost, due to their high density, viscosity and sulfur content. Conversely, conventional oils, which are low-density, relatively fluid and low in sulfur, are easier to exploit. Figures 8.16 and 8.17 present the Rock-Eval® 7S comparative analysis of a conventional crude oil (S content = 1.7 wt.%) and a heavy crude oil (S content = 5.0 wt.%). The thermograms demonstrate that the conventional oil contains a large proportion of thermally labile fractions (Figure 8.16(a)): a fraction that is vaporizable at 180°C (S1r) and a fraction that is crackable under pyrolysis at low temperature between 200°C and 450°C (S2a). Conversely, the heavy crude oil contains few labile fractions with respect to the fraction that is crackable under pyrolysis between 450°C and 600°C (S2b) and the residual fraction that burns in air (S4CO_2) (Figure 8.16(a) and (b)). Regarding sulfur, the thermograms indistinctly show, for the two oil types, that it is mainly carried by the crackable fraction at high pyrolysis temperature (S2b), and to a lesser extent by the crackable fraction at low pyrolysis temperature (S2a) (Figure 8.16(c)). However, two sulfur properties differentiate the two oil types: heavy oil has higher pyrolysis sulfur contents than conventional oil and contains a large proportion of its sulfur in the thermally refractory fraction (Residual S), unlike conventional oil, whose refractory fraction is virtually sulfur-free (Figures 8.16(d) and 8.17).

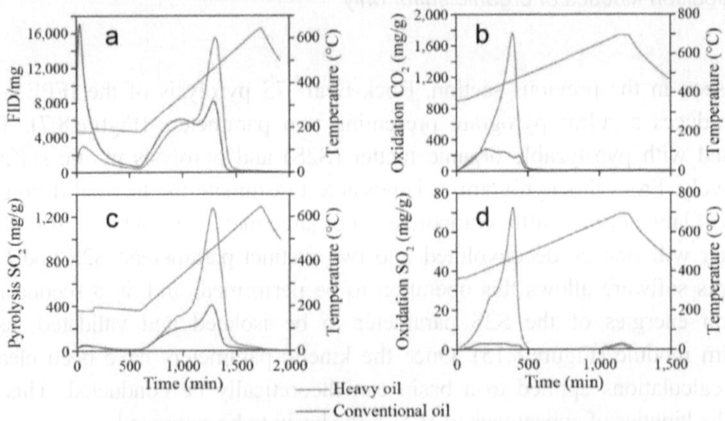

Figure 8.16. *Comparison of thermograms of heavy crude oil and conventional crude oil. (a) Pyrolysis hydrocarbons; (b) oxidation CO_2; (c) pyrolysis SO_2; and (d) oxidation SO_2. Red curve: conventional crude oil (S = 1.7%); green curve: heavy crude oil (S = 5.0%); blue curve: pyrolysis temperature (a, c) and oxidation temperature (b, d). The "Reservoir" analytical method was used. It differentiates four classes of components in oil according to their thermal reactivity: a class of compounds volatilized under pyrolysis at 180°C (S1r); a class of compounds cracked under pyrolysis between 200°C and 450°C (S2a); a class of compounds cracked under pyrolysis between 450°C and 600°C (S2b); and a class of compounds decomposed under oxidation between 300°C and 650°C (Residual S)*

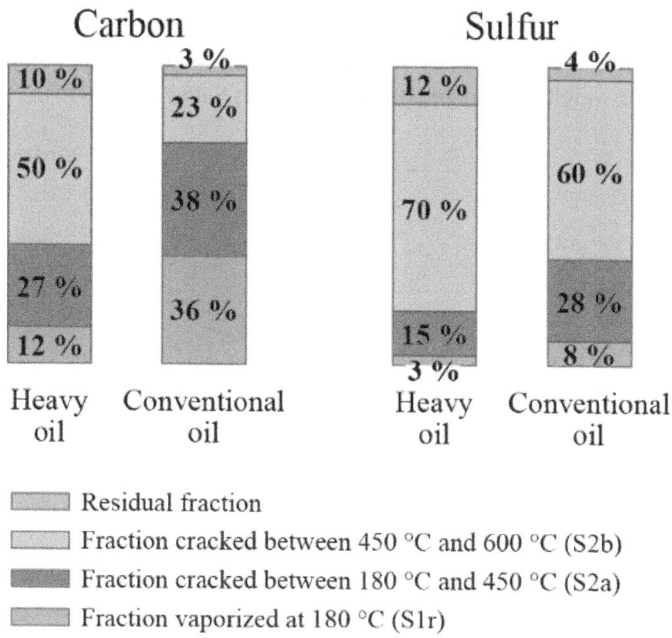

Figure 8.17. *Distribution of carbon and sulfur in fractions of a heavy crude oil and a conventional crude oil*

Information on the thermal stability of organosulfur compounds in oil can then be interpreted in terms of the genesis potential of sulfur compounds under thermal stress in oil production conditions (Lamoureux-Var and Lorant 2005). One application is the risk of H_2S emission associated with the production of oil by thermal processes, for example, with the injection of water vapor or hot air into the oil reservoir (Lamoureux-Var et al. 2011; Uzcategui et al. 2011; Preux et al. 2018). The more thermally labile organosulfur compounds the oil contains, the higher the risk of H_2S production (Lamoureux-Var and Barroux 2013). This is highlighted by the example in Figure 8.18.

Thus, the information obtained on crude oils, namely the carbon and sulfur contents, as well as the thermal reactivity classes, are useful in two respects for players in the oil industry. First, they provide knowledge of the "conventional" or "heavy" status of crude oils, which is crucial information for their exploitation. In addition, they provide data on the thermal stability of their organosulfur compounds, which is useful information for thermal production and refining. Rock-Eval® is particularly appropriate for characterizing heavy oils and bitumens with their thermal stability, as it is perfectly suited to the analysis of organic compounds with high molecular weight (Lamoureux-Var et al. 2011; Sanei et al. 2015).

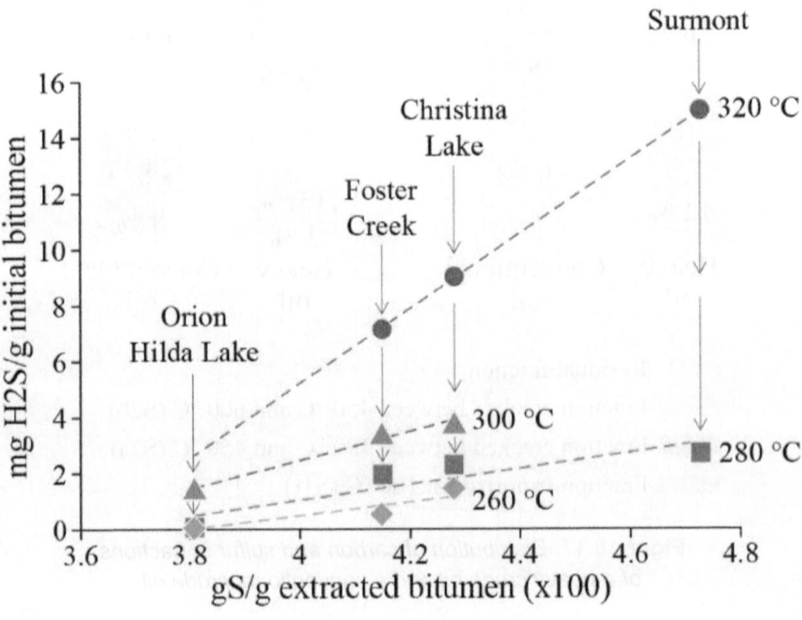

Figure 8.18. *Production of H_2S by aquathermolysis (oil/water volume ratio = 1) as a function of the pyrolysis sulfur content measured by Rock-Eval® 7S (Lamoureux-Var and Barroux 2013). The quantity of H_2S was measured upon completion of closed-system aquathermolysis experiments (oil/water volume ratio = 1) for 203 hours at a constant temperature (260°C; 280°C; 300°C or 320°C) on four oil sands. The pyrolysis sulfur content was measured by a Rock-Eval® 7S prototype on the four bitumens extracted from the original oil sands. A positive relationship can be observed between the H_2S generated by the oil sand under aquathermolysis in the presence of water and the Rock-Eval® pyrolysis sulfur content of the bitumen*

8.7. Characterization of sulfur in reservoir rocks

Reservoir rocks are sedimentary rocks in which oil and gas accumulate via natural processes. They are permeable rocks (evaporites, limestone, sandstone or sand) covered with a low-permeability geological layer (clay or salt), which blocks the oil and gas from rising during their migration from the source rock to the surface. Reservoir rocks are structured by a porous mineral matrix that contains three types of fluids in their porosity: oil, gas and water. Sulfur can be encountered there in five phases: liquid organic (oil), solid organic (pyrobitumen, for example), gaseous inorganic (mainly H_2S), solid inorganic (gypsum, for example) and finally, aqueous (in ionic form dissolved in water). Rock-Eval® 7S identifies sulfur in three phases:

– the liquid organic phase constituted by oil, where sulfur atoms form covalent bonds with carbon atoms or in dissolved elemental form;

– the solid organic phase, which can exist in different forms:

- pyrobitumens, which result from the thermal cracking of oil in reservoir rocks subjected to high temperatures; $T > 150°C$ (Mort 2004),

- bitumens, which result from advanced biodegradation of oil by sulfur-reducing bacteria, in shallow reservoir rocks subjected to low temperatures and leached by atmospheric water; $T < 80°C$ (Machel 2001),

- tar mats, which are levels in reservoir rock containing at least 50% bitumen (Wilhelms and Larter 1994; Carpentier et al. 2007); Rock-Eval® 7S is particularly suitable for identifying these solid organosulfur compounds, which are very dense and very sulfur-rich (Woods et al. 2008);

– the solid inorganic phase constituted by the porous mineral matrix in the form of:

- metal sulfides and mainly iron sulfides such as pyrite (FeS_2) (Wade et al. 1988; Hutcheon 1998; Worden et al. 2003),

- elemental sulfur (S_8),

- sulfates such as anhydrite ($CaSO_4$) or gypsum ($CaSO_4.2H_2O$), whether they are sedimented in an evaporitic environment (Worden et al. 1997) or neoformed by oxidation of pyrite (Hutcheon 1998),

- barium and calcium sulfates precipitated in the vicinity of production wells, following the combination of formation and injection waters (Mahmud et al. 2020).

The thermograms in Figure 8.19, obtained from a sample of oil sand, indicate the sulfur present in two phases: in oil, divided into different thermal stability classes, and in the form of sulfides and sulfates.

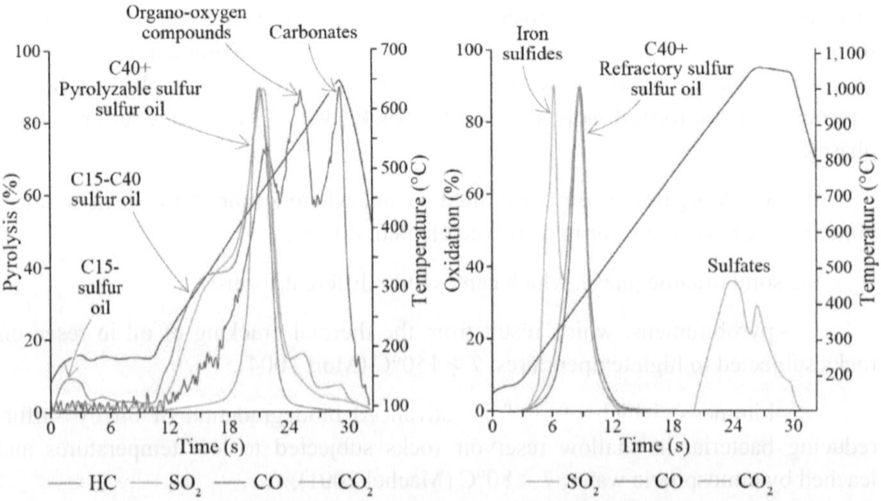

***Figure 8.19.** Thermograms resulting from Rock-Eval® 7S analysis of a sample of oil sand from the McMurray Formation, Athabasca. The pyrograms (left) highlight three thermal classes of bitumen (blue curve representing hydrocarbon classes S1r, S2a and S2b) and their associated sulfur (yellow curve representing sulfur by classes S1rS, S2aS and S2bS). The oxygrams (right) highlight a fourth class of sulfur-containing bitumen, which is thermally very refractory*

8.8. Prospects for the application of Rock-Eval® 7 Sulfur in understanding the sulfur sedimentary cycle

In a world where human activity is strongly rooted in carbon energies and the climate is significantly impacted, a better understanding of the carbon cycle is essential. Rock-Eval®, a technique for characterizing pedogenic and sedimentary carbon, naturally finds its place in the study of the global carbon cycle, which takes place in particular in the pedosphere and the geosphere, and which concerns soils, recent sediments and sedimentary rocks. Since the sedimentary cycle of sulfur is closely linked to that of carbon, the full benefit of a Rock-Eval® 7S characterizing both carbon and sulfur in soil and sedimentary organic matter can be appreciated.

The major sulfur reserves on Earth are found in the marine environment, mainly in the form of dissolved sulfates. Marine sediments are then the site of bacterial reduction of these sulfates, combined with mineralization of sedimentary organic matter, both processes being key in the sedimentary cycles of sulfur and carbon (Jørgensen 1982). They generally occur in the first sedimentary depths, beneath the oxic or suboxic zones, where sulfates are used as electron acceptors to allow bacterial respiration when the other oxidants are depleted (O_2, MnO_2, NO_3^- and Fe_2O_3) (Jørgensen 1983). Bacterial reduction of sulfates generates reduced sulfur species (H_2S, HS^-), which can then be incorporated into organic matter by forming sulfur-crosslinked polymers and organosulfur molecules with low molecular weight (Aizhenshtat et al. 1995; Amrani and Aizhenshtat 2004; Werne et al. 2008). This natural sulfurization of organic matter is deemed to effectively preserve it from biodegradation at geological time scales (Boussafir et al. 1995; Boussafir and Lallier-Vergès 1997). Reduced sulfur species can also react with reactive metals such as iron and manganese, generating metal sulfides. In the presence of reactive detrital iron, iron sulfides are formed, pyrite (FeS_2) being the most stable and widespread compound (Rickard and Luther 1997; Tribovillard et al. 2004, 2015; Shawar et al. 2018). Sulfur sequestration within sedimentary pyrite constitutes the major sink in the biogeochemical sulfur cycle, and has a significant impact on the associated carbon and oxygen cycles (Rickard and Morse 2005). Pyrite can then oxidize to sulfates, and both types of minerals are frequently encountered within the same sedimentary rock. The sulfur thus combined with organic matter and metals constitutes a marker of the sedimentary deposition conditions (marine/non-marine; oxic/anoxic/euxinic) (Berner 1984; Raiswell and Berner 1986). A third pathway of removal of sulfates from sedimentary waters to sediments is the precipitation of sulfates into evaporites, such as anhydrite ($CaSO_4$) and gypsum ($CaSO_4.2H_2O$) (Vairavamurthy et al. 1995; Worden et al. 1997).

As sediments are buried deeper over geological time, their temperature and confining pressure gradually increase, substantially altering their porosity, water saturation and chemical structure. In particular, organosulfur compounds are subjected to transformations (Sinninghe Damsté and de Leeuw 1990). As soon as the temperature passes a sufficiently high threshold, solid organic matter (kerogen) cracks to form liquid organic compounds (oil). The sulfur of this organic matter, which can reach contents by weight of 14% (Orr 1986; Calkins 1994), significantly accelerates the cracking kinetics (Orr 1986; Lewan and Ruble 2002) and enriches the generated oils with sulfur, the content by weight of which can reach 6% (Tissot and Welte 1984).

These different organic and inorganic sulfur species, which impact the carbon and sulfur cycles by participating in these biogeochemical processes, can be observed using Rock-Eval® 7S. While their identification is already acquired with this technique, their quantification is not yet fully resolved, however, and research is needed. The quantification of pyritic sulfur and organic sulfur has already been the subject of research (Aboussou 2018; Cohen-Sadon et al. 2021) and this continues to be so. The quantification of elemental sulfur and sulfates, meanwhile, remains to be resolved.

8.9. Prospects for the application of Rock-Eval® 7S in industry

In the oil production industry, bacterial sulfate reduction (BSR) is a growing phenomenon in North Sea oil reservoirs, where reservoir pressure is maintained by injecting seawater to support production. The latter contains sulfates that, under the reservoir temperature conditions, are reduced by the bacteria that integrate them into their biological cycle by generating H_2S. This production process is increasingly employed, leading to greater quantities of H_2S being emitted at the wellhead. Since this sulfur gas is both lethal from 100 ppm, and corrosive for production tools, treating it on-site is essential, inducing increased production costs. It is conceivable that tests with BSR inhibitors will one day be tested on a Rock-Eval®, in order to verify and simulate the effectiveness of such treatments before they are operated on a large scale. The Rock-Eval® 7S could then be transformed into a bacterial incubator, at different isothermal temperatures, with the SO_2 detector as a detector of microbial activity, to perform tests of biomass growth and population growth kinetics.

The process to produce oil by injecting hot water into the reservoirs, as soon as it is operated in evaporitic reservoirs, entails the risk of H_2S production. This H_2S results from the thermal reduction of sulfates (Worden et al. 2000), induced by human activity (Kowalewski et al. 2010). The use of Rock-Eval® 7S is ideal for the quick screening of a large number of reservoir rocks in order to rapidly assess the presence of pyrite and sulfates.

In the field of tire production, where sulfur is the chemical element that allows vulcanization of natural rubber, it is not difficult to imagine Rock-Eval® tests on the thermal stability of vulcanized products.

9

Study of Organic Matter in Recent Sediments

Yoann COPARD[1] and François BAUDIN[2]
[1] *M2C, Université de Rouen Normandie, Rouen, France*
[2] *ISTeP, Sorbonne Université, Paris, France*

9.1. Introduction

Sedimentary organic matter, whether of continental or marine origin, generally consists of a mixture of terrestrial sources and continental or marine aquatic sources, resulting from primary productivity. Once mobilized by weathering and continental erosion, the terrestrial particulate organic matter reaches rivers, which transfer it laterally to different continental or marine sedimentary reservoirs. Depending on the geological, anthropogenic and climatic characteristics of the drained surface, the transfer times of organic matter conveyed to the oceans can be very fast (e.g. Eel River) or rather long (e.g. Congo River). Thus, the organic matter conveyed by rivers can either be preserved or be subjected to weathering, which will be all the more probable and all the stronger, the longer the residence time in the hydrosystem. Added to this are the physicochemical properties of organic matter, which depend on the biological precursors and also on their pre-fluvial history. It is again a matter of residence time, but this time within continental surfaces (soils, continental sediments and sedimentary rocks), which, if it increases, raises the probability that this organic matter can already be weathered before it enters the hydrosystem. The journey of this organic matter reaches an end in continental reservoirs (lakes and floodplains) or marine receptacles (deltas, continental margins and turbidite

---

For a color version of all figures in this chapter, see www.iste.co.uk/baudin/rock-eval.zip.

The Rock-Eval Method,
coordinated by François BAUDIN.

systems). Further to the lateral transfer, a vertical transfer conditioned by the sedimentation rates recorded by these reservoirs is superimposed. Here again, and beyond the physicochemical characteristics of the organic matter, other parameters favor its preservation over long periods of time.

This chapter proposes to provide an overview of the contributions of the Rock-Eval® method applied to recent sediments. The associated themes concern (i) environmental changes governed by climatic and anthropogenic forcings, (ii) mass balances (source-to-sink), (iii) the global carbon cycle, (iv) the organic matter preservation conditions and lastly (v) the fate of metallic or organic contaminants. We have chosen to first describe the contributions of the Rock-Eval® method to the study of (i) pure compounds or fresh organic matter, (ii) organic matter from continental archives (including suspended particulate matter (SPM)), (iii) organic matter preserved in marine archives or present in the water column and (iv) relationships between organic matter and contaminant concentration. Here, the time scales concerned cover the event-scale to several hundred thousand years.

9.2. Reminder of the principles of analysis of recent sediments by the Rock-Eval® method

The analysis of recent sediments by the Rock-Eval® method requires precautions to be taken (see Chapter 4, section 4.1) so as to not alter the still-fresh and remineralizable organic matter that they can contain. If the samples are from a saline medium, it is necessary for them to be washed with demineralized water in order to remove salts that cause artifacts during analysis of organic matter (Baudin et al. 2015; and Chapter 4, section 4.3), as well as carbonates (Wattripont et al. 2019; and Chapter 13, section 4.3). Furthermore, the decarbonatation of these materials also alters the Rock-Eval® parameters (see Chapter 4, section 4.3). Pyrolysis should be started at a lower temperature than that applied to sedimentary rocks because recent organic matter is generally more thermolabile than kerogens. Espitalié et al. (1985b) recommended the use of cycle 3 of Rock-Eval® 2 and OSA, with an initial temperature of 180°C and a ramp of 30°C min^{-1}. However, it can be observed that the choice of analytical conditions varies according to the studies (Table 9.1), with sometimes unspecified conditions. In the face of this lack of order, Carrie et al. (2012) attempted to standardize the Rock-Eval® method for the analysis of soils and recent sediments. However, by starting pyrolysis at 300°C (Basic cycle) they caused very rapid thermovaporization of their samples (S1 parameters too high), and the oxidation conditions are not specified. Baudin et al. (2015) thus proposed a "recent sediment" cycle that uses the pyrolysis conditions recommended by Espitalié et al.

(1985) in cycle 3 of Rock-Eval® 2 and applies the oxidation conditions of the Bulk Rock method (Table 9.1).

References	Pyrolysis temperature			Oxidation temperature		
	Initial (°C)	Final (°C)	Ramp (°C/min)	Initial (°C)	Final (°C)	Ramp (°C/min)
Espitalié et al. (1985)	180	600	30	600°C - 7 min in REII and OSA		
Disnar et al. (2003)	200	650	30	450	750	?
Sanei et al. (2005)	300	650	25	400	850	20
Copard et al. (2006)	200	650	25	400	750	25
Marchand et al. (2008)	200	650	30	400	850	30
Carrie et al. (2012)	300	650	25	?	850	?
Hare et al. (2014)	100	650	25	300	850	25
Baudin et al. (2015)	**180**	**650**	**30**	**300**	**850**	**20**
Sebag et al. (2016)	200	650	25	400	850	25
Sebag et al. (2018)	200	650	25	300	850	?

Table 9.1. *Examples of several analytical conditions in the analysis of recent sediments, soils and pure organic compounds. The "recent sediments" cycle, shown in bold, is that recommended for the analysis of thermolabile organic matter*

9.3. Analysis of fresh organic matter and pure compounds

While early work focused on a description of thermograms of compounds such as sugars, cellulose and lignin (Sebag et al. 2006), the Rock-Eval® signature of leaves and woody tissues showed distinct thermograms within the same species (Figure 9.1), with generally higher HI values for leaves (Marchand et al. 2008). Despite unsuitable analytical conditions, Carrie et al. (2012) also analyzed pure organic compounds and fresh plant or animal matter. Among the pure organic compounds studied on Rock-Eval®, there are proteins, carbohydrates, lipids, various lignins, tannins, chitin and humic substances. Fresh plant matter consists of leaves and wood of various species, spruce needles, cedar bark, leaves and wood of mangrove plants, fungi, lettuce, corn kernels, coffee beans, apples, copepods and phytoplankton. In total, and although they are degraded to varying degrees during diagenetic or pedological processes, these constituents can be considered as source terms of sedimentary or pedological organic matter (Sebag et al. 2016).

The Rock-Eval® signature of these compounds can be assessed through diagrams using parameters such as S2, TOC, HI, OI and Tmax, or by crossing parameters RC/COT, S2/S3, log [(S1+S2)/S3] and HI_{400} (Carrie et al. 2012) and I- or R-index (Sebag et al. 2016; Figure 9.2 and see Chapter 3 for the definition of these indices). This collection of diagrams highlights great diversity in the values of these indicators, attributable to their chemical composition. Logically, their Tmax values are lower than those of organic matter that has been subjected to diagenetic or pedological transformations, and vice versa for the OI values.

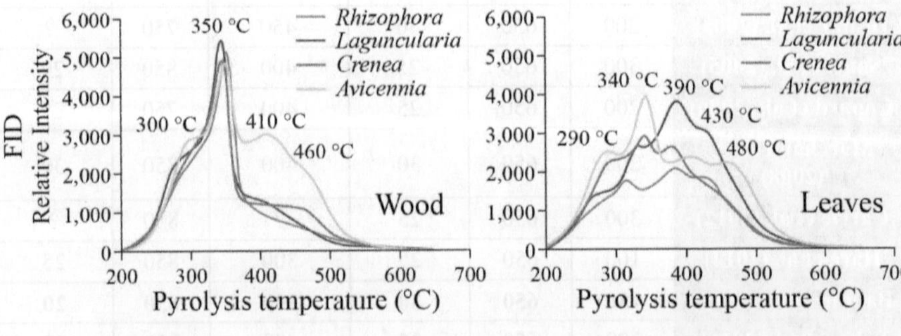

Figure 9.1. *S2 thermograms of wood and leaves of different genera of mangrove trees growing in the mangrove forest of French Guiana (according to Marchand et al. (2008))*

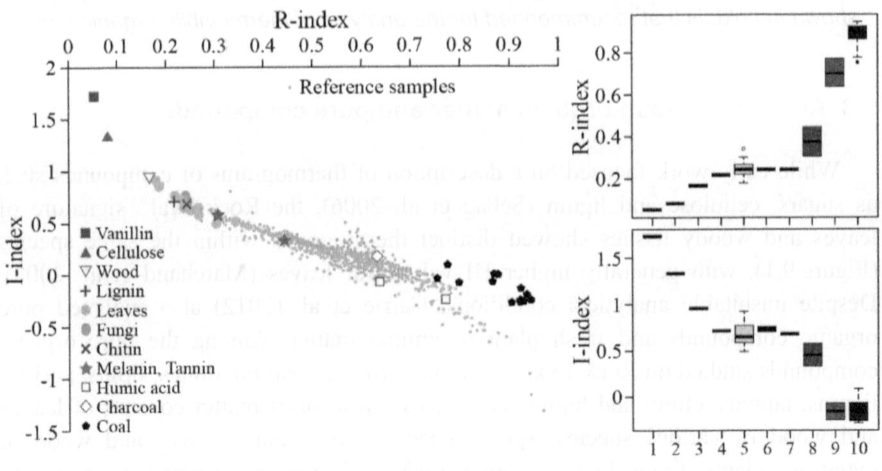

Figure 9.2. *R- and I-indices of pure compounds considered as source terms of soil organic matter (Sebag et al. 2016)*

Two limitations to the analysis of pure compounds should be noted. The first is that the TOC may differ from that determined by another method (elemental analysis; see Chapter 5). Indeed, a TOC deficit can be observed for the majority of the carbohydrates analyzed, whereas there is none for the lipids or lignins, since compounds with hydrocarbon structures rich in heteroatoms (O, N and S) are poorly detected with a FID (see section 2.3.1). It should be added that the TOC calculation, which integrates the area of the S1 and S2 parameters, considers that they contain 83% Corg. This percentage is an average determined after elemental analysis of pyrolysis effluents of kerogens enriched in Corg with respect to biomolecules following the loss of oxygenated chemical functions (alcohols, acids and ketones). A second limitation stems from the fact that Rock-Eval® analysis is rarely performed on pure organic products, but on mixtures of these with minerals resulting in matrix effects. During pyrolysis, these are marked by hydrocarbon retention and/or oxygen enrichment (see section 4.2.2) that result in HI decrease of up to 400% and OI increase of up to 600% depending on the mixture (Kanari 2018; Figure 9.3).

Ultimately, the Rock-Eval® analysis of pure organic compounds or fresh organic matter provides, to a certain extent, their "Rock-Eval® signature", but does not yet allow them to be identified and quantified when they are mixed or associated with minerals.

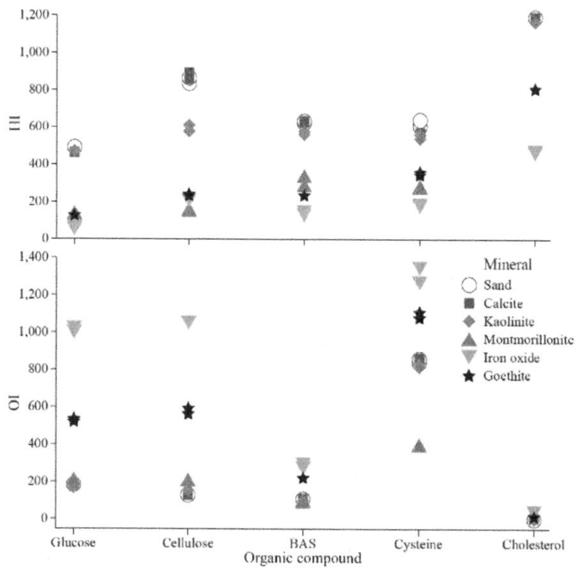

Figure 9.3. *Effect of different mineral matrices on the HI and OI parameters for pure organic compounds (according to Kanari (2018))*

9.4. Continental archives as witnesses of environmental changes

These changes, governed by climatic and anthropogenic forcings, can be deciphered through the restitution of paleoenvironments and the paleohydrological context, which modulate the contribution of sources of organic matter in sedimentary archives. They also modify the Corg trapping efficiency and therefore play a major role for the carbon cycle. A large number of works show that the organic matter quality indicators derived from the Rock-Eval® method make it possible to determine, and even evaluate, the contribution of major sources (pedological, aquatic, fossil and incomplete combustion) of organic matter that feed sediments in transit in rivers or deposited in different continental environments.

9.4.1. *Suspended particulate matter and fluvial archives*

Beyond the simple analysis of particulate organic carbon (POC) from SPM, the Rock-Eval® method is a valuable tool for determining the origin of organic matter transiting through river corridors. Thus, comparison of the ratio 1/TOC versus HI of the SPM of waterways draining the badlands of Draix in the Alps of Haute Provence (Figure 9.4) showed that their organic content essentially derives from the marls constituting the badlands. Also associated with this absence of Corg loss is the quality of the organic matter, revealed by the HI and the decomposition of the S2 signal, which remains unchanged in comparison with the organic matter of the Jurassic marls (Copard et al. 2006; Graz et al. 2011). For the Rhone drainage basin, the coupling of the ^{14}C and POC measurements of the SPM collected just before the land/sea interface reveals more mixing of different sources of organic matter, including a diversity of fossil, as well as modern, organic matter (Copard et al. 2018).

Beyond a flow of POC, close to 1.1 Mt year^{-1} delivered into the Beaufort Sea (Carrie et al. 2009) and coupled with an optical study method (maceral analysis), numerous parameters such as S1, S2, S3 and RC/TOC have provided information on the origin of organic matter conveyed by the Mackenzie River (Canada) and its main tributaries. Here, the river sediments appear to be more enriched in refractory POC (higher RC/TOC) than the tributaries draining lacustrine systems containing a large quantity of aquatic organic matter, whose signature is an enrichment in hydrocarbon compounds underlined by the high values of the S1 and S2 signals and of the HI and lower values of OI RE6 (Figure 9.5). Thus, the exploitation of these indicators provides information on the origin of the organic matter, its labile or refractory character and allows the lateral flows of POC to be specified. In the case of the

Mackenzie, the labile portion of the POC would represent 9% of exported POC, while in the case of the Rhone, the POC of fossil origin would represent 25%. Still using classical Rock-Eval® parameters coupled with other methods of studying (molecular) organic matter, Nádudvari and Fabiańska (2015) were able to highlight the importance of the fossil contribution (coal) and its different sources in the organic content of bank sediments deposited by the Bierawka River (Silesia, Poland). In the Amazon floodplain, the coupling of HI, OI, Tmax and TpeakS2 indicators and S2 signal shape with other (molecular) methods also allow large environmental changes to be shown through the fluctuation of different terrestrial and aquatic sources (Zocatteli et al. 2016).

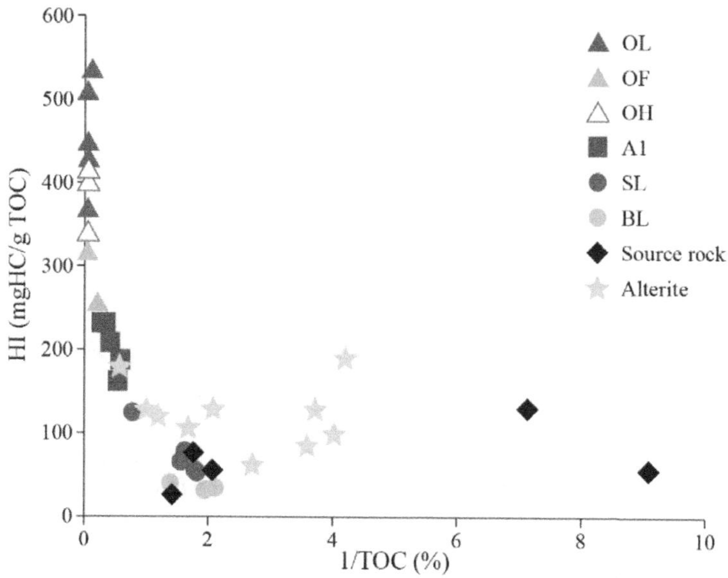

Figure 9.4. *Distribution of the main pedological sources (OL, OF, OH and A1), geological sources and surface formations (source rock and alterite) feeding the SPM of the instrumented sites of Draix according to the HI and 1/TOC values (according to Copard et al. (2006))*

However, few studies on lateral POC transfer use the Rock-Eval® method to characterize not only the Corg content of particles in transit, but also the quality of this organic matter. Given the ease with which these measurements are made, this method can only be promising, especially if high-frequency measurements in rivers are sought.

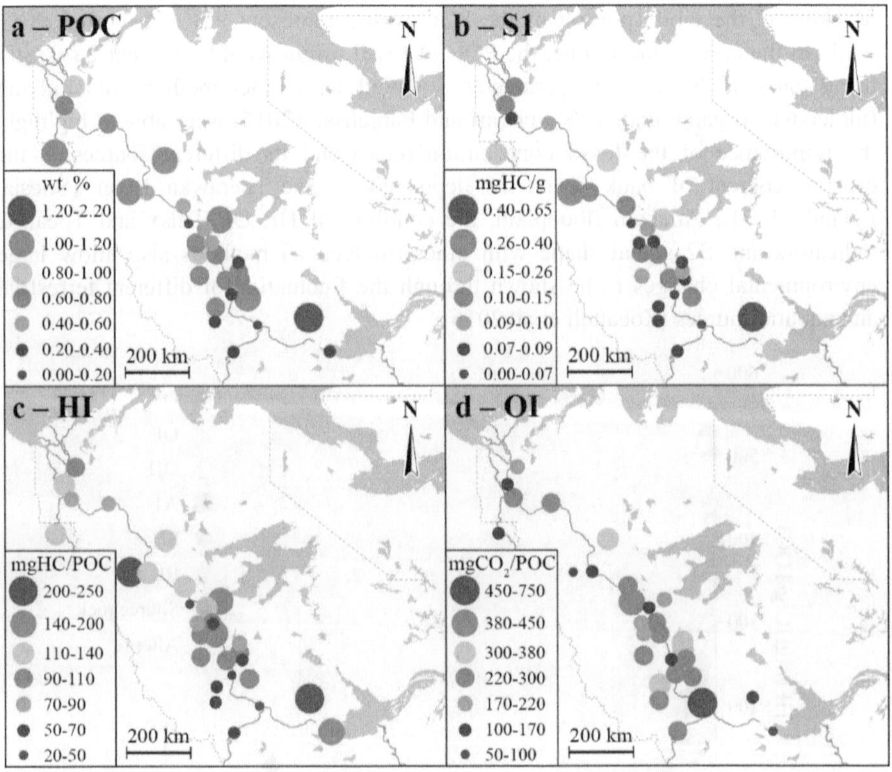

Figure 9.5. *Spatial distribution of POC, S1, HI and OI values of SPM of the Mackenzie River and its various tributaries (according to Carrie et al. (2009))*

9.4.2. *Lacustrine archives, reservoirs*

The Rock-Eval® method has belatedly been used to characterize the organic matter of recent continental sediments, particularly lacustrine, where the first works date back to the late 1980s (Talbot and Livingstone 1989; Bertrand et al. 1992; Lallier-Vergès et al. 1993). Most often, the TOC, HI, OI, S2 and Tmax parameters and more rarely R400, $S3CO_2$ and Tpeak S2 are coupled with other study methods to decipher paleoenvironmental or paleohydrological changes and their fluctuations over time (Siffedine et al. 1996, Meyers 1997, Meyers and Lallier-Vergès 1999, Di-Giovanni eSiffedine et al. 1996; Meyers 1997; Di-Giovanni et al. 1998; Meyers and Lallier-Vergès 1999; Wilkes et al. 1999; Ariztegui et al. 2001; Manalt et al. 2001; Noël et al. 2001; Jacob et al. 2004; Wohlfarth et al. 2004; Talbot et al. 2006; Millet et al. 2007; Mayr et al. 2009; Disnar et al. 2011; Giguet-Covex et al. 2011; Magny et al. 2012; Zocatelli et al. 2012b; Lavrieux et al. 2013; Leng et al. 2013;

Sebag et al. 2013; Simonneau et al. 2013; Chassiot et al. 2018; Lacey et al. 2018). For example, indicators such as R400 can provide information on the quality of organic matter stored in these environments prone to accumulate organic matter, which depends closely on hydrological fluctuations and therefore on climatic variations occurring during the Holocene (Lake Bourget, Chautagne marshes, France (Disnar et al. 2008); Figure 9.6).

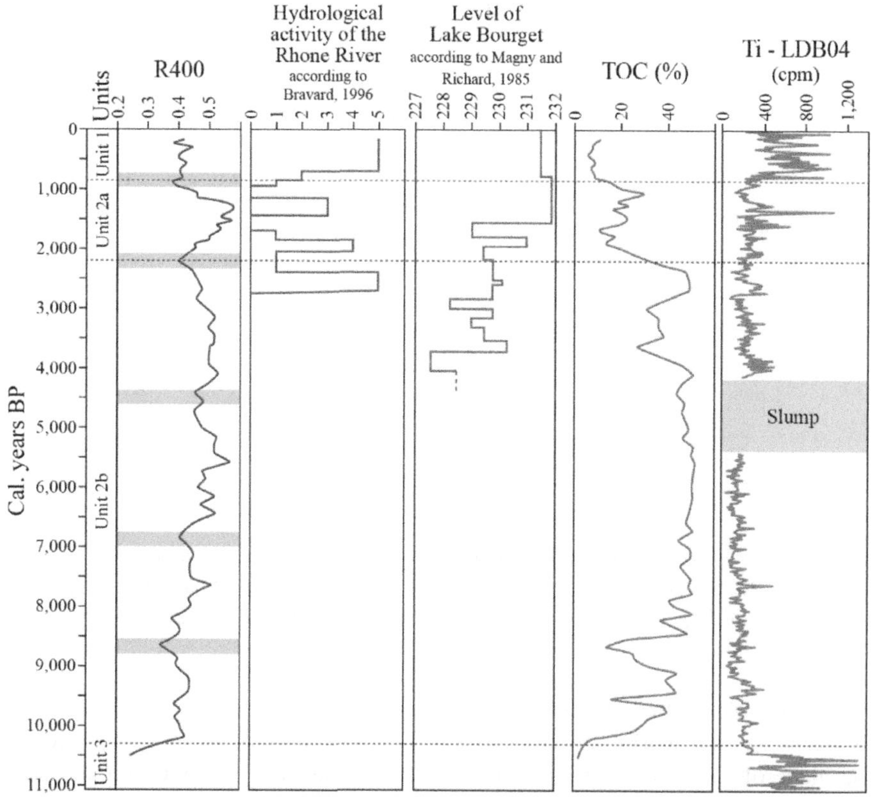

Figure 9.6. *Comparison between the evolution of the R400 and TOC parameters with markers of the hydrological activity of the Rhone River, the height of Lake Bourget and the detrital load highlighted by the titanium signal (according to Disnar et al. (2008))*

Other works link bathymetric variations, such as in Lake Caçò (Brazil), to early diagenetic processes that modify the quality of organic matter at the water/sediment interface (Boussafir et al. 2012). One example of the anthropogenic impact relates to the contributions of combustion residues resulting from deforestation that are trapped in sediments in the lakes of Saskatchewan, Canada. These are highlighted by tracking the RC/TOC parameter (Figure 9.7), which shows a notable increase toward the top of sedimentary cores (Ahad et al. 2015).

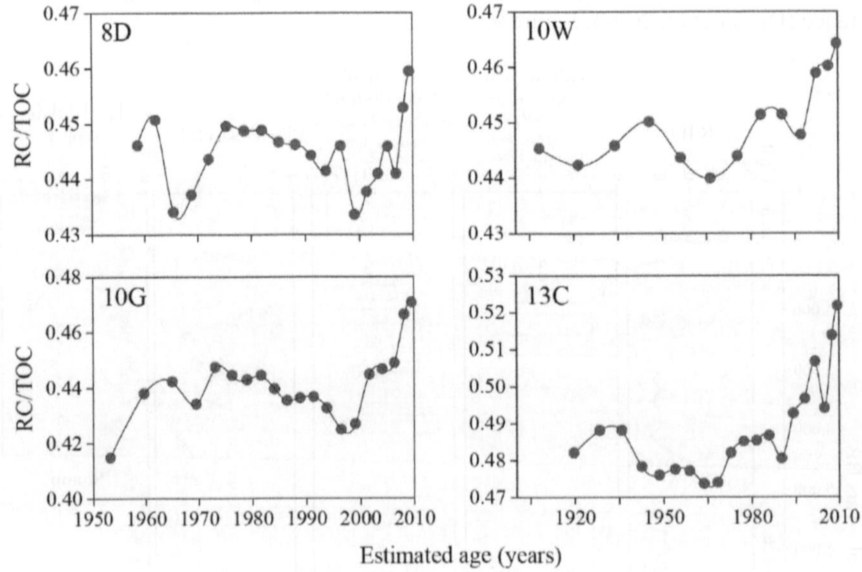

Figure 9.7. *Temporary evolution over several decades in the RC/TOC ratio values of lacustrine sediments (lakes in the state of Saskatchewan, Canada) (according to Ahad et al. (2015))*

As organic matter produced and trapped in lakes is at the origin of type I kerogens, several works have focused on the transformation of organic matter and the mode of formation of these kerogens. In sediments of Lakes Ochki and Beloe (Siberia, Russia), the HI measurements and signal S2 have thus shown that the formation of this kerogen begins as of early diagenesis from the neoformation of aliphatic macromolecules inherited from phyto- and zooplankton (Melenevskii et al. 2015), whereas macrophytes and bacteria appear to be precursors of steranes and hopanes (Melenevskii et al. 2011). Other works have been conducted on lacustrine sediments, such as those from Lake Wabamun (Alberta, Canada) to characterize organic matter contributing to the S2 and S3 parameters (Sanei et al. 2005). Thus, pyrolysis triggers the production of bitumen from 300°C, as well as the progressive appearance of pyrobitumen up to 650°C.

In sediments trapped in reservoirs, HI has been used as an indicator of aquatic primary productivity that increases with global warming in subtropical reservoirs in southern China (Duan et al. 2017). In a source-to-sink concept, parameters such as HI, OI and TOC can be used as a tracer of archived sediment sources. Thus, a simple van Krevelen pseudo-diagram made it possible to assess the detrital origin and the state of degradation of the sedimentary organic matter trapped in a reservoir (Kamech, Tunisia) (Ben Slimane et al. 2013). The modest size of the drainage basin coupled with the physicochemical conditions of the reservoir favors the preservation of the already-resistant organic matter from the agricultural soils constituting the bulk of the Kamech drainage basin.

Finally, the Rock-Eval® parameters or the exploitation of pyrolysis signals make it possible to validate the use of biomarkers, such as pentacyclic triterpene methyl ethers (Zocatelli et al. 2016) or fatty acids (Zocatelli et al. 2012a), or indeed validate markers from hyperspectral spectroscopy (Van Exem et al. 2019). For this last work, a transfer function was used to link a hyperspectral index with the TOC contents by the Rock-Eval® method in order to restore a very high-resolution TOC content in sediments of Lake Linné (Spitsbergen).

9.4.3. *Peatlands*

Peatlands store a large quantity of Corg and therefore play a significant role in the global carbon cycle. Here again, and coupled with other analytical techniques (optical, molecular, isotopic spectroscopic and geochemical), classical indicators (TOC, HI, OI and Tmax) and certain more recent indices such as R330 (Delarue et al. 2013), R400 (Gogo et al. 2011) and the I- and R-indices (Girkin et al. 2019) have been used to decipher environmental changes, in particular climate change (Van der Putten et al. 2008; Malounguila-Nganga et al. 2017). These changes are at the origin of the variability in organic contributions and the stabilization mechanisms of Corg that facilitate or decrease the efficiency of its storage (Bourdon et al. 2000; Hoş-Çebi and Korkmaz 2015; Swindles et al. 2017; Upton et al. 2018) and modulate the escape of this Corg to hydrosystems or the atmosphere, in the form of DOC, CO_2 and CH_4 (Cooper et al. 2019; Girkin et al. 2019). In turn, the modification of plant groups, of natural or anthropogenic origin (land conversion) is marked by modifications to the physicochemical properties of the organic matter that will be stored, altering or facilitating the efficiency of the C sink represented by these environments (Gogo et al. 2011; Cooper et al. 2019, Figure 9.8; Hudson et al. 2019, Figure 9.9). In these wetlands, which are very sensitive to hydrological variations, natural seasonal climate variations or environmental changes modulate the elevation of the piezometric surface, which is at the origin of fluctuations in redox conditions. For example, episodes of drought and flooding make it possible to alter or preserve the Corg (Delarue et al. 2011, 2013). In the case of a shallow aquifer, where the peatland becomes a marsh, the hydro(geo)logical context modifies the diversity of plant species and can facilitate aquatic primary production (Bourdon et al. 2000), which is easily identified with the increase in the values of the S2 parameter (Newell et al. 2016). At the land/sea interface, classical indicators can be used to trace sea-level fluctuations, as examined in the Dvina Bay (White Sea, Russia), where a continentalization marked by the establishment of peatlands from 2,800 years BP above more-marine facies was highlighted. This change is related to the spatial variations in Eurasian and Fennoscandian ice caps (Kemp et al. 2017).

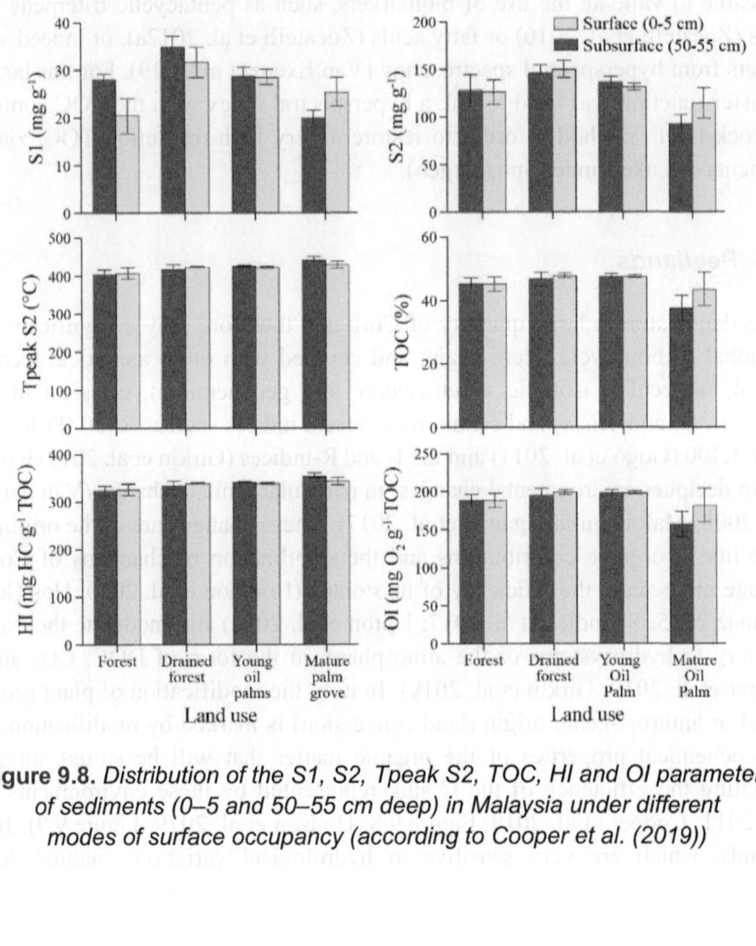

Figure 9.8. *Distribution of the S1, S2, Tpeak S2, TOC, HI and OI parameters of sediments (0–5 and 50–55 cm deep) in Malaysia under different modes of surface occupancy (according to Cooper et al. (2019))*

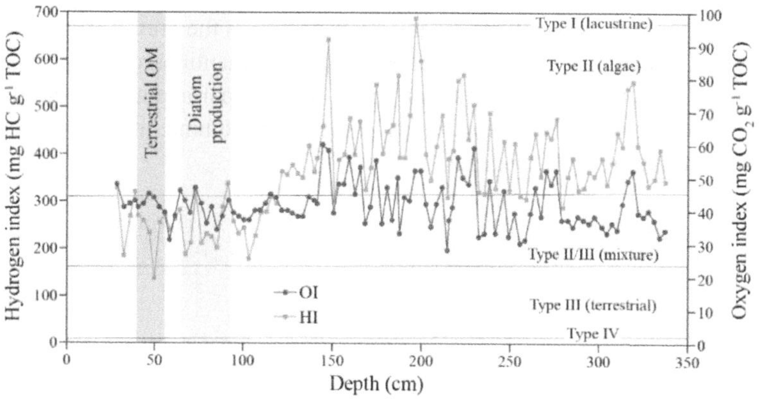

Figure 9.9. *Evolution of the HI and OI parameters of sediments of a minerotrophic peatland (Uinta Mountains, Utah, USA). The covariance, sometimes negative and sometimes positive, between these two parameters highlights different types of organic matter, terrestrial and aquatic (diatom), respectively (according to Hudson et al. (2019))*

Certain works focus more specifically on humification processes that can be traced by tracking the evolution of the values of the HI and OI parameters (Biester et al. 2014) and signals S1 and S2 with depth. These indicators are thus considered as markers of the intensity of decomposition of organic matter. Based on the concept of bio- and geopolymers defined by thermal-stability range (Sebag et al. 2006), the thermolabile compounds, whose Tpeak S2 < 400°C, assumed to be lipids, are obviously those that suffer the consequences of the degradation processes (Melenevskii et al. 2019). These thermolabile biomacromolecules, characterizing fresh organic matter, also demonstrate a strong OI and are leached during piezometry fluctuations (Delarue et al. 2011; Figure 9.10). These humification processes can also be examined in order to understand the mechanisms of stabilization and hence long-term storage of Corg, by tracking the signals of the RC, HI and OI parameters of a peat profile collected in the Northwest Territories of Canada (Zaccone et al. 2011).

In total, coupled with other study methods, the Rock-Eval® method provides information on the preservation of continental sedimentary organic matter, its origin and state of degradation, whose varying contributions depend on climatic and anthropogenic forcings. These forcings, modifying the mode of occupation of drainage-basin surfaces and hydrology, have a significant effect on the exported flows and on the physicochemistry of the detrital organic matter trapped and stored in the continental reservoirs, as well as on that of the aquatic primary production. In turn, this chemical variability of organic matter impacts the preservation potential of Corg in sediments, which also depends on (i) the quantities of Corg exported to these reservoirs, (ii) the sedimentation rates, (iii) the particle size of the deposited sediments and (iv) the redox conditions in the water column, at the water/sediment interface and during early diagenesis.

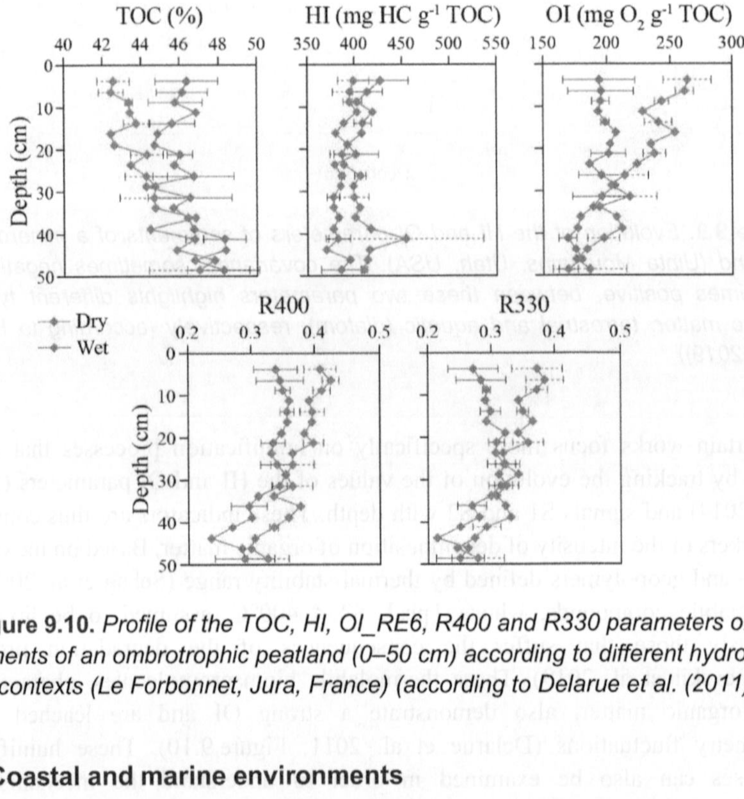

Figure 9.10. *Profile of the TOC, HI, OI_RE6, R400 and R330 parameters of the sediments of an ombrotrophic peatland (0–50 cm) according to different hydrological contexts (Le Forbonnet, Jura, France) (according to Delarue et al. (2011))*

9.5. Coastal and marine environments

It is estimated that 200 Tg year^{-1} of POC are carried out by rivers to oceans (Blair and Aller 2012). A significant proportion, between 55 and 75%, would be mineralized in the water column or at the water/sediment interface, while more than 99% of marine organic matter would be mineralized (Burdige 2007). In sediments, mineralization processes involve redox oscillations (oxic conditions), favored by bioturbation and microbial activity. In total, on a global scale, about $^{2}/_{3}$ of the Corg buried in marine sediments, particularly at deltas and continental margins, which represent 80% of Corg storage, would be of terrestrial origin (Burdige 2007). Therefore, although it undergoes intense mineralization, terrestrial organic matter is preferentially preserved. However, in open-marine environments, its concentration tends to decrease offshore (dilution process) in favor of the marine organic matter produced in situ.

Coupled with other study techniques (isotopic geochemistry and petrography), the Rock-Eval® method makes it possible to decipher the methods of preservation of sedimentary organic matter in different coastal or marine environments, which is mainly controlled by natural climate variability. Indeed, this climatic forcing impacts exchanges of water masses, which regulate the redox conditions that may or may not be favorable to the preservation of the Corg, but also the marine primary production and the contributions of terrestrial organic matter. The Rock-Eval® method has also proven to be a valuable tool to understand early diagenesis processes that modify the efficiency of Corg storage, as well as the quality of the organic matter. Some emblematic examples are provided below.

9.5.1. *Mangroves*

Coastal environments such as lagoons, fjords and mangroves have the capacity to store considerable quantities of Corg and play a key role in the carbon cycle. The parameters delivered by the Rock-Eval® method enable an understanding of the factors that influence the efficiency of the long-term preservation of C. These factors include the sedimentation rate, the redox conditions prevailing during early diagenesis, as well as the diversity of the types of organic matter (algal and microbial) constituting these biomes, as well as the allochthonous contribution, of marine or terrestrial origin. For the latter, they can either be brought directly via a land/sea continuum or be considered as "marine" contributions, but of terrestrial origin.

Thus, the analysis of sedimentary records of Guyanese mangroves reveals a limit between a pedogenetic level enriched in autochthonous organic matter and underlying marine sediments containing allochthonous organic matter, depleted in Corg and with low HI and high OI values, originating from the Amazon discharge (Marchand et al. 2008; Figure 9.11). In Senegal, the interpretation of Rock-Eval® signals demonstrates the importance of the nature of the species developing in mangroves where for genera such as *Avicennia*, the root system coupled with bioturbation facilitates suboxic conditions, conducive to the mineralization of Corg from the first 20 centimeters of sediment (Sakho et al. 2015). The indicators of the Rock-Eval® method underline the refractory nature of certain compounds such as phenols, derived from lignin, which tend to be stored in sediments, unlike much more labile carbohydrates (Marchand et al. 2003, 2008). In early diagenesis processes, the decomposition of autochthonous organic matter in soils leads to the production of dissolved inorganic carbon (DIC), which can migrate deep and favor carbonate precipitation (siderite or rhodochrosite) due to the prevailing anoxic conditions. Beneath the senescent mangrove, the high OI values would then reflect a greater quantity of these authigenic carbonates, resulting from the decomposition of higher quantities of organic matter (Marchand et al. 2003; Figure 9.12).

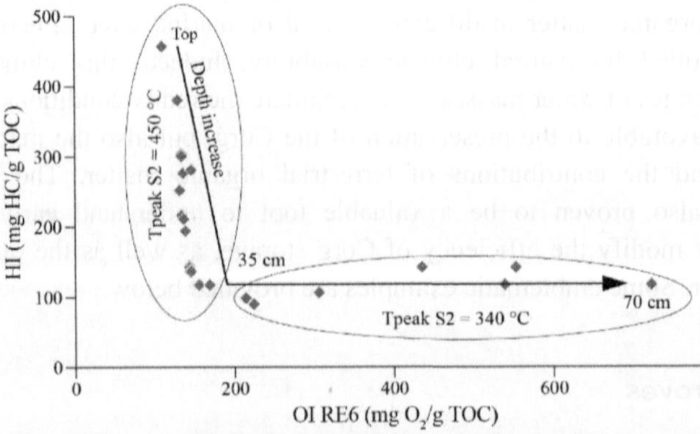

Figure 9.11. *Van Krevelen pseudo-diagram showing the evolution of the HI and OI RE6 parameters as a function of the depth of the organic horizon in a mangrove soil in French Guiana and the highly oxidized signature of the organic matter of the muddy sediments on which the mangrove develops (according to Marchand et al. (2008))*

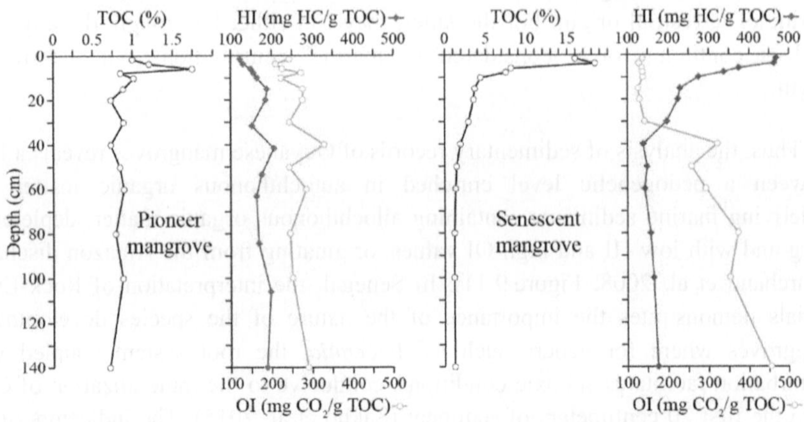

Figure 9.12. *Rock-Eval® geochemical profiles showing the evolution of the TOC, HI and OI parameters in pioneer and senescent mangrove soils in French Guiana (according to Marchand et al. (2003))*

9.5.2. *Epicontinental seas*

The evolution of the Rock-Eval® signature of Hudson Bay sediments (Canada) was studied by Hare et al. (2014). Terrigenous organic matter is subjected to intense mineralization as soon as it enters the bay, underlined by low HI values and high OI values. In contrast, at the NE of the bay, the high marine primary production is marked by slightly higher HI values, hence a higher HI/OI ratio. Coupled with the $\delta^{13}C$ of the organic matter (Figure 9.13), the use of the HI/OI ratio characterizes a mixing line between a non-woody terrigenous organic matter pole ($\delta^{13}C$ of ~ –23 ‰; HI/OI ~0.5) and a marine pole ($\delta^{13}C$ of ~ –21 ‰; HI/OI = 1). Two regions are exceptions to this logic: the outlet of a river where the fresher lignocellulosic matter is underlined by a value of $\delta^{13}C$ of ~ –26.5 ‰ and the center of the bay where the redox cycling of Mn in the sediments leads to very high OIs in organic matter with the heaviest isotopic signature.

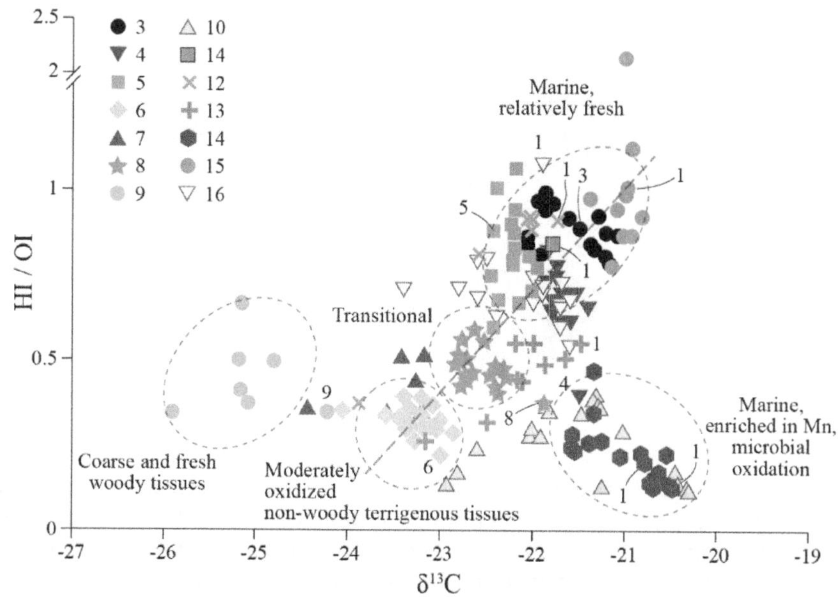

Figure 9.13. *Characterization of sources of organic matter in surface sediments of Hudson Bay (Canada) based on the carbon isotope ratio ($\delta^{13}C_{org}$) and the HI/OI ratio provided by Rock-Eval® analysis (according to Hare et al. (2014))*

9.5.3. *Deep turbidite systems*

The Congo River extends into a unique underwater canyon that cuts deep into the continental shelf. Thus, the active turbidite system is formed by a canyon/channel, levees and terminal lobes over an area of 21,300 km^2. The Congo sedimentary discharge is therefore largely channeled and descends in the form of turbid flows to a depth of 5 km, more than 800 km from the mouth (Figure 9.14). Added to this is a hemipelagic sedimentation composed of fine particles, the origin of which is both continental (plume of the river) and marine (biogenic, mineral and organic particles).

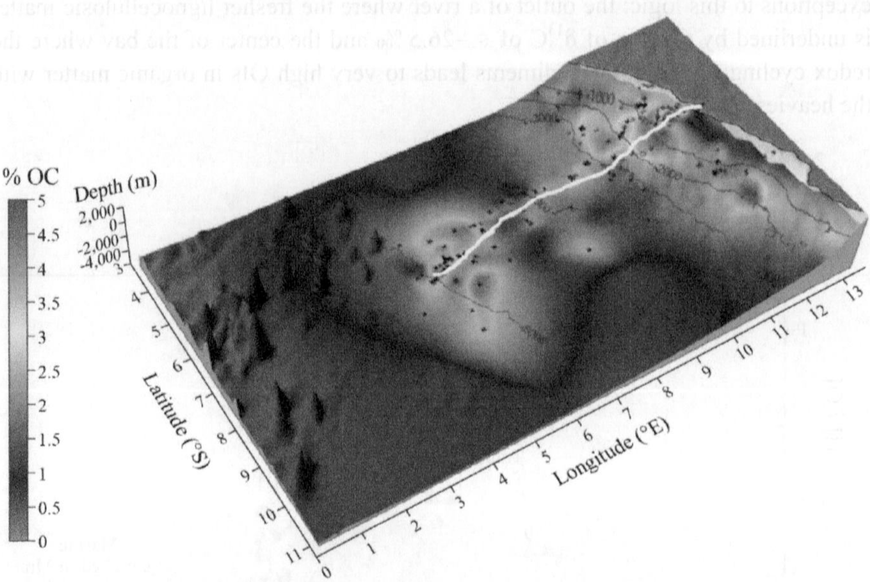

Figure 9.14. *Three-dimensional view off the mouth of the Congo River and outline (in white) of the canyon/channel of the turbidite system, which transfers sediments from the river at great depth. The map of TOC in surface sediments is based on more than 300 analysis points, and interpolation between the data points was performed by kriging. The TOC in surface sediments of the terminal lobes is 10 times higher than that outside the turbidite system (according to Baudin et al. (2020))*

Due to the TOC richness of SPM in the Congo River, the TOC contents of the terminal lobes reach 4–5%, whereas they are less than 0.5% outside the influence of the turbidite system (Figure 9.14). Terrigenous organic matter can be distinguished that is better preserved in the turbidite system than the marine organic matter deposited in hemipelagic environments, which record a much higher rate of mineralization (Figure 9.15). In the turbidite system, despite the particle size sorting performed by the currents, there is little variation in the TOC content and organic matter quality. Changes in sedimentation rates therefore have little effect on the TOC content in sediments (Baudin et al. 2010). Ultimately, the Congo turbidite system can be seen as an exceptionally efficient Corg sink where 0.74 Tg year^{-1} of Corg (Rabouille et al. 2019; Baudin et al. 2020) represents nearly 40% of the Corg stored in the South Atlantic Ocean. If the efficiency of this Corg sink is extended to the deep turbidite systems of other major world rivers (Amazon, Niger and Ganges–Brahmaputra), having functioned during glacial periods (with a lower sea level of 120 m), then these large rivers would play a new role in the global C cycle, beyond the lateral C transfer (Ciais et al. 2008).

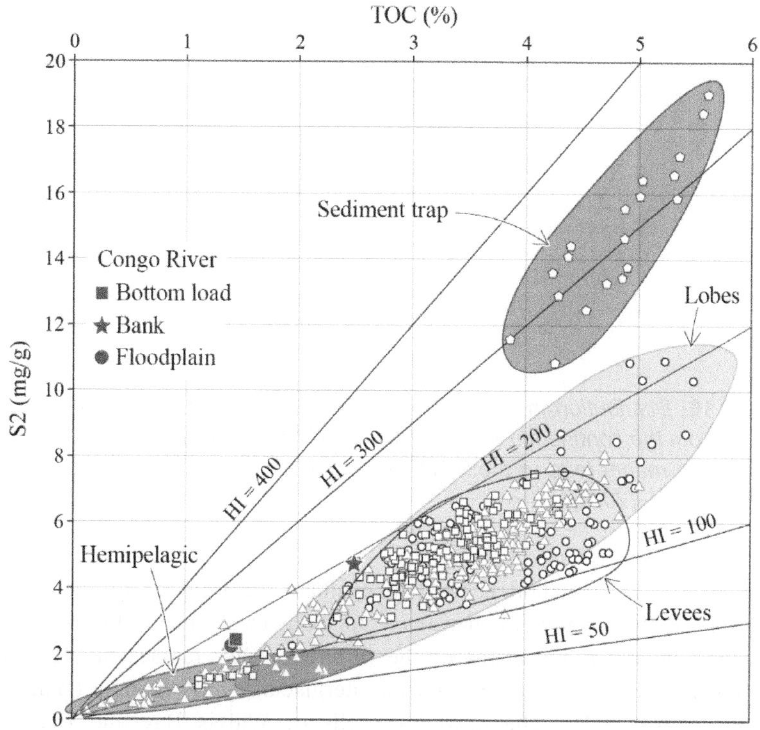

Figure 9.15. *Characterization in an S2–TOC diagram of the organic matter of the different sedimentary domains (levees, lobes and hemipelagic) associated with the turbidite system of the Congo River, compared with that at the outlet of the river (red dots) and with that originating from surface marine production (sediment traps). 95% of the organic matter deposited in the muddy and sandy turbidite facies comes directly from the river (according to Baudin et al. (2017b))*

9.5.4. *Upwellings*

Although subjected to intense mineralization in the water column and during early diagenesis, the marine organic matter that is predominant outside the influence of continental discharges can be durably stored in certain contexts such as upwellings. This is demonstrated by the example of the Benguela Current, a cold coastal current that rises along the SW African margin to join the warm South Equatorial Current. It is characterized by a succession of upwelling cells that induce the rising of cold, nutrient-rich waters stimulating primary production.

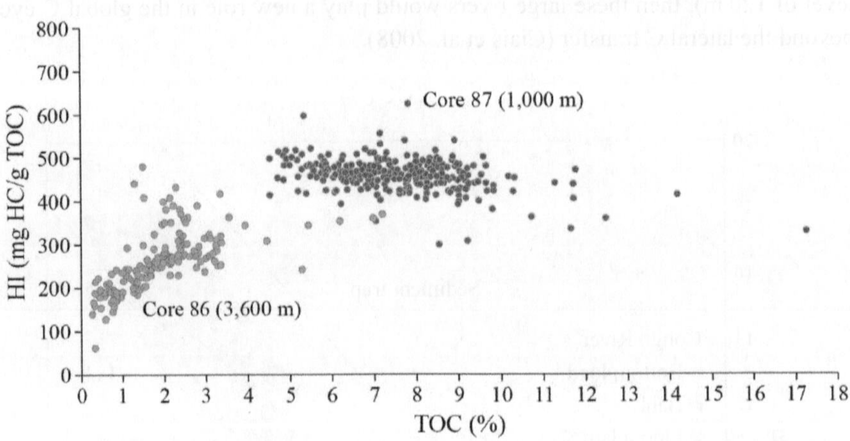

Figure 9.16. *Distribution in an HI–TOC diagram of samples of cores #86 and #87 collected from the continental slope off Namibia. Two pathways to preserve the same initial organic matter lead to these differences in Rock-Eval® parameters (according to Pichevin et al. (2004))*

The Rock-Eval® method is used to evaluate the trapping efficiency and the modalities of preservation of marine-origin Corg in sediments by also providing information on the redox states of the different water masses involved and the upwelling dynamics over several glacial/interglacial cycles (280 ka (Blanke 2004; Pichevin et al. 2004)). The upwelling cells of Walvis Bay and Lüderitz Bay (Namibia) are the most productive, with Corg contents of about 12% measured in the sediments of the shelf, as well as on the top of the continental slope. Based on the Corg contents, upwelling cells appear more active during glacial periods because flows of biogenic carbonate and detrital particles varied little between the two climatic stages (Pichevin et al. 2004). In a core located at a depth of 3,600 m (#86; Figure 9.16), the lowest TOC and HI values correspond to the interglacial periods when sediment was deposited in well-oxygenated bottom waters, while Corg flow from the surface was limited. In contrast, in a core collected at a depth of 1,000 m, the HI values are constant at approximately 450 mg HC/g TOC (#87; Figure 9.16), indicating that the water masses at the top of the slope were continuously anoxic. Lastly, coupled with optical and molecular analyses, the HI-TOC diagram highlights two types of organic matter that derive from a process of preservation by adsorption on the surface of clays or by natural sulfurization.

9.5.5. *Sapropel deposits*

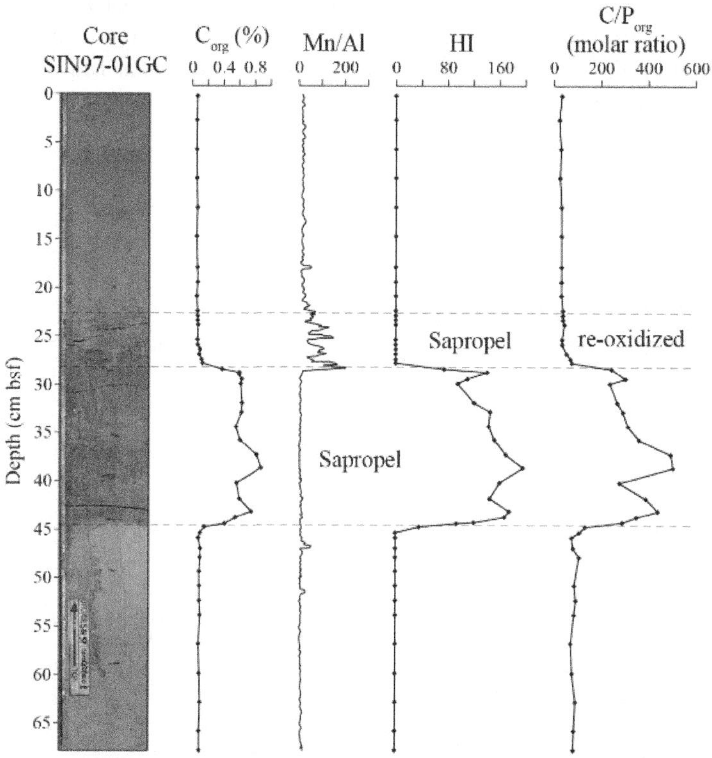

Figure 9.17. *Variations in TOC contents, HI, the Corg/phosphorus ratio and Mn contents normalized by Al in a core sampled from sapropel S1 (Aegean Sea). The upper part of the sapropel was reoxidized after its deposition (according to Gennari et al. (2009))*

The quaternary sapropel deposits in the eastern Mediterranean (Levantine Basin) coincide with periods of strengthening of the African monsoon causing a massive influx of fresh water via the Nile River. As a result, the fall in salinity of surface waters inhibits their immersion and leads to the establishment of dysoxic to anoxic conditions on the oxygen-poor bottom, favoring the preservation of organic matter mainly of marine phytoplanktonic origin in the sediments. Here, the Rock-Eval® method was coupled with analysis of chemical elements (notably P and Mn), to evaluate the significance of this trapping and the conditions of preservation of the Corg in relation to natural climatic variability (millennial to decadal solar cycles; Gennari et al. 2009). By way of example, the most recent deposition (10,000–7,000 years BP) occurred under dysoxic conditions, highlighted by the high Corg/P ratio values, favoring Corg trapping (Figure 9.17). However, the presence of Mn parameters across about 5 cm, precipitating in the form of oxyhydroxides under oxic conditions, clearly shows changes in redox state whose origin would be either an upward remobilization of Mn or episodes of downward incursion of oxygen from the bottom waters.

9.5.6. *Marine sediments of coral origin*

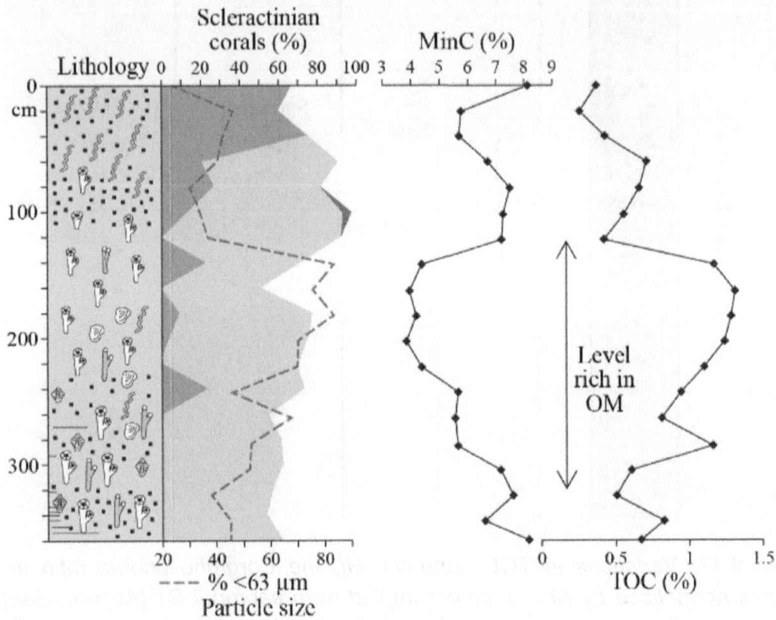

Figure 9.18. *Lithology of a core collected from a depth of 300 m in the Alboran Sea and changes in the assemblages of benthic corals and foraminifera, particle size, % mineral carbon (MinC) and organic carbon (TOC) over the past 14,000 years (according to Stalder et al. (2018))*

At a depth of approximately 300 m on a rocky promontory off Mellila (Morocco), bioconstructions, associating corals, bryozoans and benthic foraminifera, are well known in the Alboran Sea (West Mediterranean). Stalder et al. (2018) studied the dynamics of change in the coral population during the last glaciation (from 33,000 to 14,700 years) and during deglaciation at the beginning of the Holocene (since ~10,000 years; Figure 9.18). Isotopic data on corals and foraminifera, associated with Rock-Eval® data on sediment, indicate that surface waters had high marine primary production (high HI values) during the glacial period with well-ventilated bottom waters. In the interglacial period, the bottom waters, less well-ventilated, allowed the preservation of organic matter of terrestrial origin (lower HI; Figure 9.19). Changes in coral assemblages would then be controlled by the ecological needs of the species, sensitive to the availability of oxygen and nutrients, as well as to changes in bottom-water temperature.

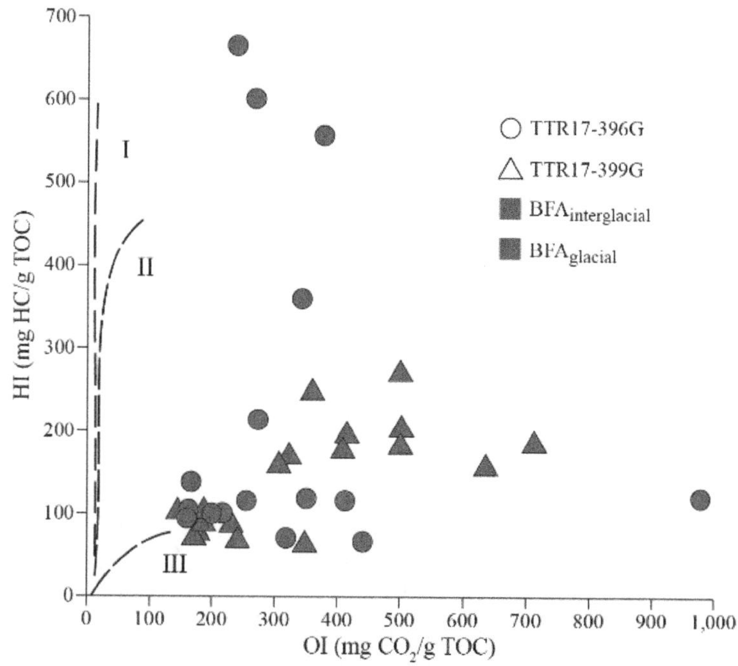

Figure 9.19. *Van Krevelen pseudo-diagram for two cores of coral-rich carbonate sediments collected from a depth of 300 m in the Alboran Sea. The transition between the glacial and interglacial periods is accompanied by a change in the nature of organic matter. The core in Figure 9.18 is TTR17-396G (according to Stalder et al. (2018))*

9.6. Relationships between organic matter and contaminant concentrations

In sediments, the relationships between organic matter and contaminants, whether organic or metallic, are well known. Indeed, sedimentary organic matter tends to be considered as a trap, making it possible to explain the different amounts of metals (Chin and Gschwend 1991) or of organic contaminants such as polycyclic aromatic hydrocarbons (Fernandez et al. 1999).

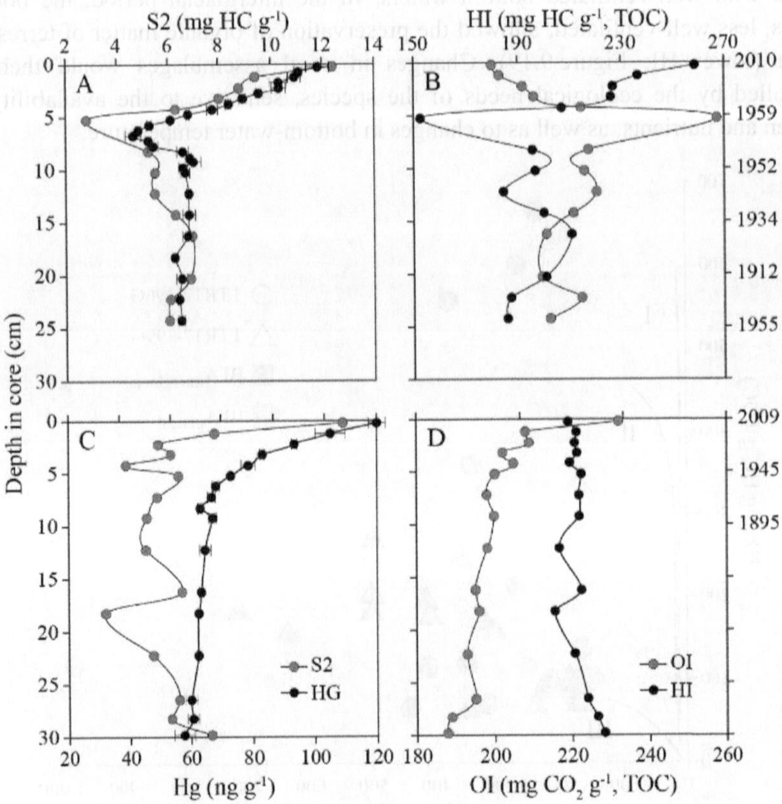

Figure 9.20. *S2, HI and OI parameter profiles compared to that of mercury in sediments from two lakes near Hudson Bay (Ontario, Canada) (according to Brazeau et al. (2013))*

In sedimentary archives from the Pearl River Delta (SE China), the highest concentrations of metallic trace elements show a correlation with the presence of algal organic matter characterized by high values of the S1 and S2 indicators (Duan et al. 2014). Similarly, it has been shown that the highest mercury concentrations were measured in the labile part, constituting the S1 parameter (Sanei and Goodarzi 2006; Wu et al. 2013) or the S2 parameter of lacustrine organic matter (Brazeau et al. 2013, Figure 9.20). This high Hg/labile organic matter affinity would be related to the physicochemical characteristics of organic matter, in particular its reactive surface and/or its sulfur content (Sanei et al. 2012). For the latter authors, interested in the mercury content prior to anthropogenic impact in a lake near the delta of the Mackenzie River (Canada), the increase in Hg concentration coincides with global warming, which stimulates lacustrine phytoplankton primary productivity. This Hg relationship with certain types of lacustrine aquatic organic matter explains why the TOC relationship of the sediment, including other types of organic (detrital) matter, with mercury is sometimes not obvious. In contrast, rates of mercury accumulation in peatlands are generally lower than in lacustrine sediments, although it is not possible to advance if humification processes are at the origin (peat collected in the Northwest Territories of Canada; Outridge and Sanei 2010). In this work, humification processes are tracked with Rock-Eval® indicators such as S2, RC, HI and OI coupled with petrographic observations. It thus appears that trace elements, notably metallic, show a strong affinity for organic matter of algal origin so that a high aquatic primary productivity, boosted by climate change, ultimately allows these contaminants to be concentrated in the sediments. Lastly, changes in the composition of lacustrine organic matter (terrestrial vs. aquatic) can also be seen as witnesses to the response of aquatic ecosystems to fluctuations in sources of continental contributions rich in trace metals attributable to anthropogenic activity (Vidy Bay, Lake Geneva; Gascón Díez et al. 2017). Such studies can be considered as decision support tools for defining sustainable development strategies aiming to maintain aquatic ecosystems in good ecological condition. In sediments trapped upstream of a dam (Eure, France), polycyclic aromatic hydrocarbons (PAHs) appear bound to the refractory fraction of the organic matter defined by the RC/TOC ratio (Gardes et al. 2020; Figure 9.21).

However, other studies conducted on lacustrine sediments in China show that PAHs with low molecular weight have a greater affinity for the labile fraction of algal organic matter (Wu et al. 2012). This relationship between the total PAH concentration and algal contribution to organic matter is observed to a lesser or greater extent in sediments trapped in dam lakes (Duan et al. 2015; Figure 9.22).

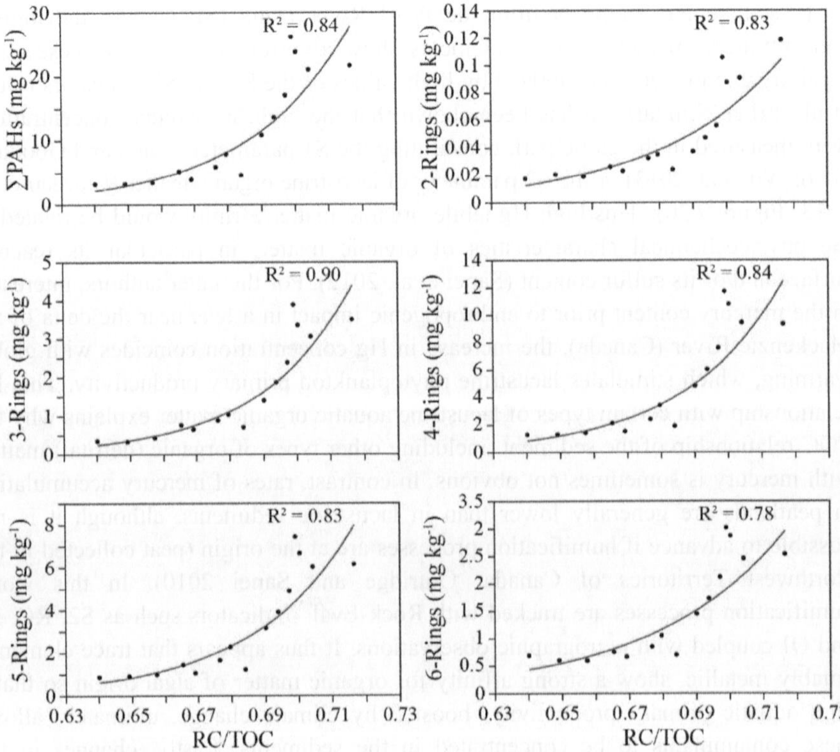

Figure 9.21. *Positive correlation between the RC/TOC ratio and different PAHs determined by the number of aromatic nuclei of river sediments trapped upstream of a dam (Eure, France) (according to Gardes et al. (2020))*

Although little used, the Rock-Eval® method seems to be a good tool for tracing organic matter versus contaminant relationships, particularly in terms of quality indicators. The labile organic fraction shows an affinity for certain metallic or organic contaminants, and in the case of a remobilization of sediments or degradation of labile organic matter, the Rock-Eval® method can evaluate the ability to restore these contaminants in aquatic ecosystems. Conversely, an affinity with the refractory fraction indicates that the contaminant can be lastingly trapped in sediments. However, these relationships are not yet clear and are worth investigating further. The fact remains that the Rock-Eval® method could be applied in the context of ecotoxicology-related issues.

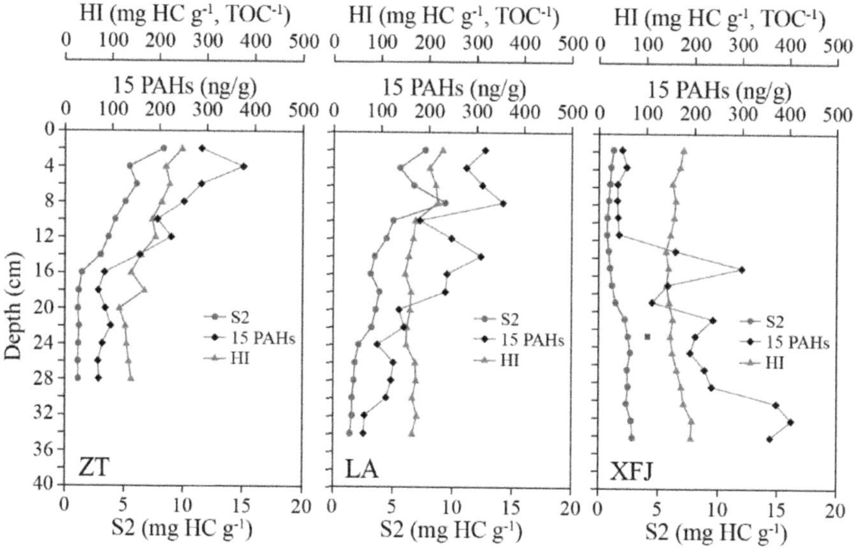

Figure 9.22. *Profile of the HI and S2 parameters compared to those of 15 PAHs and total neutral sugars from three sedimentary archives collected from dam lakes (Pearl River drainage basin, China) (according to Duan et al. (2015))*

10

Characterization and Evaluation of the Stability of Soil Organic Matter

Pierre BARRÉ[1], Lauric CÉCILLON[2] and Eva KANARI[1,3]

[1] *Géologie ENS, CNRS, Paris, France*

[2] *ENS/Université de Rouen Normandie, INRAE, Rouen, France*

[3] *Sorbonne Université, Paris, France*

10.1. Introduction

Soil organic matter (SOM) is particularly important for soil functioning (Koch et al. 2013; Keesstra et al. 2016). Thus, the water retention capacity of a soil and its ability to purify water increase with its organic matter content, and a soil rich in organic matter will be less subject to erosion. SOM is also the main source of matter and energy for soil organisms, notably microorganisms. Moreover, at the global scale, soils contain, in the form of SOM, about three times more carbon than the atmosphere and positive or negative variations in soil organic carbon stocks can significantly influence the atmospheric CO_2 concentration over short time periods (year/decades) (Eglin et al. 2010). Due to its importance for soil quality and climate regulation, SOM has been extensively studied in recent decades. These studies focus mainly on SOM characterization, as well as on its dynamics. Indeed, being able to know the residence time of carbon in the soil is essential in order to be able to predict the evolution of organic carbon stock in soils.

The Rock-Eval Method,
coordinated by François BAUDIN.

SOM consists of a mixture of various compounds of plant and microbial origin at various stages of decomposition, which interact to a lesser or greater extent with the mineral matrix of the soil (Kögel-Knabner 2002). This organic matter is difficult to characterize, with its analysis generally being carried out using complementary techniques such as elemental analysis (CHNS dry combustion), nuclear magnetic resonance spectroscopy (NMR ^{13}C or ^{15}N), mid-infrared spectroscopy, X-ray absorption spectroscopy or pyrolysis coupled with gas chromatography and mass spectrometry (GC-MS), each having its advantages and disadvantages (Kögel-Knabner 2000). Since the late 1990s, a growing number of studies have sought to evaluate the capacity of Rock-Eval® thermal analysis to characterize the SOM and evaluate its dynamics. Compared to most other SOM characterization techniques, Rock-Eval® presents the advantage of being fast (approximately 1 h per sample), inexpensive (excluding depreciation of the device) and requiring only a small quantity of matter to perform an analysis (a few tens of milligrams of finely ground soil).

This chapter provides a summary of the scientific works using Rock-Eval® thermal analysis to study SOM. It will explore the advances enabled by the Rock-Eval® method for the characterization of SOM and evaluation of its dynamics. It will mention some gray areas or precautions to be taken and will discuss the broad prospects opened up by the use of Rock-Eval® in studying organic carbon dynamics in soils.

10.2. Relevance and general teachings of Rock-Eval® analysis of soil samples

10.2.1. *Analyzing soils with Rock-Eval®, a reasonable diversion*

Rock-Eval® was of course not originally developed to analyze SOM. The first works on SOM date back to the late 1990s (Di-Giovanni et al. 1998, 2000) and show quite clearly that Rock-Eval® thermal analysis of soil samples is possible and can be scientifically enlightening. To analyze SOM with Rock-Eval®, however, it is preferable to start the pyrolysis phase at a lower temperature than for rock samples because organic matter in soils is more thermolabile than that of kerogens that is several tens to hundreds of millions of years old (Disnar et al. 2003; Hetényi et al. 2005). The thermograms obtained are fairly similar to those observed for source rocks or sediments. There are, however, several notable differences. In the absence of soil contamination by organic pollutants with low molecular weight (see

Chapter 12; Lafargue et al. 1998a), the S1 parameter is often negligible for soil samples (Disnar et al. 2003). For most soil samples, the S1 parameter likely does not correspond to free hydrocarbons, but rather to the thermal cracking of weak chemical bonds (Hetényi et al. 2005). Moreover, the molecules produced during the pyrolysis phase are probably richer in oxygen than the source-rock hydrocarbons. Since the FID is less sensitive to oxygenated compounds, such as polysaccharides (Disnar et al. 2003; see section 2.3.1), the quantity of hydrocarbons determined with the S2 parameter is probably underestimated. In addition, the carbon content in the compounds measured by the S2 signal has been estimated at 83% in the case of source rocks (Espitalié et al. 1977a). If the compounds measured in the S2 signal are richer in heteroatoms (O as well as N), their carbon content is likely less than 83%. It is thus possible that, when analyzing soil samples, the quantity of S2 may be underestimated, but its C content is slightly overestimated.

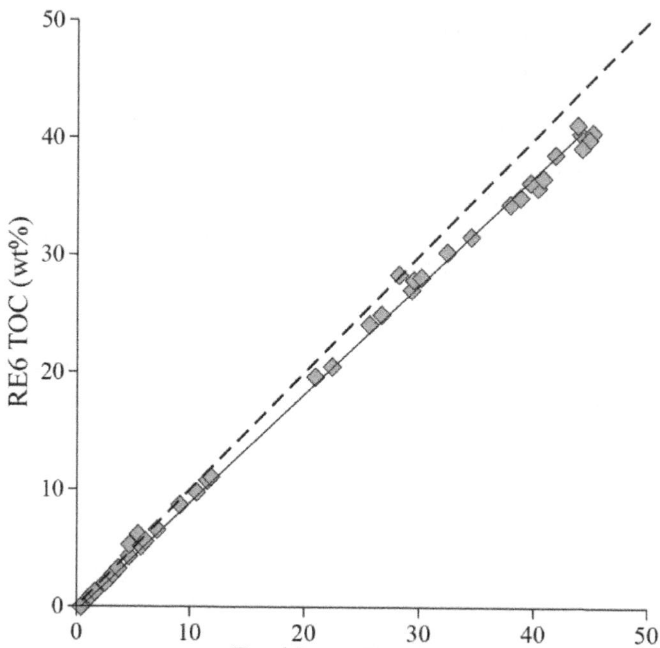

Figure 10.1. *Values of TOC contents measured by Rock-Eval® (RE6 TOC) as a function of the values of TOC contents measured by elemental analysis (dry combustion) for approximately 100 samples from a variety of pedoclimatic contexts. Figure from Disnar et al. (2003). For a color version of this figure, see www.iste.co.uk/baudin/rock-eval.zip*

Although these possible inaccuracies should be kept in mind, they remain relatively minor as the soil organic carbon (SOC) content measured by Rock-Eval® (RE6 TOC) is always close (±5% to ±10%) to that determined by elemental analysis in dry combustion. Figure 10.1 shows the TOC measured by elemental analysis as a function of the TOC measured by Rock-Eval® for soil and litter samples. Although the correlation between the two values is excellent, Rock-Eval® slightly but systematically underestimates the TOC, with underestimation generally increasing with the TOC content. This example is fairly representative, and similar results have been observed by several authors (notably Saenger et al. (2013)). Disnar et al. (2003) suggest that due to the systematic nature of TOC underestimation, it is possible to correct it. However, as this is low (of the order of magnitude of the TOC measurement accuracy), generally no correction is applied in the majority of published scientific works. In non-carbonate soils, by adding the value of the MinC parameter (non-zero despite the absence of mineral carbon) to that of the TOC, the correlation between [TOC + MinC] and TOC measured by elemental analysis in dry combustion is improved (Volland-Tuduri 2005).

Despite the slight underestimation of TOC discussed above, Rock-Eval® appears to be a very relevant method for the analysis of SOC content. In addition, it enables information to be obtained on soil organic carbon without pretreatment in carbonate soil samples and also enables the inorganic carbon of the soil to be quantified (MinC parameter). Beyond quantifying soil organic and inorganic carbon contents, Rock-Eval® thermal analysis provides information on elemental stoichiometry: H, via the hydrogen index (HI); O via the oxygen index (OI) and on the thermal stability of the SOM (Espitalié et al. 1977a; Disnar et al. 2003). This explains its use in a significant number of works focusing on SOM, although, as already highlighted by Feller et al. (2010), Rock-Eval® thermal analysis is still used by only a limited number of Soil Science research teams.

Reference	Pyrolysis start and end temp.	Pyrolysis start isotherm	Pyrolysis ramp	Oxidation start and end temp.	Oxidation start isotherm	Oxidation ramp	Isotherm at the end of oxidation
Disnar et al. (2003)	200–650°C	?	30°C min^{-1}	400–750°C	?	?	?
Hetenyi et al. (2005)	300–600°C	4 min at 300°C	25°C min^{-1}	Plateau at –600°C	Not applicable	Constant temp.	Not applicable
Sebag et al. (2006)	200–650°C	?	30°C min^{-1}	300–850°C	?	?	?
Saenger et al. (2013)	200–650°C	?	30°C min^{-1}	400–750°C	?	?	?
Delarue et al. (2013)	200–650°C	?	30°C min^{-1}	?	?	?	?
Gregorich et al. (2015)	300–650°C	300°C, duration?	25°C min^{-1}	300–850°C	300°C, duration?	20°C min^{-1}	?
Albrecht et al. (2015)	200–650°C	3 min at 200°C	?	400–850°C	3 min at 400°C	?	?
Barré et al. (2016)	200–650°C	?	?	300–850°C	?	?	?
Cécillon et al. (2018)	200–650°C	?	30°C min^{-1}	300–850°C	?	20°C min^{-1}	?
Soucémarianadin et al. (2018)	200–650°C	3 min at 200°C	30°C min^{-1}	300–850°C	?	20°C min^{-1}	?
Cécillon et al. (2018)	200–650°C	3 min at 200°C	30°C min^{-1}	300–850°C	1 min at 300°C	20°C min^{-1}	5 min at 850°C

Table 10.1. *Main acquisition parameters used in various studies on SOM characterization by Rock-Eval® thermal analysis*

Unfortunately, these different works do not always use the same acquisition parameters of the different Rock-Eval® thermograms, as illustrated in Table 10.1, which presents the main ones, used in several studies on the characterization of SOM by Rock-Eval®. Thus, Disnar et al. (2003) proposed starting pyrolysis at 200°C and ending it at 650°C with a temperature ramp of 30°C/min, and they limited oxidation to the interval 400–750°C with an unspecified temperature ramp. Gregorich et al. (2015), meanwhile, used very different acquisition parameters. Their temperature at the start of pyrolysis is 300°C and their temperature ramp in pyrolysis is 25°C/min (Table 10.1). It is also notable that the temperature ramp used in oxidation is not indicated in several works, and that the presence of an isothermal phase at the start of pyrolysis and oxidation, as well as at the end of the oxidation ramp is rarely mentioned. When they are indicated, the acquisition parameters of the oxidation phase most often correspond to the Rock-Eval® Bulk Rock method (300–850°C at a rate of 20°C/min, preceded and followed by short isothermal phases at these temperatures). The Bulk Rock method is also recommended for the oxidation phase because it enables the determination of the inorganic carbon (MinC, see Chapter 13), which is particularly significant for carbonate soils. While it would be desirable to harmonize the acquisition parameters for Rock-Eval® analysis of SOM, the acquisition parameters used by the different research teams are nevertheless close enough that most published results can be compared.

10.2.2. *A marked evolution in Rock-Eval® parameters with soil depth*

10.2.2.1. *Rock-Eval® parameters related to the bulk chemistry of soil samples*

Following the pioneering work of Di-Giovanni et al. (1998, 2000), Disnar et al. (2003) and then Sebag et al. (2006) used Rock-Eval® to rapidly characterize the SOM of a wide variety of soils. Disnar et al. (2003) thus present the results of the analysis of around a hundred litter samples and soil organomineral horizons from 24 sites collected in different regions of the world. A subsequent study also synthesized Rock-Eval® data from 263 sites worldwide (more than 1,300 soil samples; Sebag et al. 2016).

The results confirm the clear influence of soil depth on the values of the Rock-Eval® parameters of SOM across a wide pedological diversity, which was identified as of the first, more occasional, studies (Di-Giovanni et al. 1998, 2000). This is particularly notable for the HI and the OI. Indeed, SOM tends to dehydrogenate (decrease in HI) and oxidize (increase in OI) with soil depth (Figure 10.2), resulting in a frequent decrease in the HI/OI ratio with soil depth (Disnar et al. 2003; Hetényi et al. 2005). These changes in the HI and OI with soil depth have also been noted in the organomineral horizons of forest and grassland soils of the French Alps, the

Massif Central and the Swiss Jura (Lavrieux et al. 2012; Hasinger et al. 2015; Bajard et al. 2017), as well as in the sites of several forest monitoring networks in Metropolitan France (RENECOFOR, GNB and ALTER). Cécillon et al. (2017), Soucémarianadin et al. (2018b) or Collet et al. (2019) thus observed that soil depth is one of the factors that best explains the variability of HI and OI in these forest soils.

Within the organic litter layer, the HI values of intact litter horizons (OL; > 300 mg HC g^{-1} TOC; Figure 10.2) correspond to what is expected for the material of plant origin slightly transformed by soil microorganisms. The HI value clearly decreases during the process of microbial decomposition of litter (from intact litter (OL) horizons to fragmentation (OF) and humified (OH) horizons) and wood debris buried in the soil (Disnar et al. 2003; Sebag et al. 2006, 2016; Fujisaki et al. 2015). These same studies show that the OI tends to increase during the litter decomposition process (Figure 10.2), although this tendency is less obvious within the same humiferous horizon and is not observed during the decomposition of wood debris buried in the soil.

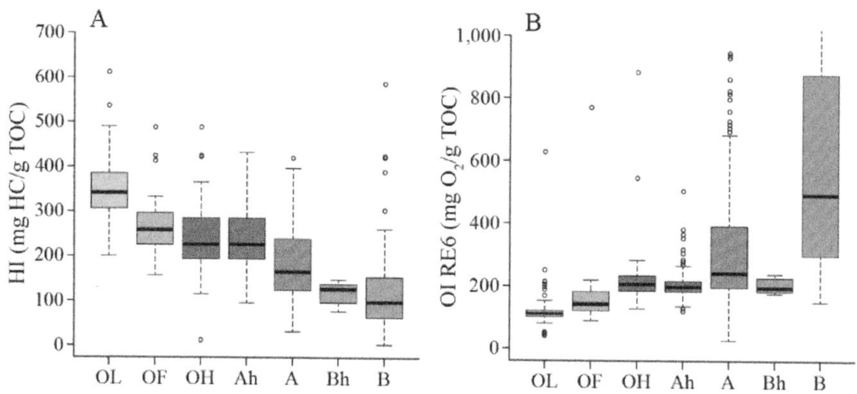

Figure 10.2. *Evolution of Rock-Eval® HI and OI indicators with soil depth, represented here by the succession of litter (OL, OF and OH) and soil (Ah, A, Bh and B) horizons. The data come from 1,385 samples collected from different regions of the world. Figures from Sebag et al. (2016). For a color version of this figure, see www.iste.co.uk/baudin/rock-eval.zip*

The link between HI, OI and soil depth is more complex for deeper soil cores (up to several meters) and in peatlands (Histosols) (Upton et al. 2018; Cooper et al. 2019). In this case, the evolutions in the OI and HI are interpreted as differences in the degree of preservation of organic matter and are then often used as indicators of evolutions of paleoenvironmental conditions (Van der Putten et al. 2008; Zaccone et al. 2011; Biester et al. 2014; Newell et al. 2016).

10.2.2.2. *Rock-Eval® parameters related to the thermal stability of soil samples*

In addition to the parameters related to the bulk chemistry of the sample (elemental stoichiometry; HI and OI), several studies have also reported the evolution in parameters relating to SOM thermal stability with soil depth. Di-Giovanni et al. (1998, 2000) were the first to observe an increase in the Tmax parameter of the S2 signal with the depth of the soil. Disnar et al. (2003) confirmed these results over their wider pedological range and observed the same for the Tpeak parameter of the S2 signal (see Behar et al. (2001) for details on the relationships between Tmax and Tpeak S2, or section 2.4.3 of this book). The low Tpeak values observed (320–390°C) in litter horizons are typical of the temperatures at which biological compounds such as polysaccharides or lignin crack in pyrolysis. These results were confirmed on the even more extensive dataset collected by Sebag et al. (2016). The increase in the thermal stability of SOM (S2 signal) with soil depth is also observable in the surface parts of peatland organic soils (Histosols; see, for example, Delarue et al. (2011)) in the 0–50 cm layer.

The variation in the shape of the S2 signal with soil depth was exploited to propose new Rock-Eval® indices. R400 (Disnar et al. 2003) and the R-index (Sebag et al. 2016), directly derived from R400, then the I-index (Sebag et al. 2006, 2016) and T-50 (Gregorich et al. 2015; Soucémarianadin et al. 2018b; see Chapter 3 on Rock-Eval® indicators) have been proposed, for example. The I-index increases with the quantity of thermolabile hydrocarbons, whereas the R-index is at a maximum when the proportion of thermally resistant hydrocarbons (> 400°C) increases. Sebag et al. (2016), Matteodo et al. (2018) and Soucémarianadin et al. (2018b) clearly observe that the I-index decreases with soil depth, while the R-index and T-50 of the S2 signal increase. It should be noted, however, that contradictory results on the evolution of the I and R indicators with soil depth were reported recently in Senegalese Arenosols by Malou et al. (2020).

Variations in the shape of Rock-Eval® thermograms other than the S2 signal with soil depth were generally less investigated. Nevertheless, Soucémarianadin et al. (2018b) and Collet et al. (2019) observed an increase in the T-50 parameter of the S4 signal of the SOM of forest soils with depth (see Chapter 3 on Rock-Eval® indicators).

10.2.3. *Rock-Eval® signal specificities of certain soil types, soil horizons and soil aggregates*

While soil depth strongly influences the values of most Rock-Eval® parameters, the characterization of SOM by Rock-Eval® also provided other interesting results, with several authors noting the effect of soil type on the SOM characteristics observed by Rock-Eval®. Disnar et al. (2003) noted the highest oxidation of organic matter in Podzosols, which was not necessarily found within forest sites of mainland France (strong HIs, low OIs in Podzosols compared to other soil types; Cécillon et al. 2017; Soucémarianadin et al. 2018b). Saenger et al. (2013) observed higher HI values in the surface horizons of anthroposols marked by sheep pastoral activity, relative to the other alpine soils that they studied. In a work conducted on 53 French forest sites, Soucémarianadin et al. (2018b) observed that while the thermal stability of SOM of eutric Cambisols is lower at the surface (0–10 cm) than that of Podzosols and dystric Cambisols, the opposite is observed for deep layers (40–100 cm). The effect of soil type and lithology on SOM thermal stability has also been observed in soils of the Swiss Alps, notably for deep soil layers (Matteodo et al. 2018). Climate also seems to influence the SOM Rock-Eval® signature. Soucémarianadin et al. (2018b) thus observed that SOM of mountain forests has a lower thermal stability on average than SOM of lowland forests in mainland France.

Large differences can be observed in Rock-Eval® parameter values for soils at the same depths. Disnar et al. (2003) attribute these differences notably to differences in mineral matrix and land cover. Indeed, cultivated soils appear to have lower HI values than forest or grassland soils. Zocatelli et al. (2012a) highlighted that SOM of forest Andosols is often characterized by a higher OI than Andosols covered by grasslands, within a watershed in the Massif Central. An increase in surface SOM linked to exogenous contributions or soil non-tillage increases the values of the HI (Malou et al. 2020) and S2 indicators (Cooper et al. 2021). However, it still remains for a sound demonstration of the effect of land use, vegetation cover and management practices to be performed.

The influence of several physicochemical properties of soils on the Rock-Eval® signature of SOM has been studied. Soil texture appears to influence SOM thermal stability. Romanens et al. (2019) showed lower R-index values in tropical Arenosols, with a very sandy texture, compared to other soil types in the Chobe region of Botswana. However, the effect of texture on SOM thermal stability depends on the soil depth; thus, it was no longer observable when soil samples collected at different depths in forest soils were combined (Cécillon et al. 2017; Soucémarianadin et al. 2018b). On the other hand, soil pH strongly influences the Rock-Eval® signature of SOM; the OI increases, the HI decreases and the thermal stability of SOM increases with soil pH (Cécillon et al. 2017; Soucémarianadin et al. 2018b). In these same studies of forest soils in mainland France, the C/N ratio of SOM was related to its thermal stability evaluated by various Rock-Eval® indicators (T-50 of the S2 and S4 signals): the thermal stability of SOM decreases when the C/N ratio increases (Cécillon et al. 2017; Soucémarianadin et al. 2018b).

The presence of "geological" or "fossil" organic carbon or pyrogenic carbon in the soil influences the Rock-Eval® signature of SOM. Indeed, a soil sample enriched in organic carbon of pyrogenic origin will have lower HI and OI values and will often present higher Tmax or Tpeak S2 values, while a sample rich in organic carbon of geological origin will often present low OI values, and variable HI values, depending on the source of the fossil carbon (Di-Giovanni et al. 2000; Bodineau et al. 2002; Volland-Tuduri 2005; Copard et al. 2006; Hetényi et al. 2006; Poot et al. 2009; Graz et al. 2010, 2011, 2012; Vindušková et al. 2015; Chassé et al. 2021).

Finally, studies suggest that Rock-Eval® thermal analysis could identify a specific SOM signature of biogenic structures produced by earthworms, compared to other soil or compost aggregates (Schomburg et al. 2018; Bottinelli et al. 2020; Le Mer et al. 2020). However, the effect of these soil organisms on the Rock-Eval® I and R parameters strongly depends on the physicochemical properties of the soil (Schomburg et al. 2019). Current results suggest that the activity of ecosystem engineering organisms, such as earthworms or termites, decreases the thermal stability of SOM, represented by the RC/PC ratio or by the T-50 indicators of the Rock-Eval® S3 and S4 signals (Romanens et al. 2019; Bottinelli et al. 2020).

10.3. Difficulties in linking the Rock-Eval® signature and chemical composition of soil organic matter

10.3.1. *Contributions of Rock-Eval® analysis of pure compounds*

Rock-Eval® analyses of pure compounds were carried out in an attempt to interpret the results obtained on soil samples in greater detail. Carrie et al. (2012) probably carried out the most complete study on the subject using about 15 pure compounds, analyzed in a sand matrix. Unfortunately, their results are not fully comparable to the results of other works as they used a protocol with a pyrolysis start temperature of 300°C and not 200°C, as is classically done for Rock-Eval® soil analysis (Disnar et al. 2003; see Table 10.1). Nevertheless, the work of Carrie et al. (2012) provides some interesting information.

They notably observe that the S2 signal of pure compounds is generally bimodal. The proportion of carbon associated with the S2 signal is much higher for lipids than for the other compounds. The lowest quantities of S2 are measured for lignins and carbohydrates, the proteins presenting intermediate values on average (Figure 10.3). It is also notable that the quantity of hydrocarbons released before 400°C is higher for lipids, in proportion to the S2 signal. Carbohydrates and lignins, on the other hand, contribute more strongly to the S3 signal. This is also the case for plant debris, which makes it regrettable that a significant part of works on soils essentially only consider the S2 signal.

Sebag et al. (2016) showed that the analysis of pure organic compounds (vanillin, cellulose and lignin) and natural samples (plants, charcoal and geological charcoal) makes it possible to limit the values of the Rock-Eval® I and R parameters of the SOM of a wide range of soil samples.

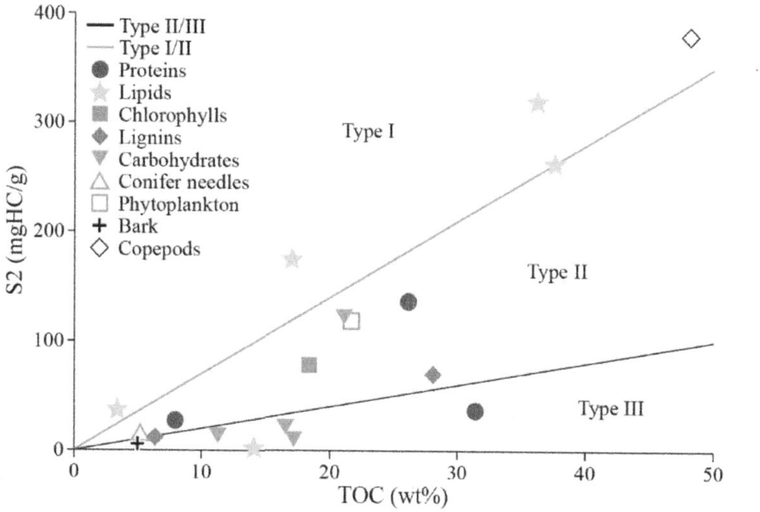

Figure 10.3. *Rock-Eval® S2 signal as a function of the TOC parameter for different pure compounds and natural materials. The majority of lipids and copepods that are rich in lipids have the strongest S2 signals for a given TOC, while carbohydrates and plant materials have the weakest S2 signals (according to Carrie et al. (2012)). For a color version of this figure, see www.iste.co.uk/baudin/rock-eval.zip*

The results obtained on pure compounds or natural samples of plants or coals are illuminating, but they do not take into account the full complexity of SOM and above all, they do not consider the effect that the mineral matrix can have on the Rock-Eval® signals of SOM. Indeed, the mineral matrix of the soil can modify the Rock-Eval® signals by catalyzing reactions or by retaining certain effluents (Zegouagh et al. 2004; Volland-Tuduri 2005). Disnar et al. (2003) and Zegouagh et al. (2004) thus pointed out that the mineral matrix of the soil can lead to the retention of significant proportions of pyrolysis products. Interactions between organic molecules and mineral surfaces also likely modify the temperature at which the molecules crack (see section 9.3). Nevertheless, Disnar et al. (2003) note that for a wide range of soils, the S2 versus TOC correlation curve tends toward the origin of the axes, suggesting that the mineral matrix has no significant effect and there is therefore no or very minimal alteration of the HI values for SOM-poor samples. Information on the behavior of pure compounds in the presence of different mineral matrices representative of those found in soils show that the matrix effect is highly variable according to the compound being considered and the nature of the mineral matrix (see section 9.3).

10.3.2. *Links between chemical composition of soil organic matter determined by various methods and Rock-Eval® signature*

Certain works have sought to link the chemical composition of SOM evaluated using spectroscopic techniques, selective extractions or GC-MS pyrolysis to the Rock-Eval® signature of this organic matter. It should be noted that several of these works have been carried out on pure organic matter (composts or particulate organic matter) or soil samples very rich in SOM (Histosols, within peatlands), and not on organomineral soils. Nevertheless, several works on soil organomineral samples have carried out multi-technique characterizations of SOM, including Rock-Eval® thermal analysis.

Within two types of peatlands (Histosols), Delarue et al. (2013) observed a strong positive correlation between the OI and the main absorption band attributed to polysaccharides (1,050 cm^{-1}) in mid-infrared spectroscopy, as well as a negative correlation between the HI and the band at 1,710 cm^{-1} attributed to C=O (C-carboxyl). They also observed a negative correlation between R400 (proportion of the area under the curve of the S2 signal before 400°C; see Chapter 3 on Rock-Eval® indicators) and the relative proportions of the bands attributable to C-alkyls (2,918 and 2,850 cm^{-1}) and aromatics (1,611 and 1,513 cm^{-1}). Also in Histosols, Disnar et al. (2008) found a strong positive correlation between R400 and the quantity of carbohydrates determined by selective extraction. The relationships between R400 and the quantities of lignin and fatty acids determined by GC-MS pyrolysis in these peatland samples were less clear. Within other Histosols, Biester et al. (2014) noted a positive correlation between the quantity of aliphatic compounds determined by GC-MS pyrolysis and the Rock-Eval® HI parameter.

When analyzing particulate organic matter extracted from organomineral layers of forest soils in mainland France, Soucémarianadin et al. (2019) observed a negative correlation between the HI and the proportion of the signal obtained by mid-infrared spectroscopy attributable to aromatic compounds. In contrast, the HI was positively correlated with the quantity of aliphatic compounds. In a ^{13}C NMR study of composts of different maturities, Albrecht et al. (2015) noted no significant correlation between R400 and the chemical functions observed by NMR. The authors did, however, observe a positive correlation between C-alkyl and the proportion of S2 evolved before 340°C, as well as a positive correlation between aromatic C and the proportion of S2 evolved between 460 and 550°C. However, these correlations remain low (R^2 of approximately 0.2). Again, with regard to composts (and vermicomposts) of different maturities, Le Mer et al. (2020) also observed an increase in the thermal stability of the SOM with the increase in the relative proportion of aromatic compounds determined by ^{13}C NMR.

Among the Rock-Eval® studies of mineral soils that have integrated another technique for characterizing the chemical composition of SOM, data published by Vindušková et al. (2015; mining soils of the Czech Republic containing high proportions of fossil carbon rich in aliphatic compounds) showed that the Rock-Eval® HI parameter is positively correlated with the proportion of aliphatic carbon estimated by ^{13}C NMR (C-alkyl; $R^2 = 0.31$; $n = 14$). In this study, the HI is negatively correlated with O-alkyl-C compounds ($R^2 = 0.32$) and aromatic compounds ($R^2 = 0.19$), and uncorrelated with the carboxylic groups of soil organic carbon. The Rock-Eval® OI parameter is strongly correlated with O-alkyl-C compounds (positive correlation; $R^2 = 0.66$) and C-alkyl compounds (negative correlation; $R^2 = 0.65$). The OI is positively correlated with aromatic compounds ($R^2 = 0.34$) and, to a lesser extent, with C-carboxyl compounds ($R^2 = 0.13$). In this study, the RC/PC ratio is negatively correlated with C-alkyl compounds ($R^2 = 0.27$), positively correlated with O–C-alkyl compounds ($R^2 = 0.29$) and aromatic compounds ($R^2 = 0.19$), and uncorrelated with C-carboxyl compounds (Vindušková et al. 2015). Gillespie et al. (2014) showed, in frozen forest soils in northern Canada, that the increase in the thermal stability of HC could be related to the concomitant increase in carboxyl-C:ketone, carbohydrate:ketone or phenol:ketone ratios, quantified by NEXAFS spectroscopy at the carbon edge. Using the same X-ray absorption spectroscopic technique at both the carbon and nitrogen edges, Gregorich et al. (2015) showed that the Rock-Eval® parameters were well correlated with the chemical composition of the SOM of agricultural mineral and bare-fallow soils (see the explanation in section 10.4.1) in New Zealand. The works of Barré et al. (2016, 2018) conducted on long-term bare-fallow soils in Europe show that in most of the studied sites, the decrease in the HI, concomitant with the increase in the thermal stability of the SOM (T-50 of the S4 signal) with bare-fallow duration, is generally accompanied by a decrease in the signal intensity of aromatic compounds and C-carbonyl, while C-carboxyl compounds tend to increase (measurements obtained by NEXAFS spectroscopy at the carbon edge). GC-MS pyrolysis analysis of the same soil samples indicates that the decrease in the HI and the increase in thermal stability of the SOM with bare-fallow duration is associated with the relative increase in organic compounds of microbial origin containing nitrogen, while certain markers of organic matter of plant origin persist on a multi-decadal scale in the soil (Barré et al. 2018). Lastly, the recent works by Khedim et al. (2021) conducted on soils in formation on glacial margins of different regions of the world show that the constitution of the organic carbon stock of a soil is accompanied by a decrease in thermal stability and an increase in the hydrogen richness of the SOM, and that these evolutions observed in Rock-Eval® are associated with an increase in the C/N ratio of the SOM and the intensity of the signal corresponding to the functional groups of the C=C carbon, and with a decrease in the functional groups C–H and C=O, as quantified by mid-infrared spectroscopy. Taken as a whole, the works comparing Rock-Eval® results and the chemical composition of SOM suggest that the decrease in the proportion of carbohydrates goes hand in hand with an increase in the thermal stability of SOM. An increase in the quantity of aliphatic compounds appears to lead to an increase in the HI; albeit that there are exceptions, especially when the soil contains pyrogenic carbon (Barré et al. 2016). These results should nevertheless be considered with caution as the chemical functions have often not been characterized using the same techniques and the Rock-Eval® signals have not been treated and perhaps even obtained in an identical manner in all of the studies cited. In addition, an important part of this work has been carried out on pure organic matter and not on organomineral assemblages. Further work would therefore be needed to link the chemical characteristics and Rock-Eval® signature of SOM.

10.4. Evaluation of soil organic carbon stability by Rock-Eval®

10.4.1. *What is the biogeochemical stability of soil organic carbon?*

Organic carbon can persist in soil from several minutes to several centuries or even millenia. These highly contrasting residence times are explained by the different mechanisms which protect organic carbon against microorganisms operating in soils. Thus, a C atom will have a lower probability of being mineralized rapidly if it is located in a molecule that is more difficult to degrade, which is notably the case for pyrogenic carbon. Similarly, a C atom of a molecule protected from the degrading action of microorganisms by a mineral matrix will probably be mineralized after a longer period than a C atom located within an unprotected molecule (see, for example, von Lützow et al. (2006) for more details).

Over several decades, numerous authors have developed methods to evaluate the stability of soil organic C, i.e. the difficulty for soil microorganisms to mineralize it. Several direct measurements of soil C stability exist, but these are only possible in very specific sites, such as long-term bare fallows (Rühlmann 1999) or chronosequences of C3/C4 vegetation changes (Cerri et al. 1985; Balesdent et al. 1987). Bare fallows are sites in which soils have been kept free of vegetation for decades, thus making C inputs negligible, while microbial decomposition of organic C continues in these soils. Thanks to these sites, it is possible to directly track over several years (up to 90 years for the Versailles bare fallow, which is the longest), the mineralization of the soil organic C (Barré et al. 2010). Tracking the decline of organic C stocks in bare-fallow soils directly determines the evolution of the biogeochemical stability of the soil organic C. In C3/C4 chronosequences, the change at a given time from C3 photosynthetic vegetation (e.g. wheat) to C4 photosynthetic vegetation (e.g. maize) enables monitoring, using isotopic analyses, the decline of the soil C stock derived from C3 vegetation over time. The rate of this decline directly provides information on the stability of organic C of C3 origin. However, with the exception of these very useful and interesting long-term experiments, the vast majority of soils are neither bare-fallow nor included in C3/C4 chronosequences. Direct measurements of biogeochemical stability of soil organic C thus being impossible, indirect measurements or biogeochemical stability markers were developed.

The most frequently used methods are physical fractionation (granulometric, densimetric or granulodensimetric) or chemical fractionation of SOM and the tracking of soil organic C mineralization in incubation flasks under controlled conditions. For physical fractionation methods, certain works have shown that the organic C found in fine fractions (clays and fine silts) or heavy fractions (organic matter associated with minerals) has a longer mean residence time than the organic C of coarse and/or light fractions (Balesdent 1996; Poeplau et al. 2018). For incubations, the lower the proportion of mineralized organic C during the incubation period (generally a few months), the more the soil organic C is considered to be stable.

In order to predict the evolution of the organic C stock of a soil, evaluating its biogeochemical stability is crucial. Indeed, an unstable C stock is likely to decrease rapidly if soil C mineralization is not compensated for by significant C inputs. Conversely, a soil with a stable C stock is one in which it will be possible to increase the C stock with little effort because a low quantity of C inputs is necessary to compensate for the low mineralization of the C stock. The methods used to evaluate the biogeochemical stability of the soil C can either be used at very specific sites (bare fallow and chronosequence) or take a fairly long time to implement (fractionations and incubations). If a Rock-Eval® measurement only requiring about an hour provided information on the C biogeochemical stability, Rock-Eval® would then have an undeniable advantage in establishing itself as an effective method for evaluating the C biogeochemical stability in soils. In the following sections, we will see whether Rock-Eval® effectively provides information on the stability of C in the soil and how the Rock-Eval® results compare with the results of the usual techniques for assessing the biogeochemical stability of C in the soil.

10.4.2. *Links between Rock-Eval® results and biogeochemical stability of soil organic matter observed in the field*

It was the pioneering works of Di-Giovanni et al. (1998, 2000) and then Disnar et al. (2003) that first introduced the possibility that Rock-Eval® results could provide information on the biogeochemical stability of SOM. Indeed, many studies have shown that the biogeochemical stability of soil organic C, on a multi-decadal or even multi-secular scale, increases with soil depth (see, for example, Balesdent et al. (2018)). However, as indicated above, these studies have shown a clear evolution in the Rock-Eval® signature of SOM with soil depth. The trends that they observed with soil depth (decrease in the HI and increase in the OI and the thermal stability of the soil organic C) were confirmed by all subsequent studies.

The results of Albrecht et al. (2015), obtained on composted organic matter, suggest that the thermal stability of SOM during the pyrolysis phase of Rock-Eval® analysis provides information on the biogeochemical stability of SOM in the short term (time scale from one week to one quarter). Indeed, they observed that the Rock-Eval® I and R parameters evolve as the compost matures. Since the least-stable organic matter is mineralized the fastest, the biogeochemical stability of compost increases with its maturation time. The results of Albrecht et al. (2015) show that the thermal stability in the pyrolysis phase, expressed using the I- and R-indicators, increases with the biogeochemical stability of compost.

In mineral soils, Barré et al. (2016) clearly established the link between the long-term biogeochemical stability of SOM (multi-decadal scale) and its Rock-Eval® signature. To do so, they studied the evolution of the Rock-Eval® signature of SOM with the duration of bare fallow (soils kept artificially free of vegetation) in five sites in Europe (Figure 10.4). The biogeochemical stability of soil organic C increases with bare-fallow duration. Indeed, the most labile C is mineralized first and the bare-fallow soil is thus gradually enriched in stable C (in proportion). This can be observed directly on the evolution of the soil organic C content with bare-fallow duration. Indeed, the decline in C content reduces with bare-fallow duration, even appearing to approach a plateau. The results show that the HI decreases and that thermal stability during the oxidation phase (T-50 of the S4 signal; see Chapter 3) increases with bare-fallow duration. These results confirm those of Gregorich et al. (2015), obtained on a New Zealand soil maintained in bare fallow for 11 years: decrease of the HI and the T-50 index of the S2 signal of the bare-fallow SOM compared to adjacent grassland and crop soils. They were further supported by a recent study in which different-sized fractions of SOM from the Versailles bare-fallow trial were analyzed using Rock-Eval® (France; Chassé et al. 2021).

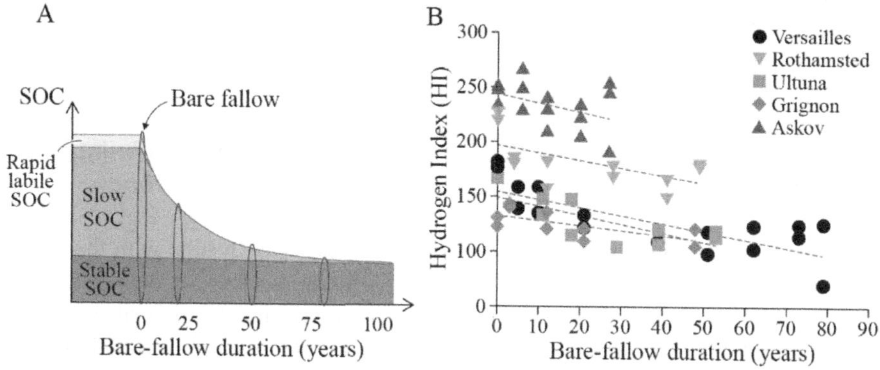

Figure 10.4. *(A) Evolution of the SOC content and its biogeochemical stability with bare-fallow duration; (B) evolution of HI with bare-fallow duration. The more the SOM is enriched with stable organic matter, the lower the HI. Versailles (FR), Rothamsted (GB), Ultuna (SW), Grignon (FR) and Askov (DK) correspond to the experimental sites. Figures from Barré et al. (2016). For a color version of this figure, see www.iste.co.uk/baudin/rock-eval.zip*

Poeplau et al. (2019) obtained fairly similar results in soils of an Icelandic grassland subjected to intense warming. Following an earthquake, geothermal pipes near the surface shifted and imposed warming on soils that were not subjected to the action of these geothermal pipes prior to the earthquake. Poeplau et al. (2019) analyzed soils that were subjected to heating between 0.6°C and 80°C for six years.

Warming triggered mineralization of the SOC, and as a result soils subjected to intense warming came to be enriched in stable C. Poeplau et al. (2019) observed that the HI decreased with the intensity of the warming (and thus with the decrease in the SOC content and the simultaneous increase in the proportion of stable C) in contrast to the OI and the thermal stability of the SOC both in pyrolysis and in oxidation, which each increased.

Taken as a whole, these different results show that the Rock-Eval® signature provides information on the biogeochemical stability of SOM. The stable soil organic C has a higher thermal stability in pyrolysis and oxidation and a lower HI value than the less-stable organic C. These results seem valid for periods ranging from one week to several decades (Albrecht et al. 2015; Gregorich et al. 2015; Barré et al. 2016), but also concerning the stability of soil organic C to warming (Poeplau et al. 2019). These results could be explained by the fact that stable SOM contains fewer high-energy C–H bonds possibly highly coveted by soil microorganisms (Barré et al. 2016). It is also possible that the higher thermal stability of stable C can be explained by the fact that it is more associated with minerals, which would delay its thermal degradation.

10.4.3. *Links between Rock-Eval® results and the usual techniques for quantifying the biogeochemical stability of soil organic carbon*

Several studies have sought to compare the Rock-Eval® results with the results from physical or chemical fractionations of SOM or incubations of soil samples under controlled conditions. Copard et al. (2006) showed that the thermal stability of the S2 signal of the SOM of soils and black marls increases after incubation under controlled conditions. Saenger et al. (2015) observed, in alpine soils, that the organic matter associated with minerals has a lower HI value than particulate organic matter. Gregorich et al. (2015) observed that organic matter associated with sands has a higher HI value and a lower thermal stability in pyrolysis (T-50 of the S2 signal) than organic matter associated with clays.

Based on these results, Soucémarianadin et al. (2018a) attempted to correlate the values of different Rock-Eval® parameters with the results of physical fractionation of SOM and soil incubation under controlled conditions on approximately 100 soil samples from forests in mainland France. The results show that several Rock-Eval® parameters (HI, OI, T-50 of the S2 and S4 signals, I- and R-index; see Chapter 3 for the definition of the parameters) are strongly correlated with the proportion of particulate organic matter (Table 10.2). On the other hand, Rock-Eval® parameters are less well correlated with the proportion of mineralized C in incubation flasks. These results suggest that the Rock-Eval® signature can provide information on the stability of soil organic C at the decade or century scale, but more imprecisely on the very short-term stability of the soil C, as observed with incubations under controlled conditions.

	Biogeochemical stability indicators	
	Breathed-C	POM-C
$T_{50\_HC\_PYR}$	-0.26**	-0.73***
$T_{50\_CO2\_OX}$	-0.16	-0.56***
HI	0.06	0.67***
OI_{RE6}	-0.02	-0.76***
I-index	0.32**	0.35***
R-index	-0.31**	-0.69***

Table 10.2. *Correlations between the percentage of C mineralized in 10 weeks in an incubation bottle (breathed-C) or the proportion of C in the form of particulate organic matter (POM-C) and different Rock-Eval® parameters. The data come from 53 surface layers (0–10 cm) and 46 depth layers (40–80 cm) of French forest soils. Significance levels are indicated as follows: ***, $p < 0.001$; **, $p < 0.01$. Data from Soucémarianadin et al. (2018a)*

Other studies used chemical methods of SOM fractionation to isolate an insoluble or non-hydrolyzable fraction (ROM), or a fraction resistant to NaOCl or H_2O_2 oxidation, and then analyzed these chemical fractions of SOM using Rock-Eval® (Hetényi et al. 2006; Czirbus et al. 2016; Poeplau et al. 2019). While neither the ROM fraction nor that resistant to H_2O_2 oxidation are recommended for studying the biogeochemical stability of carbon (Poeplau et al. 2019), the fraction resistant to NaOCl oxidation is sometimes recommended for this purpose (Poeplau et al. 2018), although it struggles to isolate the biogeochemically stable SOM at the multi-decadal scale (Lutfalla et al. 2014). The HI index of the oxidation-resistant fraction is greatly reduced, while its OI index is greatly increased (Poeplau et al. 2019) with respect to the initial soil. However, the Rock-Eval® signature of oxidation-resistant chemical fractions differs greatly from that of soils naturally enriched in biogeochemically stable organic carbon, particularly with regard to the OI or thermal-stability indicators (Poeplau et al. 2019).

The results obtained by comparing Rock-Eval® thermal analysis with usual techniques for estimating the biogeochemical stability of soil organic carbon, in particular those that are relatively robust, based on the physical fractionation of SOM, are consistent with the results relating the Rock-Eval® signature to the biogeochemical stability of the SOM observed in the field. Taken as a whole, this shows that the Rock-Eval® signature of SOM can be used as a marker to qualitatively assess its soil organic C biogeochemical stability in the short (one week to one quarter), medium and long terms (multi-decadal scale).

However, it is important to note that precise knowledge of biogeochemical stability, i.e. knowledge of the mineralization dynamics of organic carbon present in the soil at a given moment, is only slightly enlightened by the wide variety of qualitative indicators derived from Rock-Eval® thermal analysis (HI, OI, I- or R-index, T-50 of the different signals, etc.) proposed by different research teams (Hetényi and Nyilas 2014; Gregorich et al. 2015; Barré et al. 2016; Sebag et al. 2016; Thoumazeau et al. 2020; Khedim et al. 2021). All these indicators can only provide a qualitative assessment of the protection of SOM C from microbial decomposition. Indeed, thermally stable or hydrogen-poor soil organic carbon remains a mixture of active and stable carbon pools at the century scale (Cécillon et al. 2021). Only by robustly calibrating the different Rock-Eval® indicators could they be made quantitative with regard to the biogeochemical stability of soil organic carbon, and therefore useful for improving knowledge of soil organic C dynamics, and even for the development of operational tools to track soil organic C dynamics. An example of the use of Rock-Eval® thermal analysis to quantitatively evaluate the biogeochemical stability of organic C in soils at the multi-decadal scale is discussed in the next section.

10.5. Quantifying soil organic carbon stability using Rock-Eval® with a view to improving soil carbon dynamics models

10.5.1. *Soil organic carbon dynamics models*

In most soil organic C dynamics models, the C stock is divided into several kinetic compartments in which the C has different residence times (Jenkinson 1990). Conventionally, three kinetic compartments are distinguished between: (1) a "labile" compartment with a residence time of the order of a year, (2) an "intermediate" compartment with a residence time of the order of several decades (3) and a "slow" compartment with a residence time of the order of several centuries. The "labile" compartment represents a small proportion of the total stock (1–5%) and is sometimes not represented. In certain models, the "slow" compartment is considered to be inert on the simulation time period.

The AMG model (Andriulo et al. 1999) is a widely used, efficient model to reproduce the evolution of C stocks in agricultural soils, particularly in France (Clivot et al. 2019). In this model, the soil organic C stock is divided into two kinetic compartments: an "active" compartment (Ca) with a residence time of the order of several decades, and a "stable" compartment (Cs), inert at the scale of the duration of the simulations, which last several decades at most. When this model is used, the total quantity of soil organic C (TOC) is thus distributed into the two compartments such that: TOC = Ca + Cs, where "Ca" is the quantity of C in the active compartment and "Cs" is the quantity of C in the stable compartment.

As with all soil organic C dynamics compartment models, the initialization of the kinetic compartment size at the beginning of the simulation is problematic. Indeed, since no technique exists to measure the quantity or proportion of C to be assigned in the different compartments, initialization is always uncertain. This question is all the more significant as the distribution of C within the different compartments is a variable that is particularly sensitive for the result of the simulations (Luo et al. 2016). In the case of the AMG model, the model considers 65% of the soil organic C to be "stable" (Cs = 0.65 x TOC) if the soil has been cultivated for a long time, and that only 40% of the C is "stable" (Cs = 0.40 x TOC) if it is a permanent grassland soil. This initialization of the model by default does not give optimal results (Clivot et al. 2019). Developing a method to improve the initialization of soil organic C dynamics models would thus improve the accuracy of their simulations.

10.5.2. *Construction of the stable carbon model at the century scale*

Cécillon et al. (2018, 2021) wanted to test whether it was possible to predict the quantity of stable C at the century scale in a soil (hereinafter referred to as "stable C") from Rock-Eval® measurements made on unknown soil samples. As we highlighted in the previous section, such an ambition requires prior calibration of the different Rock-Eval® parameters to the biogeochemical stability observed in the field, for example, via a machine learning model. To achieve this, a set of soil samples must first be available, for which the quantity of stable C at a specific time horizon is known, and these samples were analyzed in Rock-Eval®.

Soil samples from long-term bare-fallow trials offer a unique opportunity for this. Indeed, the decrease of TOC in bare-fallow soils follows a trend that can be modeled by a decreasing exponential function and a constant (Barré et al. 2010; Figure 10.5). It is possible to determine through optimization the value of the different parameters and in particular the value of the constant that determines the quantity of TOC corresponding to the "compartment of C whose mineralization cannot be seen at the scale of several decades" (Barré et al. 2010; Franko and Merbach 2017; Cécillon et al. 2018). Using this value, the proportion of stable C in each soil sample can be simply determined by dividing the value of the constant by the TOC value (Figure 10.5). As soil samples collected from the bare fallows were stored dry and in the dark (which blocks the evolution of SOM), bare-fallow tests provide a set of soil samples that can be analyzed in Rock-Eval® and whose proportion of stable C can be estimated fairly accurately.

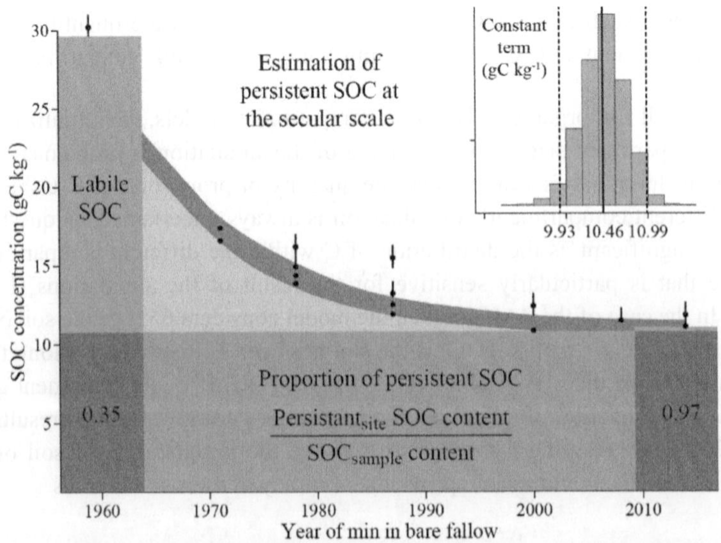

Figure 10.5. *Example of calculating the proportion of stable C at the century scale in a soil sample of a long-term bare-fallow test. The size of the stable kinetic compartment of the soil C, which, by definition, does not vary between the start of bare fallow and the last measurement point is estimated by optimization. At each measuring point, the ratio between the stable C content of the soil and the TOC content measured by elemental analysis corresponds to the proportion of stable C in the soil (Cécillon et al. 2018). For a color version of this figure, see www.iste.co.uk/baudin/rock-eval.zip*

Cécillon et al. (2018, 2021) began by performing Rock-Eval® analysis of the soil samples from six European sites and one Colombian site, in which the proportion of stable C can be estimated, and they correlated these proportions with different Rock-Eval® indicators. Table 10.3 shows that certain parameters correlate very strongly, positively or negatively, with the proportion of stable C. These include the HI, the OI and the thermal stability parameters involving CO_2, in the pyrolysis phase (S3 signal) and oxidation phase (S4 signal).

In a second step, Cécillon et al. (2018, 2021) divided the set of soil samples into a learning set and a validation set. They then built a "random forest" machine learning model using the learning set and looked at the predictions of this model on the validation set. The result, presented in Figure 10.6, shows that the machine learning model, entitled $PARTY_{SOC}$, very satisfactorily predicts the proportion of stable C on the soil samples in the validation set. This result shows that it is possible to predict the quantity of stable C for the bare-fallow sites. The next steps are to determine whether this model allows determination of the quantity of stable C for other sites, in particular sites that are not in bare fallow, and whether it thus improves the initialization of the AMG model and therefore the accuracy of its simulations.

T_{10}_HC_PYR	0.38	<0.0001
T_{30}_HC_PYR	0.46	<0.0001
T_{50}_HC_PYR	0.46	<0.0001
T_{70}_HC_PYR	**0.54**	**<0.0001**
T_{90}_HC_PYR	**0.81**	**<0.0001**
T_{10}_CO_PYR	0.40	<0.0001
T_{30}_CO_PYR	0.36	<0.001
T_{50}_CO_PYR	0.33	<0.001
T_{70}_CO_PYR	0.31	<0.001
T_{90}_CO_PYR	0.31	<0.001
T_{10}_CO_2_PYR	0.35	<0.001
T_{30}_CO_2_PYR	**0.56**	**<0.0001**
T_{50}_CO_2_PYR	**0.55**	**<0.0001**
T_{70}_CO_2_PYR	**0.55**	**<0.0001**
T_{90}_CO_2_PYR	**0.58**	**<0.0001**
T_{10}_CO_OX	0.31	<0.001
T_{30}_CO_OX	0.41	<0.001
T_{50}_CO_OX	0.49	<0.0001
T_{70}_CO_OX	**0.58**	**<0.0001**
T_{90}_CO_OX	0.33	<0.0001
T_{10}_CO_2_OX	0.09	0.33
T_{30}_CO_2_OX	0.39	<0.0001
T_{50}_CO_2_OX	**0.63**	**<0.0001**
T_{70}_CO_2_OX	**0.70**	**<0.0001**
T_{90}_CO_2_OX	**0.60**	**<0.0001**
pseudoS1	**-0.50**	**<0.0001**
pseudoS1/PC	0.28	<0.01

pseudoS1/TOCre6	-0.05	0.57
S2	**-0.85**	**<0.0001**
S2/PC	**-0.70**	**<0.0001**
HI	**-0.72**	**<0.0001**
OIre6	-0.09	0.35
HI/OIre6	**-0.68**	**<0.0001**
PC/TOCre6		
TOCre6	**-0.71**	**<0.0001**
I-index	**-0.55**	**<0.0001**
R-index	-0.40	<0.0001
	0.47	<0.0001

Table 10.3. *Correlation between the proportion of stable C in a sample and different Rock-Eval parameters. From Cécillon et al. (2021)*

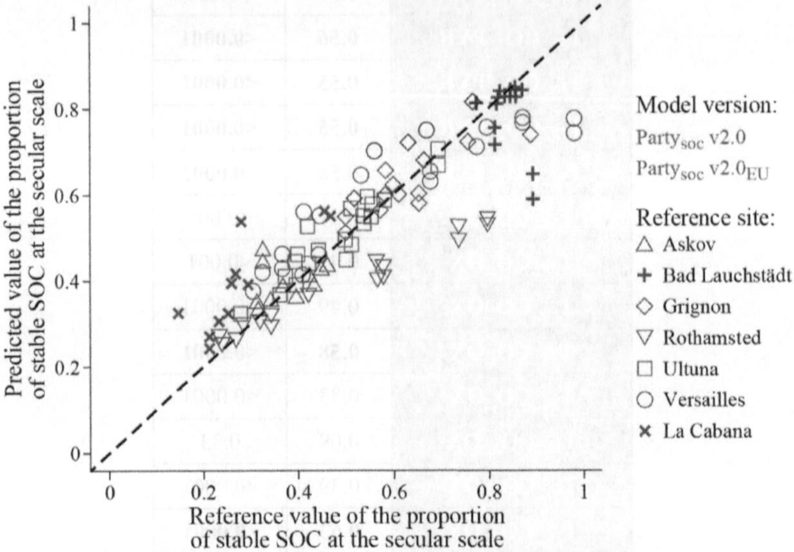

Figure 10.6. *Performance of the random-forest machine learning model to predict the proportion of stable C at the secular scale. Figure from Cécillon et al. (2021). For a color version of this figure, see www.iste.co.uk/baudin/rock-eval.zip*

10.5.3. *Using the PARTY$_{SOC}$ machine learning model to estimate the size of the stable kinetic compartment of the AMG soil C dynamics model*

The default estimation of the proportion of Cs in the AMG model is imperfect. Clivot et al. (2019) showed by optimization that in several well-characterized long-term agronomic trials, proportions of Cs different from the default values can improve the accuracy of SOC stock evolution simulations. The PARTY$_{SOC}$ machine learning model developed by Cécillon et al. (2018, 2021) enables the proportion of stable C at the secular scale to be estimated. It is therefore possible that this model can effectively predict the proportion of Cs in the AMG model. To test this, Kanari et al. (2022) performed a Rock-Eval® analysis of soil samples from long-term agronomic experiments in France (initial samples from archival collections, with a median test duration of 21 years). They then compared the values of Cs optimized by Clivot et al. (2019) with the secular-scale stable-C values determined by the PARTY$_{SOC}$ machine learning model based on Rock-Eval® thermal analysis. The results presented in Figure 10.7 show a very close fit between these values.

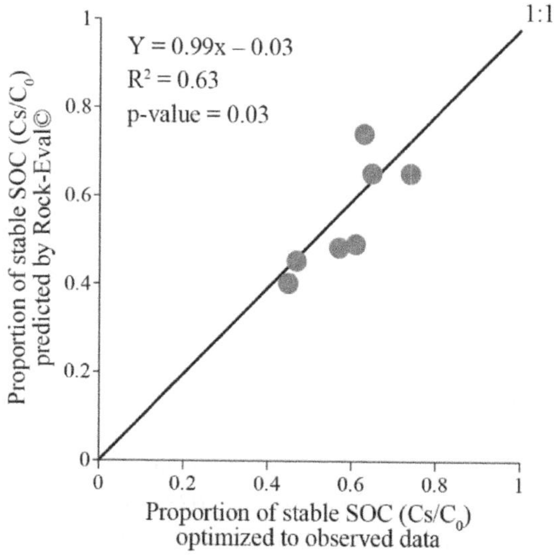

Figure 10.7. *Proportion of stable C at the century scale in soil, predicted using the PARTY$_{SOC}$ machine learning model (Cécillon et al. 2021) using Rock-Eval® soil sample data as a function of the proportion of Cs of the AMG model, optimized to obtain the best fit between model simulation and data. The optimized proportions of Cs are different from the default values (0.65 for cultivated soils) and very close to the proportions determined using the PARTY$_{SOC}$ machine learning model. Figure from Kanari et al. (2022). For a color version of this figure, see www.iste.co.uk/baudin/rock-eval.zip*

In a second step, Kanari et al. (2022) compared the AMG simulations performed with the default values to the AMG simulations performed by initializing the value of Cs using Rock-Eval® data and the $PARTY_{SOC}$ machine learning model. The results show that, on average, Rock-Eval® initialization improves the accuracy of AMG model predictions and that this improvement is particularly strong for certain sites: sites with a complex history in terms of land use or agricultural management practices (Figure 10.8).

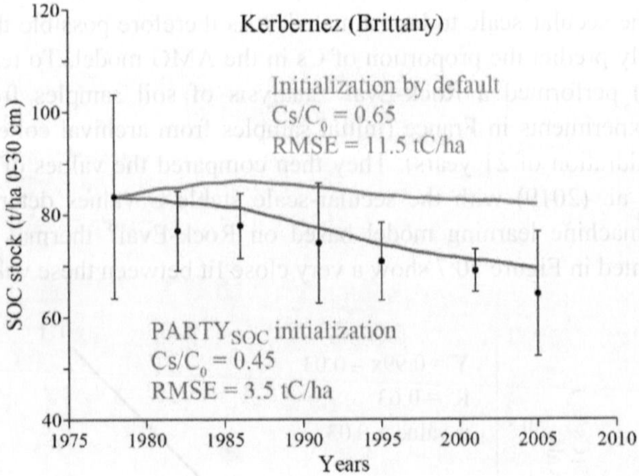

Figure 10.8. *Simulations of the evolution of SOC stocks at the Kerbernez site (Brittany) using the AMG model initialized by default or using the $PARTY_{SOC}$ machine learning model (Cécillon et al. 2021). Figure from Kanari et al. (2022). For a color version of this figure, see www.iste.co.uk/baudin/rock-eval.zip*

These results show that Rock-Eval® thermal analysis, once properly calibrated via the $PARTY_{SOC}$ machine learning model, improves the initialization of the AMG soil organic carbon dynamics model, and therefore the accuracy of these simulations at the multi-decadal timescale (Kanari et al. 2022). These results open up broad prospects. The $PARTY_{SOC}$ machine learning model based on Rock-Eval® thermal analysis presented by Cécillon et al. (2018, 2021) is continuously being improved. It is therefore set to evolve in the coming years. It is possible that the same approach can be used for other soil organic C dynamics models (Century, Roth-C, etc.) and extended to modeling organic C dynamics in grassland and forest soils.

10.6. Conclusion

Since the pioneering works of Di-Giovanni et al. (1998, 2000) and Disnar et al. (2003), SOM has been studied regularly by Rock-Eval® thermal analysis. Although the majority of these works use fairly similar Rock-Eval® signal acquisition parameters, it is important to use the exact same SOM analysis routine. This would allow for a more robust comparison of the results obtained by the different scientific groups. Nevertheless, the corpus of available results demonstrates that while the possible links between Rock-Eval® results and the chemical composition of SOM remain rather obscure, several works clearly show that Rock-Eval® analysis provides qualitative information on the biogeochemical stability of soil organic carbon, and even quantitative when the Rock-Eval® parameters are correctly calibrated to the observation data, as is the case in the $PARTY_{SOC}$ machine learning model.

These recent results show that Rock-Eval® thermal analysis enables rapid determination (~1 hour), at low cost, reproducibly and quantitatively, of soil organic and inorganic carbon content and the biogeochemical stability of soil organic carbon. This makes the Rock-Eval® technique particularly attractive for research applications, as well as for more-operational applications aimed at determining and monitoring soil quality. For both of these reasons, it is likely that the use of Rock-Eval® for soil sample characterization will develop rapidly in the coming years.

11

Study of Natural and Anthropogenic Events

Yoann COPARD[1] and François BAUDIN[2]

[1] M2C, Université de Rouen Normandie, Rouen, France

[2] ISTeP, Sorbonne Université, Paris, France

11.1. Introduction

As seen in previous chapters, the Rock-Eval® method facilitates, quickly and at low analytical costs, the characterization of the Corg content of a sample, to provide quality indicators (parameters derived from the S2 to S4 signals) and also to evaluate the degree of thermal evolution reached during classical diagenetic processes specific to sedimentary basins. Thus, this method has primarily been of interest in research on fossil resources (oil, gas and coal) insofar as the "diagenetic and energetic position" of a given sample in oil and gas windows, its potential to release its hydrocarbons and its origin, are rapidly defined through a classical evolution of its main Rock-Eval® parameters. Nevertheless, the evolution of the values of these parameters can be disturbed by diagenetic to post-diagenetic events of thermal order (hydrothermal circulation, metamorphism or fire) and/or oxidative events (supergene alteration or weathering), which generally occur under subsurface or surface conditions.

Thus, this chapter will only cover events that alter the parameters of the Rock-Eval® method, i.e. those that occur when the classical diagenetic thermal evolution has been modified locally by a hot event, or when it has been stopped. In particular, the stopping of this thermal evolution generally takes place when a geodynamic trajectory of a sedimentary basin leads to geothermal flows of lesser importance, or during exhumation of the rocks. Concerning thermal events, which can exhibit oxidizing characteristics for organic matter, they can be considered as being of geological origin (metamorphic aureole, hydrothermal circulations and lava flow) or climate-anthropogenic (fire), and concern a variable time scale (seconds to several tens of millions of years). Low-temperature oxidative events in equilibrium at the atmosphere/pedosphere/biosphere/lithosphere interface, which can be grouped together under the term "weathering", are related more to processes occurring on a time scale that can reach several thousand years.

This chapter therefore proposes to first explore the modifications to the parameters of the Rock-Eval® method of organic matter impacted by contact with a magma (dyke and sill) and by hydrothermal circulations that allow the establishment of an exploitable deposit. For this latter point, the temperature of the hydrothermal circulations may be lower than that recorded by organic matter during diagenesis, and they may also be oxidizing to a lesser or greater degree. Keeping to the same thermal register, the impact of fires with the production of incomplete combustion residues (soot charcoal) and their Rock-Eval® signature will then be examined. An overview will also be given of the impact of weathering, as well as that of sensu lato oxidation carried out in the laboratory, on the parameters of the Rock-Eval® method.

The interest of this chapter is thus twofold: for a geological- to petroleum-level issue, it intends to provide warnings on the analysis of surface samples for determining the oil potential of a future deposit or detecting indications of a uranium deposit; for a surficial geoscience or environmental issue, it shows how altering the parameters of the Rock-Eval® method can erase the origin of organic matter increasingly used as a tracer in a source-to-sink concept.

11.2. Events of geological origin

These events cover both the impact of hydrothermal circulations, magmatic occurrences (lava flow), and the impact of metamorphism on the parameters of the Rock-Eval® method. With regard to exploitable deposits, this section will not deal with the mechanisms aimed at concentrating elements in subsurface environments involving organic matter and that have therefore triggered the evolution of Rock-Eval® parameters; as such, we invite the reader to refer to the review by Disnar and Sureau (1990).

11.2.1. *Hydrothermal circulations*

In general, these circulations, capable of generating exploitable deposits (e.g. Kupferschiefer, U and MVT), disturb the expected or modeled thermal evolution of organic matter in a given sedimentary basin. Thus, in the case of shale diapirs (Yinggehai Tertiary basin, China), the evolution in depth of Tmax can emphasize a delay in thermal maturation linked to either overpressure processes and the presence of excess water, or overmaturation caused by the excess temperature of a geothermal upwelling; blocked by a caprock with conductive properties (Hao et al. 1998, 2000; Figure 11.1).

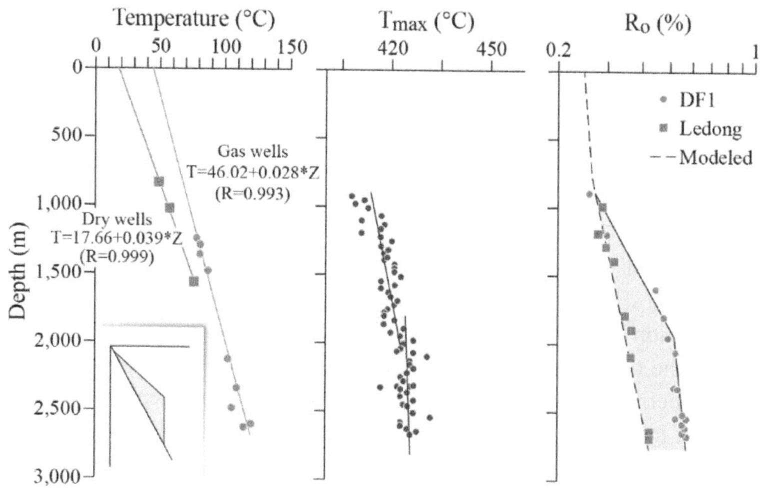

Figure 11.1. *Evolution of Tmax, vitrinite reflectance and geothermal upwelling temperature (Yinggehai basin, SW China (Hao et al. 2000)). For a color version of this figure, see www.iste.co.uk/baudin/rock-eval.zip*

For Kupferschiefer-type deposits, which occur in sedimentary basins composed of shale (Sangerhausen Basin, Germany), the organic matter serves as a proton donor during thermal sulfate reduction (TSR) to facilitate the formation of H_2S and enable the precipitation of copper-enriched sulfides. This hydrogen loss is thus visible with the collapse of the HI values, an increase in the OI and a significant gain in Tmax (Sun et al. 1995; Table 11.1). These values are consistent with other usual chemical (molecular or isotopic) or petrographic analyses showing that the remaining organic residue corresponds to a pyrobitumen.

Sample	HI	OI	Tmax
7	100	38	434
6	125	40	427
5	81	28	432
4	68	49	431
3	72	47	434
2	15	24	442
1	7	52	467

Table 11.1. *Evolution of the HI, OI and Tmax parameters of organic matter collected from a Kupferschiefer-type deposit (Sun et al. 1995)*

In the case of the Au deposit, as in the Witwatersrand (South Africa), indicators such as the HI or OI of bitumen, resulting from a secondary phase, reflect thermal alteration of hydrothermal origin to a greater extent than the degree of maturity acquired by these carbonaceous materials. The Tmax then indicates recorded temperatures well above 200°C (Spangenberg and Frimmel 2001). In Guizhou Province (China, Bao et al. 2005), the relationships between the S2 parameter and the concentration of Au and As show that this deposit is closely related to the reactivity of the organic matter (organic acids) involved in the reduction of sulfates. Other hydrothermal circulations, not generating exploitable deposits, impacted the parameters of the Rock-Eval® method. As such, in the Qikou Sag basin (East China; Yu et al. 2016), circulations of mantle origin either triggered the classical increase in the thermal evolution of organic matter or, less typically, caused a decrease in Tmax and an increase in the indicators marking hydrogen (HI and S2). For the latter process, the hypothesis retained would be that these volatile-rich supercritical fluids, with a high dissolution and diffusion ability in the host rocks, would cause the accumulation of hydrocarbons in the fluid by extraction of the hydrocarbons already produced, or by locally heating the organic matter present and affected by this fluid (Figure 11.2).

One of the least obvious examples concerning the evolution of the Tmax parameter compared to its optical counterpart, vitrinite reflectance (Ro), faced with a hydrothermal circulation, relates to Niger coals collected in the vicinity of the Arlit Fault (Harouna et al. 1993), which played a major role in the genesis of the Arlit and Akouta uranium deposits (Forbes et al. 1988). Here, only Tmax recorded the thermal effects of burial diagenesis in contrast to Ro, which emphasizes via the model of Barker and Pawlewicz (1994), a temperature close to 195°C, in accordance with the microthermometric data deduced from the analysis of fluid inclusions (175–180°C;

Forbes et al. 1988). In addition, as indicated by the low values of the O/C atomic ratio of the samples, an oxidative process should be avoided. These coals were therefore impacted by fluid circulations recorded solely by the Ro (and the H/C ratio). The proposed hypothesis is to consider an early hydrothermal event impacting these coals and producing bitumen, which, during burial diagenesis, would undergo polycondensation reactions such that the Tmax values would reflect only a simple diagenetic evolution. The difference in the degree of maturity, provided by Ro and Tmax, would then seem to come from a differential kinetic evolution of these two parameters. This example offers a good illustration of the many factors involved in fluid/organic matter interactions and the difficulty of exploiting Rock-Eval® method data without comparing them with other study methods, whether optical or geochemical. Thus, the interpretations deduced from the modifications generated by such hydrothermal fluids are not necessarily immediate, and are sometimes even contradictory with other studies, in particular those of Disnar et al. (1997), who showed that a class of organic matter had recorded an increase in Tmax without an increase in reflectance under the action of a possible hydrothermal occurrence.

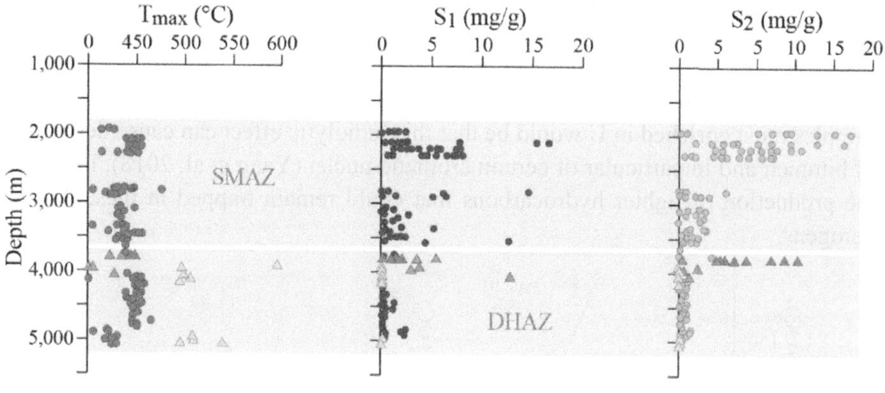

Figure 11.2. *Evolution of the Tmax, S1 and S2 parameters of organic matter with depth in contact with the mantle circulation. Tmax either increases (yellow triangle) or decreases (green and orange triangles) and vice versa for the S1 and S2 signals. The blue, red and green circles represent sinks unaffected by volcanic rocks and hydrothermal occurrences, respectively. For a color version of this figure, see www.iste.co.uk/baudin/rock-eval.zip*

11.2.2. *Radiolysis and other U/organic matter relationships*

A succession of oxidation processes, thermal and radiolytic, often coupled, affects the organic matter associated with U deposits and can be traced by the variation in the Rock-Eval® parameters (Landais 1996). For example, the effects of biodegradation such as BSR (Bacterial Sulfate Reduction) may facilitate the establishment of Roll Front-type U deposits and is at the origin of the high OI and low HI values of a highly degraded kerogen initially Type III to IV in the Erlian basin (NE China; Bonnetti et al. 2015). While the effects of organic matter oxidation are classical, the organic matter from the most U-enriched zones can later be subjected to a radiolytic process that modifies the Rock-Eval® parameters. Thus, in the Akouta deposit in Niger, these radiolytic effects cause a strong increase in Tmax (from 448 to 498°C) on Type III organic matter, coupled with an increase in oxygen and a significant loss of hydrogen following the mineralization of organic matter in the U-enriched zone (Forbes et al. 1988; Figure 11.3). In contrast, outside the deposit, a process, described by the authors as the coupled effects of (syndiagenetic) oxidation and maturation, affects the organic matter with the consequence of a slight gain in Tmax of less than 10°C, a loss of hydrogen, as well as a slight gain in oxygen (Figure 11.3). However, this thermal and oxidative coupling may mimic simple weathering if it is not compared with other molecular organic geochemistry methods. Taking the example of the U deposit of Cambrian Alum shales (Sweden), the evolution of Tmax is subject to controversy because irradiation from U decay can also cause a decrease in Tmax (Yang et al. 2018), and the IP parameter derived from the Rock-Eval® method (Dahl et al. 1988; see Chapter 3 of this book). One of the hypotheses put forward to explain this decrease in thermal maturity of the samples most enriched in U would be that this radiolytic effect can cause destruction of bitumen and in particular of certain aromatic nuclei (Yang et al. 2018), leading to the production of lighter hydrocarbons that could remain trapped in these Type II kerogens.

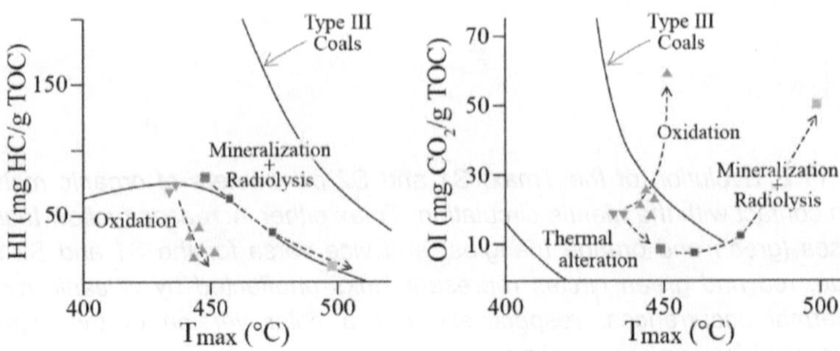

Figure 11.3. *Evolution of the HI versus Tmax and OI versus Tmax parameters of the samples (in black) subjected to U-rich hydrothermal fluids. The symbols correspond either to reference samples or to oxidized samples (triangle with point facing upwards) (according to Forbes et al. (1988)). For a color version of this figure, see www.iste.co.uk/baudin/rock-eval.zip*

11.2.3. *Magmatic and metamorphic events*

Other geological processes, which can be described as events for the geological time scale and occurring under surface to subsurface conditions, can disturb the values of the Rock-Eval® parameters. Lava flows, dykes or sills, are characterized by excessively high temperatures with respect to the environment allowing local simulation of a natural pyrolysis of organic matter, leading to a loss of hydrogen (decrease in HI), a gain in Tmax, and OIs that can remain unchanged or even decrease. In the case of a dyke affecting Late Cretaceous shales (Pierre Shales, Colorado, USA), a significant decrease in the HI and TOC can be clearly noted for a distance of 80 cm from the dyke (Clayton and Bostick 1986). It is of course to be expected that the impact of a magmatic body in spatial terms (distance from contact) and in terms of evolution of the parameters examined differ according to its temperature and size. Another example presenting the same evolutions is given for siltites and oil shales from Scotland, this time including the evolution of Tmax (Midland Valley; George et al. 1992). While a gain in Tmax and a decrease in the HI values appear systematically, the evolution of the OIs appears less clear, as shown in Figure 11.4, which represents the evolution of these indices for a series of intrusive dykes in Jurassic black shales (Kutch rift basin, India; Aroa et al. 2017), during intrusions of sills into the Basque-Cantabrian basin (Spain; Agirrezabala et al. 2014) or in the Karoo basin (South Africa; Aarnes et al. 2011).

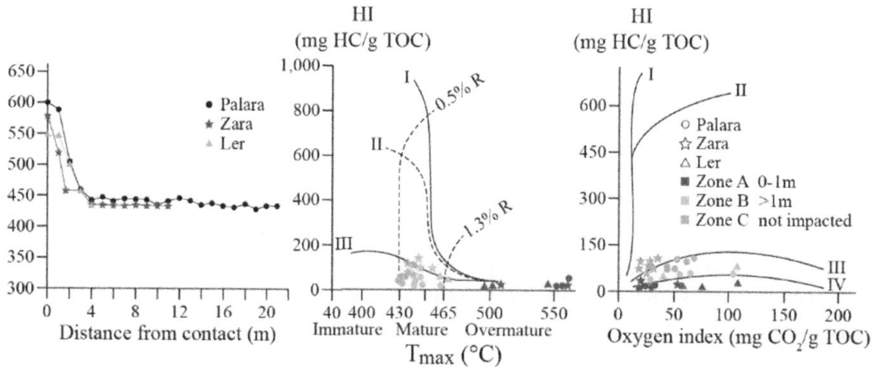

Figure 11.4. *Evolution of the Tmax, HI and OI parameters in contact with a dyke (Aroa et al. 2017). Zone A corresponds to distances between 0 and 1 m, zone B > 1 m and zone C corresponds to samples not impacted by the dyke. For a color version of this figure, see www.iste.co.uk/baudin/rock-eval.zip*

On the contrary, the fall in the OI is severe for a layer of coal in contact with an igneous intrusion (Illinois, USA; Rahman and Rimmer 2014). While the TOC will generally decrease as the organic matter/magmatic body contact diminishes, it is sometimes known to increase when the sample is very close to contact (Agirrezabala et al. 2014; Aroa et al. 2017). One of the explanations put forward consists of the production of graphitic materials that remain fixed in contact with the body due to a sudden rise in the temperature of this natural pyrolysis, as was suspected for a coal layer affected by a sill (Vermejo formation, Colorado-New Mexico, USA; Meyers and Simonet 1999). Ultimately, given the thermal process involved, all studies naturally converge toward a loss of Corg, except for samples very close to the contact. Also to be noted are a loss of hydrogen, an increase in the refractory character of the organic matter (increase in maturity underlined by Tmax) and the more versatile behavior of the OI, which can remain unchanged or even decrease. Regarding the validation of the increase in maturity, many studies conventionally compare Tmax with the vitrinite reflectance. We will see below why it is necessary to do so.

Longer-lasting geological processes can locally to regionally affect the thermal evolution of organic matter expected in a sedimentary basin. This is the case, in particular, for metamorphic aureoles, which can be syndiagenetic or post-diagenetic. The Carboniferous basin of Graissessac (Massif Central, France; Copard et al. 2000) is a good example of this, where an increase in Tmax values is to be noted from East to West, linked to a Late Variscan metamorphism in the SW of the basin. Unfortunately, these surface samples are also affected by weathering, which tends to mimic thermal alteration (Copard et al. 2002). Thus, it was necessary to compare Tmax with a parameter that is a little less versatile in the face of weathering, Ro, which highlights the degree of thermal maturity recorded by organic matter. Nevertheless, it is highly likely that samples, undisturbed by weathering, would have seen their Tmax increase. Another example from a metamorphic unit (Khushaym Matruck, Jordan; Elie et al. 2007) shows that the coupling between oxidative circulations and thermal degradation are capable of significantly increasing Tmax, decreasing the HI and locally increasing the OI of samples collected at different distances from this metamorphic unit, without, however, significantly modifying the Corg content.

11.3. Fires and signature of incomplete combustion residues

As a first approximation, the idea of characterizing refractory carbon and in particular black carbon in charcoal and soot using the Rock-Eval® method may seem risky because the combustion temperatures at the origin of their formation are often higher than or equal to that used during pyrolysis. Nevertheless, it is possible to propose the RC (Residual Carbon) parameter as a marker of condensed carbons specific to PAH with four aromatic rings showing variable desorption kinetics (fluoranthene, pyrene and chrysene; Poot et al. 2014). One study compared the Rock-Eval® method with other classical black-carbon extraction methods, namely chemical oxidation, thermal oxidation, organic geochemical markers and an optical method on a series of reference samples presenting zero (melanoidin) to high (soot) degrees of combustion (Poot et al. 2009). In this continuum of combustion, the RC appears to correlate again with materials that have recorded "low" combustion (the fraction known as *char*), while the T50_oxi parameter, the temperature at which 50% of organic matter is oxidized, correlates more with the fraction that has recorded more severe combustion (soot fraction). However, even if these conclusions appear convincing, several gray areas still remain that do not allow this method to be validated for detecting this continuum of combustion in soil or sediment samples. The first problem cited by the authors is that with the Rock-Eval® method the melanoidin sample would also contain black carbon, which is not possible. The second is that kerogens are also capable of generating such a signal (Figure 11.5).

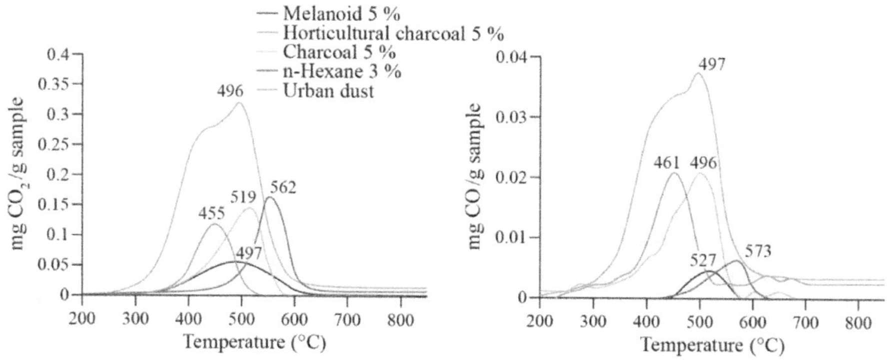

Figure 11.5. *CO_2 and CO signals of the oxidation phase of reference samples of the International Steering Committee of Black Carbon reference materials (Poot et al. 2009). For a color version of this figure, see www.iste.co.uk/baudin/rock-eval.zip*

11.4. Weathering versus syndiagenetic oxidation

Whether it is old or modern, weathering occurs naturally as soon as the organic matter is found in surface to sub-surface conditions. In general, all studies agree on the fact that weathering tends to reduce the hydrocarbon fraction (decrease in the S1, S2 and HI parameters) and increase the OI (Littke et al. 1991; Lo and Cardott 1995; Copard et al. 2002; Georgiev et al. 2012; Meinhold et al. 2013; Emmings et al. 2017; Stock et al. 2017; Kuske et al. 2019; Philp and DeGarmo 2020).

Only Tmax may be subject to controversy. Indeed, it can increase significantly, as has been shown for an outcrop of low-volatile bituminous coals of the Stephanian age (Alès, SE of the Massif Central; Copard et al. 2002; Figure 11.6).

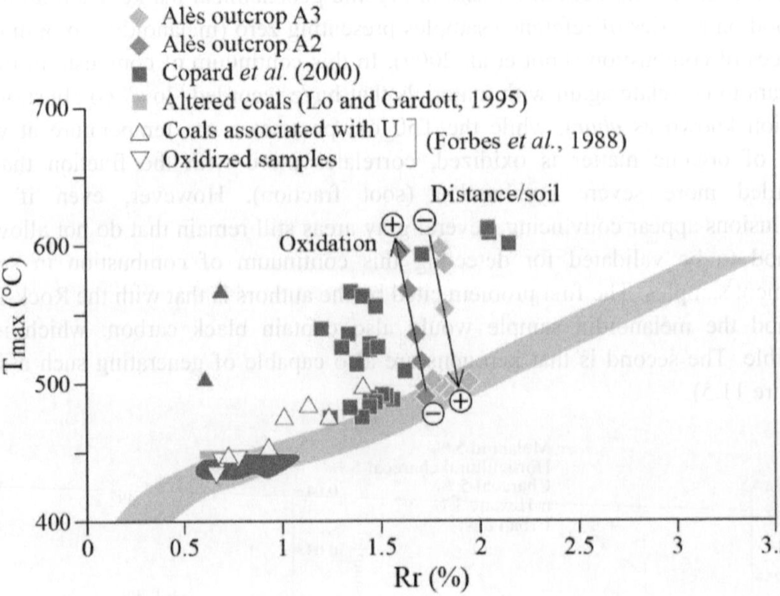

Figure 11.6. *Evolution of Tmax and the vitrinite reflectance of the samples of the Alès coal basin (Copard et al. 2002) and less-evolved coals examined by Lo and Cardott (1995). Note that the least-evolved samples do not show a significant increase in Tmax. For a color version of this figure, see www.iste.co.uk/baudin/rock-eval.zip*

However, for less-evolved coals of Middle Pennsylvanian age from the McAlester formation, Woodford shales (Oklahoma, USA, Lo and Cardott 1995) and less-mature coals of Stephanian age (Massif Central, France, Copard 2002), Tmax remains unchanged in the face of weathering (Figure 11.6). Yet, it would appear that its evolution, mimicking a classical thermal alteration, can emphasize a lesser or greater degree of oxidation, as shown in Figure 11.7, comparing Tmax and the OI (Copard et al. 2002).

The hypothesis envisaged would then be that oxidation favors the preferential degradation of the aliphatic and aromatic groups, which constitute the most thermolabile components of the hydrocarbon fraction shaping the S2 signal. As such, weathering of mature to very-mature organic matter very rapidly favors the displacement of the S2 parameter toward a higher pyrolysis temperature and thus triggers a rise in Tmax, as this labile fraction has already been significantly released during previous thermal diagenesis. In contrast, for matter that is less thermally evolved, this oxy- and thermolabile fraction is still strongly present, and weathering therefore has to be intense in order to consume this fraction and favor this displacement of Tmax (Figure 11.8; Copard et al. 2000). Tmax can therefore be considered here not as a marker of thermal evolution, as it is conventionally considered, but rather as a marker of low-temperature oxidation. This is the reason why, in order to ascertain the degree of thermal evolution achieved by samples at the outcrop, it is necessary to compare the Rock-Eval© parameters with other indicators less sensitive to oxidation, such as vitrinite reflectance (Copard et al. 2002; Yang and Horsfield 2020). Overall, these Rock-Eval© parameters thus form part of a general scheme of the weathering process of organic matter, such as coals, which make it possible to explain the set of geochemical, spectroscopic and optical signatures obtained on such outcrop samples (Figure 11.9; Copard 2002).

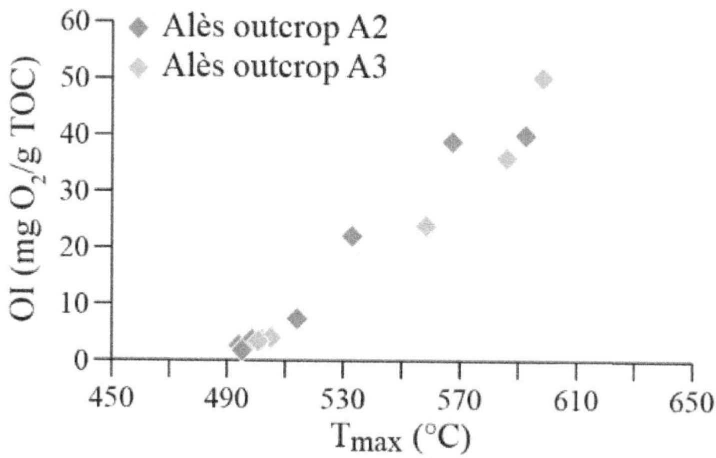

Figure 11.7. *Relationship between the Tmax values and those of the RE6 OI of the two Alès outcrops (Copard et al. 2002). For a color version of this figure, see www.iste.co.uk/baudin/rock-eval.zip*

Similarly, at low temperatures, the processes of syndiagenetic oxidation, sometimes related to uranium deposits (section 11.2.2), trigger the same trends as weathering, namely a decrease in the TOC content, the HI, an increase in the OI, as well as in certain cases, a significant increase, more than 100°C in extreme cases, in Tmax (Figure 11.10; Landais et al. 1984). For the latter authors, the gain in thermal stability marked by the increase in Tmax would be explained by chemical bonds involving oxygen, which form between the molecules already present, thus consolidating the chemical structure.

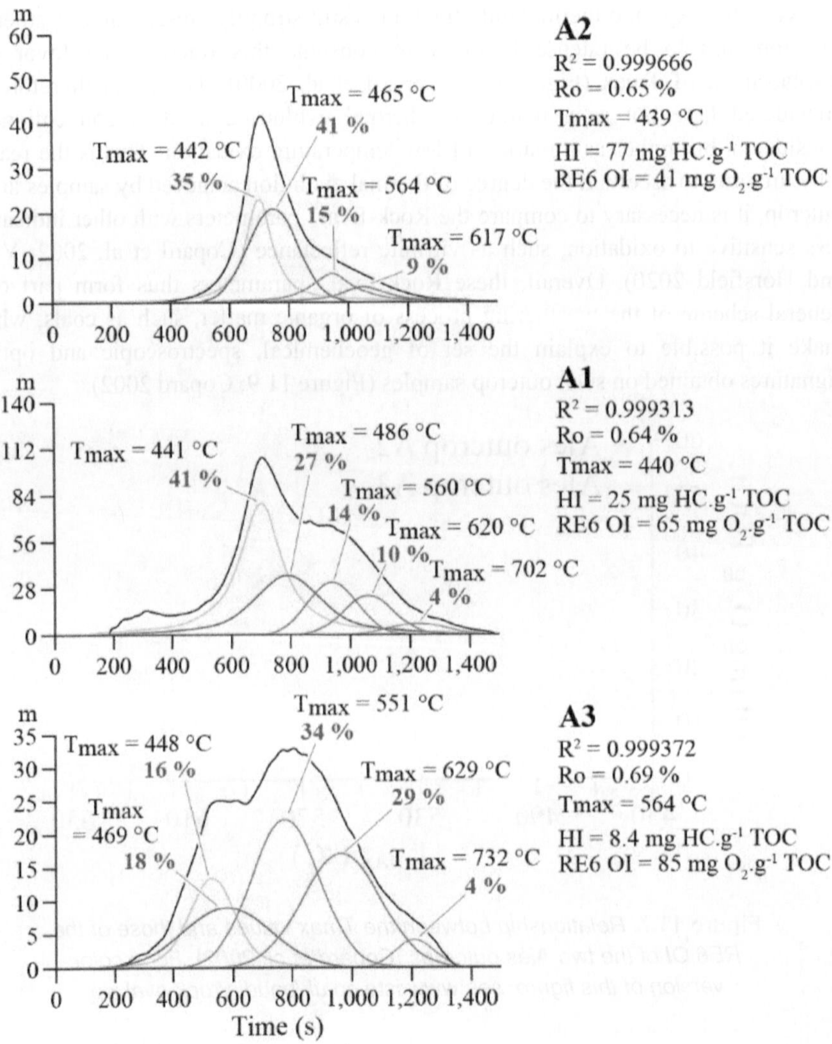

Figure 11.8. *Deconvolution of pyrograms of three low-maturity coal samples exhibiting increasing degrees of oxidation of the Argentat coal basin (Massif Central, France). Note the decrease in the signal of the most thermolabile fraction. Modified from Copard (2002). For a color version of this figure, see www.iste.co.uk/baudin/rock-eval.zip*

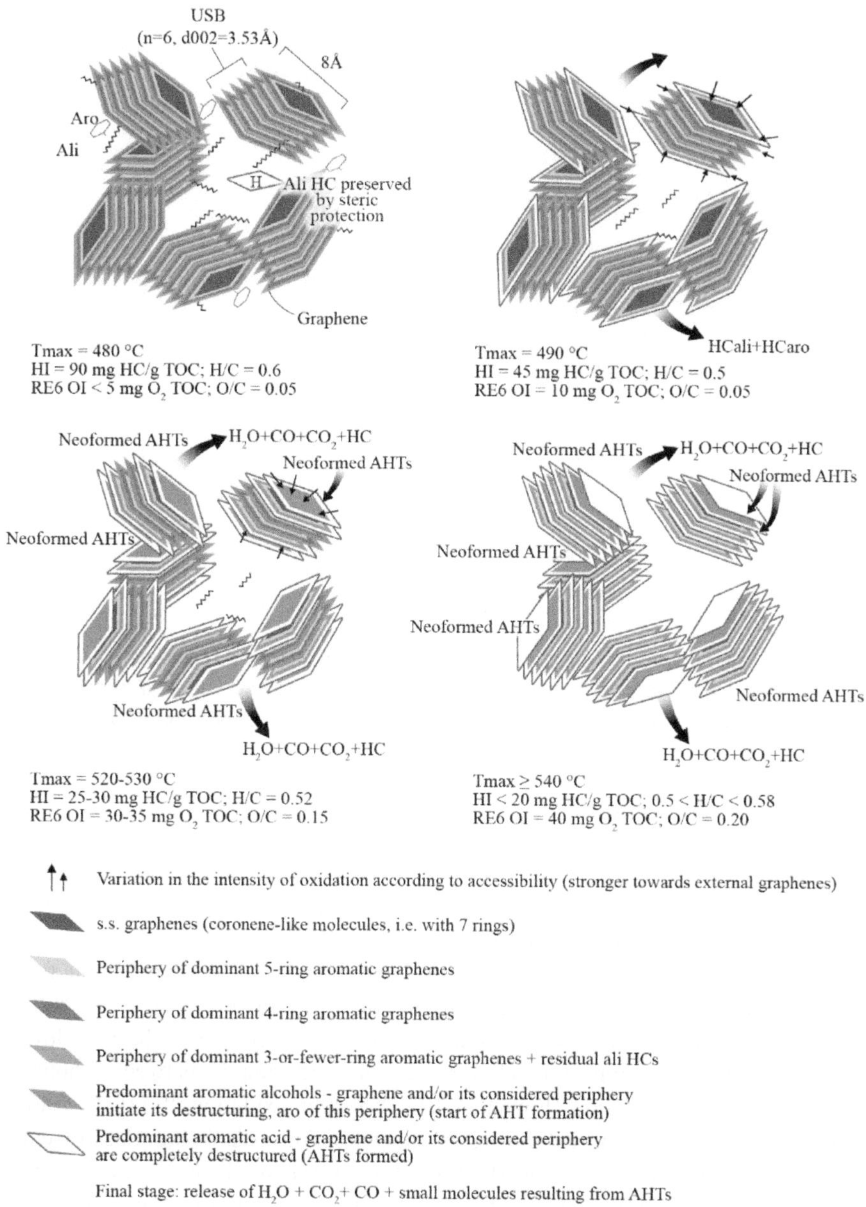

Figure 11.9. *Summary diagram of the weathering of a coal with a vitrinite reflectance > 1.5%, incorporating the overall geochemical (RE6 and atomic ratio), molecular and spectroscopic (DRX and FTIR) dataset. Modified from Copard (2002). For a color version of this figure, see www.iste.co.uk/baudin/rock-eval.zip*

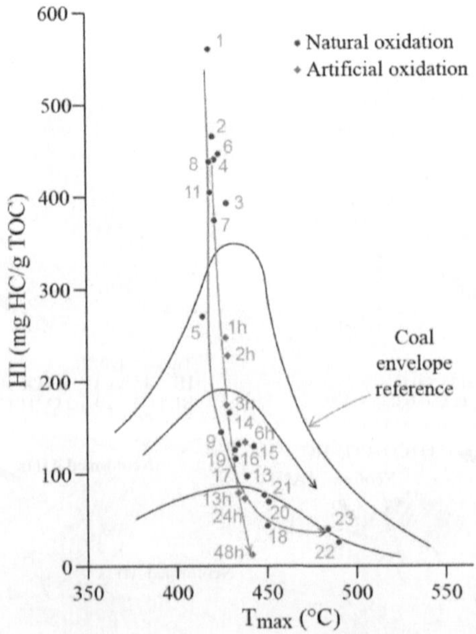

Figure 11.10. *HI versus Tmax diagram of Salt Wash samples (Landais et al. 1984); showing the effects of increasing supergene weathering compared to artifical oxidation. For a color version of this figure, see www.iste.co.uk/baudin/rock-eval.zip*

11.5. Artificial oxidation

Few studies have focused on low-temperature oxidative alteration conducted on laboratory samples. The first works, dating back to the 1980s (Landais et al. 1984), focus on a series of low-maturity coals (Tmax in the region of 427°C) originating from the Morrisson Formation (Salt Wash, Utah, USA). Fairly logically, the results show the trend observed during strict weathering for the TOC, HI and OI parameters. Tmax also seems to increase slightly, but, because of the very low HI values, the Tmax data unfortunately concern only the samples between 1 hour and 48 hours of aerial oxidation (Figure 11.10). A longer oxidation at 110°C, a lower temperature than that recorded during the diagenesis of coals of the Massif Central, was conducted for up to 278 days (Copard et al. 2004). While an increase in the OI and a decrease in the HI, overall stable TOC values (except during the experiment at 150°C) and an increase in Tmax can be noted, this study helped highlight two stages of the oxidation process. The first is accompanied by a sudden drop in hydrogen with a significant gain in oxygen, then, when the hydrogen integrated into the aliphatic molecules is entirely consumed, a progressive gain in oxygen can be noted. The increase in Tmax appears to occur between the first and second stages of this oxidation process. These two stages (Figure 11.11) concur with those described by Joseph and Oberlin (1983) based on atomic ratio data (H/C and O/C), which lead to the progressive formation of an oxy-coal via crosslinking reactions.

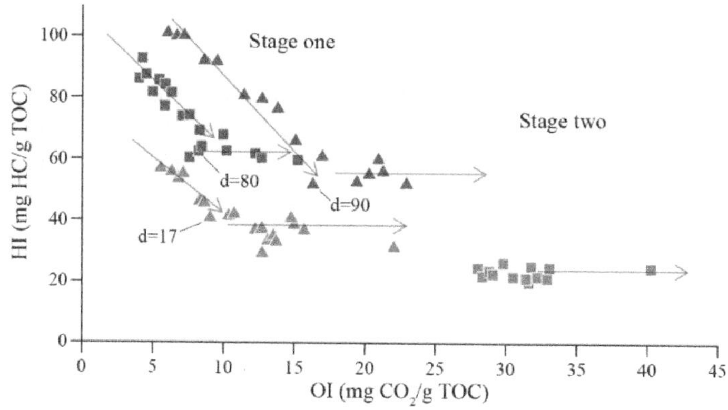

Figure 11.11. *HI versus OI diagram of four carbons artificially oxidized at 110°C, showing the two stages observed during the increase in exposure to oxidation up to 278 days. The square sample is the most oxidized and does not present any stages (Copard et al. 2004). For a color version of this figure, see www.iste.co.uk/baudin/rock-eval.zip*

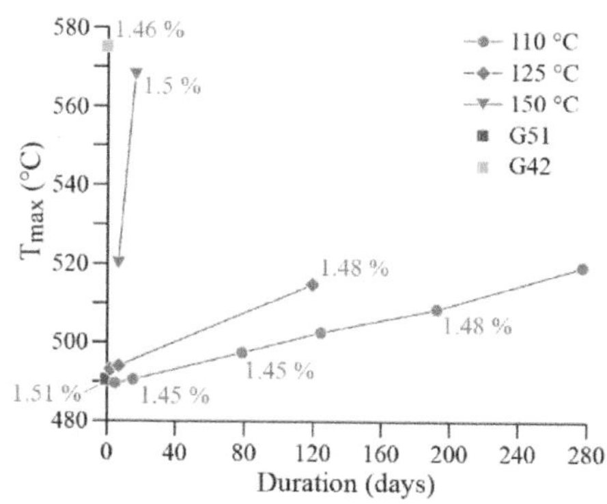

Figure 11.12. *Evolution of Tmax under an increasing oxidation temperature between 110 and 150°C for variable exposure times (Copard et al. 2004). For a color version of this figure, see www.iste.co.uk/baudin/rock-eval.zip*

It should also be noted that the initially altered sample shows no significant evolution in hydrogen and oxygen (except for one point showing a value close to 40 mg CO_2/gTOC) because its oxygen capture capacity seems to have already been reached (Figure 11.11). Finally, this gain in Tmax seems to be constrained by kinetic reactions, as shown by the change in slope in a diagram of Tmax versus oxidation duration at different temperatures (110–150°C) for the same sample (Figure 11.12; Copard et al. 2004).

This change in Tmax thus explains why altered samples with low thermal evolution do not show a significant increase in Tmax when all of the hydrogen incorporated into the aliphatic hydrocarbons is mobilized. Here again, focusing only on Tmax and the HI, weathering tends to mimic a classical thermal alteration.

12

Detection and Monitoring of Oil Pollution in the Environment

Yves BENOIT
IFP Energies nouvelles, Rueil-Malmaison, France

12.1. Introduction

In France, as in other countries of the European Union, the conservation of the state of the subsoil and water resources represents a major environmental challenge for the coming years. The preservation of the "soil resource" and its ecosystem functions is at the heart of numerous research projects. According to the latest inventories of polluted sites and soils in France, conducted by the French Agency for Ecological Transition (ADEME), the number of sites identified and potentially polluted amounts to more than 400,000 (representing approximately 100,000 hectares), some of which are in wasteland condition. This environmental liability can be explained by the numerous industrial activities developed since the mid-19th century. More than half of the soils recorded are contaminated with hydrocarbons and associated solvents. Crude petroleum cuts or raffinates are detected, as well as chlorinated solvents and heavy metals. These records cannot be exhaustive and should be commented on with caution since they only take into account industrial sites that have already been identified as polluted. In addition, they consider only part of former gas plants and abandoned gas stations.

Within this context of identifying polluted sites, the professionals in charge of the initial phases of diagnosis and risk assessment call on a number of standardized analysis methods (GC, HPLC, IR, fluorescence/UV), many of which are carried out ex situ and implemented in certified laboratories. They all need to meet the same objective: to identify and quantify the pollutants in place, and evaluate their extension. This diagnostic stage is the prerequisite for assessing the environmental impact of the existing pollutants. As industrial wasteland rehabilitation projects are subject to significant real-estate pressures and greatly constrained in terms of implementation, design offices are seeking rapid analytical techniques that can possibly be implemented in situ, with multiple objectives: improve the time taken to obtain results, optimize the costs of the initial diagnostic phases and more rapidly access the nature of pollutants, their concentrations and the extension of contaminated zones in order to best assess remediation costs.

Building on the experience gained in the oil industry over several decades for the characterization of source rocks using the Rock-Eval® method, in the early 1990s, IFPEN together with Vinci Technologies developed the Pollut-Eval® analyzer, dedicated to the characterization of soils contaminated by hydrocarbon pollutants: refined and unrefined oil products, such as fuels, coal tars and lubricants. The Pollut-Eval® method can therefore be considered as an extension of the Rock-Eval® method. Initially, this application was implemented within a laboratory device of a configuration similar to a Rock-Eval® analyzer. Nevertheless, it differed from it in terms of the equipment and functionalities adapted to this new application, which tends to vaporize highly volatile organic compounds: the initial temperature of the pyrolysis oven was consequently lowered significantly (20°C) and refrigeration of the sampler was put in place (4°C); lastly, new treatment methods, specific to the analysis of light distillates (gasoline and kerosene) were developed and qualified. For organic pollutants consisting of 5 to 20 carbon atoms and presenting high vapor pressures at room temperature, these adaptations enable an accurate carbon balance of the pollutants present within the contaminated soil sample to be obtained.

As developed later in this chapter, the Pollut-Eval® method, by adapted treatment of the FID and IR signals, also enables (1) evaluation of the moisture of the solid matrix analyzed and (2) evaluation of the proportion of signals related to the cracking of natural organic matter and those generated by hydrocarbon pollutants.

A market study then led to the development of a portable version of this device, named *Pollut-Eval®t*; its involvement within R&D projects, led by design offices, enabled the validation of rapid characterization methods (overall analysis time of 30 minutes), suitable for real-time on-site diagnosis. It was used for the in situ monitoring of drilling campaigns and sampling, as well as for the monitoring of depollution programs (Poland, Azerbaijan and Denmark) dedicated to evaluating the effectiveness of remediation methods (thermal desorption, venting and hydraulic fracturing). These stages were carried out within the framework of different partnership projects with national, European or international funding.

These studies were conducted in a mobile laboratory on wasteland and former orphan industrial sites. These campaigns led to the creation of databases consisting of characteristic thermograms of petroleum cuts, examples of which are presented in the following section. The analysis of contaminated soils using the Pollut-Eval® method was then extended to other contaminated solid matrices encountered in various environments (drilling and dredging muds, sedimentary piles at the foot of offshore drilling platforms, oily waste, burn pits, bitumen and coal tars from former gas plants).

Due to its rapid response and the reduction in the cost of the analysis, the Pollut-Eval® method has proven complementary to standardized laboratory analysis methods (GC, GC-MS and IR) by optimizing drilling and sampling operations in real time.

12.2. The Pollut-Eval® method

12.2.1. *Principle and equipment*

As mentioned in the previous section, the principle of the Pollut-Eval® method is derived from the Rock-Eval® method. We will not detail it here since its principle was described in Chapter 2. It differs from this method in terms of certain equipment and the treatment of the acquired data, which consist of releasing, quantifying and differentiating the pollutant organic carbon, often very volatile compared to Rock-Eval® classical applications, and the natural organic matter in the soil (cellulose, hemicellulose and lignin). In its laboratory version, the equipment resembles Rock-Eval® 6 Turbo (see Chapter 2) and the analysis method is similar to the Rock-Eval® Bulk Rock or multi-heating rate methods.

Due to its initial temperature, programmable from 20°C, the first pyrolysis stage of Pollut-Eval® has been adapted to thermovaporize hydrocarbons presenting high vapor pressures (gasoline, kerosene, diesel and lubricating oils) and then to pyrolyze the heaviest fractions (asphaltenes, bitumen, coal tars and heavy fuel oil) by thermal cracking. The method referred to as "standard" makes it possible, in an analysis time of about 1 hour, to successively release the different pollutant petroleum cuts present. The quantification of hydrocarbons is carried out by calibrating the response of the FID (1) either by a kerogen rock, like for Rock-Eval®, (2) or by the polluting petroleum cut, if the latter is available. Again, in its laboratory version, two infrared (IR) detectors allow the quantification of releases of CO and CO_2 originating either from carbonates or from the decomposition of the natural organic matter possibly present in the analyzed soil (humus, peat, leaves, fulvic and humic acids, etc.). The oxidation step, available only in the laboratory version, enables the total organic carbon of the soil and the rest of its mineral content to be evaluated.

It is important to note that these thermal treatments all lead to mass losses within the crucible containing the sample; these losses are continuously recorded and quantified by the different detectors. The only uncontrolled mass loss during analysis is that related to the evaporation of the water initially contained in the sample. Many soils present surface water contents of up to 40 %, particularly in the event of drilling in horizons close to aquifers and subject to variations of the water table. For comparison purposes, as design offices need to evaluate pollutant concentrations on dry soil, specific treatment of the signals of the Pollut-Eval® method has been developed to meet this need in order to determine the dry matter of the analyzed samples. In order to minimize the losses of light pollutants when they are on standby within the sampler, the samples to be analyzed are deposited between two layers of silica reactivated at 400°C. This protocol complements the refrigeration of the samples at 4°C within the sampler.

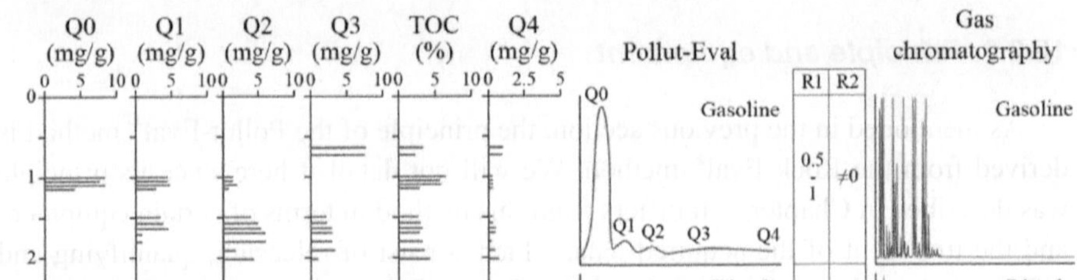

Figure 12.1. *Example of a log of pollutant concentrations encountered during drilling at a depth of 6.5 m. The discrete fractions Q0 to Q4 and the TOC representing the total hydrocarbons detected are represented. Each bar corresponds to one analysis. Shown on the right are the three polluting cuts identified on the site by GC analysis and Pollut-Eval® analysis. For a color version of this figure, see www.iste.co.uk/baudin/rock-eval.zip*

A graphical application developed directly in the *Rockint* software is used to represent the distribution of total hydrocarbons (and fractions Q0 to Q4, see Chapter 2) as a function of depth. As shown in Figure 12.1, these logs allow a two-dimensional contamination profile of the subsurface to be obtained, corresponding to each sampling point and to inform the work operators in real time regarding the zones to be excavated (or not) in order to optimize the sampling campaign duration and costs.

12.2.2. *Quantification and characterization of pollutants*

Quantification and identification of the pollutant or pollutants required the definition, within the pyrogram, of characteristic zones specific to each petroleum cut. Each zone is delimited by cursors corresponding to vaporization temperatures specific to each distillate and consequently to a precise composition of the pollutants in terms of carbon number. The surface delimited by two cursors defines an integration zone (identified by the letters Q in pyrolysis and Q' in oxidation). The integration zones (n), adapted to the analysis of hydrocarbon-type pollutants, are four in number in pyrolysis (n from 0 to 3) and three in oxidation (n from 0 to 2) (Figure 12.2). For comparison of the analysis results of different contaminated soils, the identification and quantification of pollutants requires fixed positioning of the cursors for reproducibility purposes. The main values used for calculations within each integration zone are as follows:

– the quantities of carbon measured (***Qn*** in pyrolysis and ***Q'n*** in oxidation);

– the relative values of the FID signal (***Rn***) within each zone;

– the CO_2/CO ratios (***Kn*** in pyrolysis and ***K'n*** in oxidation).

Furthermore, as in the Rock-Eval® method, the following characteristic parameters are defined:

– pyrolyzable carbon (***PC***), the mass of organic carbon released during pyrolysis, relative to the sample mass;

– residual carbon (***RC***), the mass of organic carbon released during oxidation, relative to the sample mass;

– mineral carbon (***MinC***), the mass of mineral carbon released during pyrolysis then oxidation, relative to the sample mass.

These calculations and indices make it possible to produce a complete assessment of the carbon present in the analyzed soil. The sum of pyrolyzable carbon (PC) and residual carbon (RC) defines the total organic carbon (TOC), the mass concentration of organic carbon (pollutant and/or natural). The temperature programming and the cycles were optimized after the analysis of 20 petroleum cuts representing cases of frequently encountered pollution (gasoline, diesel, kerosene, lamp oil, white spirit, heavy fuel oil and crude oils). These parameters led to the best results in terms of characterization and quantification of the polluting fractions most-frequently encountered in the environment. These pollutants of compositions known by GC/FID have been deposited on different soil types.

The *Pollut-Eval®t* treatment software enables graphical modification of the position of the integration cursors, according to the vaporization/pyrolysis temperatures of the pollutant petroleum cuts identified. In general, the light fraction of the petroleum cuts is quantified in zones Q0 and Q1, the middle distillates are quantified in zone Q2 and the heavy fraction is quantified in zone Q3 (see examples of pyrograms below). Zones Q0, Q1 and Q2 essentially correspond to the thermovaporization of hydrocarbons and zone Q3 to the cracking of the heaviest products, consisting of more than 35 carbon atoms, as explained below.

In the oxidation phase, the sample is gradually heated from 350 to 850°C at a rate of 20°C per minute and is then maintained at 850°C for 4 minutes. Two cursors are placed at 490 and 670°C, respectively (default values).

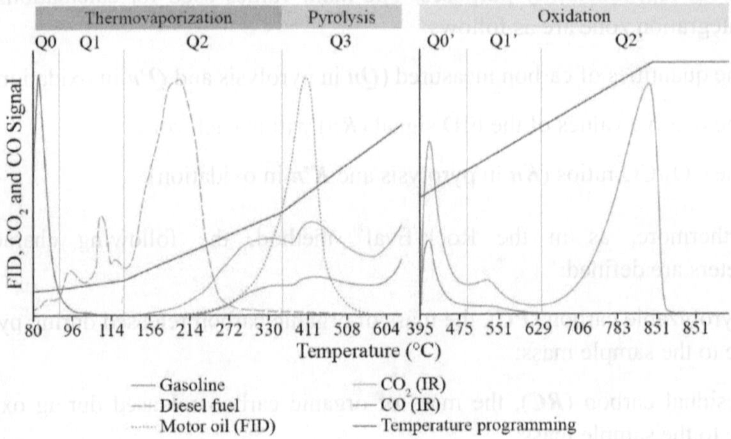

Figure 12.2. *Superposition of FID and IR signals observed during the pyrolysis and oxidation phase for a gasoline cut, a diesel cut and a motor oil. For a color version of this figure, see www.iste.co.uk/baudin/rock-eval.zip*

The thermograms shown in Figure 12.2 show the superposition of the vaporization and/or cracking curves obtained for gasoline, diesel and motor oil. GC/FID analysis reveals that each integration zone corresponds to a carbon composition and therefore to a precise petroleum cut. For example, it can be seen that gasoline is entirely vaporized at the end of zone Q0 (80–90°C) and that diesel is entirely vaporized at around 300 °C, before the end of zone Q2 (120–360°C). The motor oil begins its thermal cracking at around 300°C in Q2 and ends at 550°C in zone Q3.

12.2.3. *Characteristic pyrograms of reference hydrocarbon cuts*

During the method's metrological qualification stage, the hydrocarbons analyzed by the Pollut-Eval® method were also analyzed by gas chromatography. As indicated above, the composition and the number of carbon atoms of each constituent of the petroleum cuts studied are therefore precisely known. This enables the definition of a number of carbon atoms corresponding to each Pollut-Eval® integration zone. The pyrolysis oven of a Pollut-Eval®, subjected to a linear temperature gradient, can thus be assimilated to a distillation column with a single theoretical shelf. Comparisons of chromatographic and Pollut-Eval® analyses thus showed that the limit between zones Q0 and Q1 corresponded to C9–C10 hydrocarbons and that the limit between zones Q1 and Q2 corresponded to C12–C13 hydrocarbons. Regarding the limit between zones Q2 and Q3, it is more difficult to precisely determine the correspondence; it lies between C25 and C35 depending on the soil matrices.

It is important to note that in the presence of clay, hydrocarbons can be retained more by the soil and thus lead to a delay in vaporization (see section 12.3 on matrix effects). This can cause a shift in the peaks toward higher temperatures (mainly for the initial vaporization zones, Q0 and Q1). A database comprising several hundred pyrograms is available in its entirety in the Pollut-Eval® method manual.

12.2.4. *Two characteristic examples extracted from the reference database*

Figures 12.3 and 12.4 show the analyses of a kerosene cut and a diesel cut, two of the references in the Pollut-Eval® database with their reference indices.

12.2.4.1. *Kerosene cut: pyrogram of Pollut-Eval® analysis and GC/FID profile*

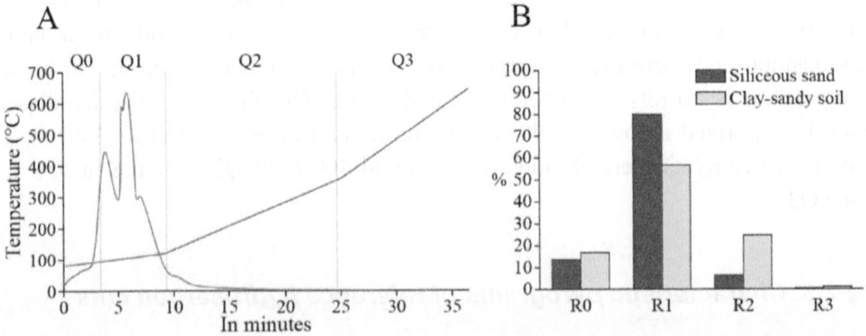

Figure 12.3. *On the left, pyrogram with the standard method of a kerosene-type oil distillate. On the right, the distribution (as a %) of the FID signal within zones Q0 to Q3 on two different soils. For a color version of this figure, see www.iste.co.uk/baudin/rock-eval.zip*

The graph on the right in Figure 12.3 represents the relative values (expressed as % m/m) of each peak detected during the pyrolysis phase of the kerosene cut presented on the left; the values correspond to the deposition of the same cut on two types of soil, siliceous and clayey. It can be observed that the pyrolysis temperature of the main peak is 103°C on siliceous sand and 110°C on clay-sandy soil. Kerosene is recognizable by the predominance of the Q1 peak. Whatever the type of soil, this represents between 55 and 85 % of the total hydrocarbons. The Q1 peak is in fact composed of two to three secondary peaks and essentially contains C11–C13 hydrocarbons.

The gas chromatogram of the kerosene cut reveals light hydrocarbons detected from 8 carbon atoms (C8) and heavy hydrocarbons present up to C15, with a maximum proportion centered around C11–C13.

12.2.4.2. *Diesel cut (domestic fuel): Pyrogram of Pollut-Eval® analysis and GC/FID profile*

The graph on the right in Figure 12.4 represents the relative values of each peak, detected during the pyrolysis phase of the same diesel cut deposited on two different soil types. It can be observed that the pyrolysis temperatures of the main peak are 193°C on siliceous sand and 220°C on clay-sandy soil. Diesel is recognizable by the predominance of the Q2 peak. Gas oil and domestic fuel are refined petroleum cuts with a maximum between C14 and C17. This type of cut generates a significant peak (60 to 70 % of the total hydrocarbons) in the beginning of zone Q2, zone Q1 being characterized by two to three secondary peaks. The gas chromatogram of the diesel cut reveals light hydrocarbons detected from C8 and heavy hydrocarbons present up to C24, with a maximum proportion centered around C14–C17.

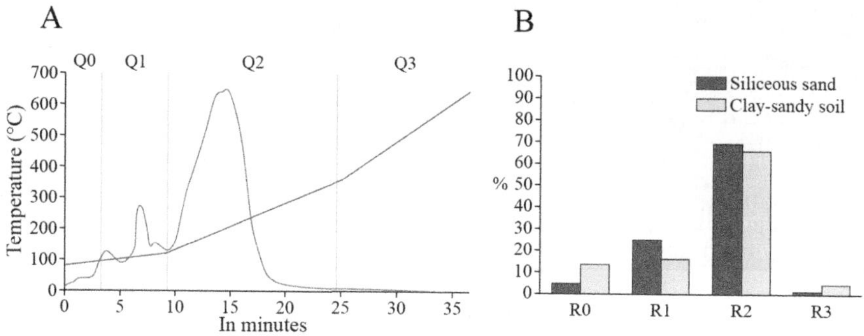

Figure 12.4. *On the left, pyrogram with the standard method of a diesel-type oil distillate. On the right, the distribution (as a %) of the FID signal within zones Q0 to Q3. For a color version of this figure, see www.iste.co.uk/baudin/rock-eval.zip*

12.2.5. *Examples of characteristic ratios for the diagnosis of contaminated soils*

As shown in the above two examples taken from the Pollut-Eval® method database, it can be noted that the shape of the pyrogram and its distribution can differ slightly depending on the soil type and that the R0 and R1 values can be subject to variations. In contrast, the relative value of the sum of the Q0 and Q1 (R0+R1) peaks is influenced little by the soil type. Ratios R2 and R3, meanwhile, undergo little variation; however, when the peak is at the zone limit, a slight shift can modify the values (this is the case for motor oil). While the shape of the pyrogram indicates that it is indeed the same peak, it is necessary to consider the sum of the two zones for more rigorous identification.

Table 12.1 shows several examples of characteristic vaporization and/or cracking ratios, measured for a sandy soil, with the values obtained from a clayey soil shown in brackets. The use of the relative values (R) allows an opinion to be formed regarding the nature of the petroleum cut present. The values obtained in each integration zone are reproducible, making it possible to assess the repeatability of the results.

Pollutant type	R0 as a %	R1 as a %	R2 as a %	R3 as a %	Average vaporization temperature
Gasoline	99.2 (> 90 %)	0 (< 10 %)	0.3 (~0)	0.5(~0)	84.7
Domestic fuel	12.2 (5–15 %)	18.7 (25–15 %)	69.1 (~70 %)	0(~0)	182.7
Kerosene	10 to 15 %	80 (55 %)	10 (30 %)	0	110
Diesel fuel	8.9 (5–15 %)	15.8 (25–15 %)	75.4 (~70 %)	0(~0)	191.0
Engine lubricating oil	0.5 (~0)	0.5 (~0)	31.8 (25-50 %)	67.2 (75–50 %)	420.3

Table 12.1. *Characteristic indices of several oil distillates obtained on two soil types*

12.3. Influence of the natural organic matrix of soils on the observed responses

Certain hydrocarbon-polluted soils may contain decomposing natural organic matter of varying maturity (forest or agricultural soils, garden soils, potting soil, humus, peats, etc.). Organic matter consists of complex compounds containing oxygen functions, such as hydroxyl and carboxyl. During its degradation under the effect of pyrolysis, the organic matter releases, along with CO/CO_2, compounds detected by FID in the same way as pollutants; this natural organic material can therefore generate a significant signal in pyrolysis, observed in zone Q2 and above all Q3.

The main function of Pollut-Eval® is to quantify and identify pollutant hydrocarbons within a soil. If this soil naturally contains thermopyrolyzable compounds, their responses will be superimposed and interfere with those of the pollutant petroleum cuts. This parasitic response can qualitatively and quantitatively alter the conclusions of the diagnosis. A study was therefore undertaken to define the existing relationships between the different signals generated during analysis; the objective was to highlight the possible existence of characteristic ratios that would make it possible to evaluate the proportion of the FID signal due to natural organic matter and that specific to pollutant hydrocarbons.

In this study, a significant factor was taken into account as certain carbonates, such as sodium carbonate and siderite, degrade as soon as the pyrolysis phase is completed (see Chapter 13). The degradation of carbonates therefore leads in turn to a release of CO/CO_2 of strictly mineral origin, which will also impact the relationships existing between the FID signal and the CO and CO_2 signals during the pyrolysis phase.

The analyses were carried out in triplicate on 20 types of soil representative of the different organic matrices that Pollut-Eval® would likely encounter during diagnostic phases (garden, forest, humus, potting soil, clays, peat, etc.). Characteristic solid matrices (rendzines or rendosols, peat and pelosols) were added to these.

12.3.1. *Types of samples and natural soils analyzed*

The analyses were carried out on the soil types presented in Table 12.2. All of the samples were dried (made possible by the absence of pollutant volatile compounds), nine of them were finely ground and the other 11 were sieved in order to assess the significance of these pretreatments in the reproducibility of the results.

Sample type	Number of samples	Sifted	Powdered
Garden soil	2	X	
Forest soil	4	X	
Clay-sandy soil	2	X	
Green clay	1	X	
Potting soil	2	X	
Peat	1	X	
Calcareous brown soil	1		X
Pelosol, its clay fraction dominant from the surface	1		X
Sandy soil	2		X
Rendzine - soil rich in humus and organic matter	2		X
Various soils (referenced 971, 973 and 976)	3		X

Table 12.2. *List of soils studied and analyzed by the Pollut-Eval® method*

12.3.2. *Correlation between the different pyrolysis and oxidation signals*

The observed values (Qt FID, CO and CO_2 in pyrolysis and oxidation) presented in this section correspond to the average integrations of three to six successive analysis results.

Designation	Number of analyses	Qt FID	Qt CO	Qt CO_2	Total Q0' + Q1'
Limestone	3	2.42	0.92	7.15	17.13
Pelosol	4	2.25	0.84	4.88	14.07
Various1	3	1.39	0.69	3.77	9.40
Randzine1	4	2.66	0.78	4.56	13.60
Various2	2	2.09	0.99	3.61	14.41
Sandy1	5	2.86	0.90	4.08	14.28
Randzine2	4	7.73	2.02	10.48	34.68
Various3	4	7.94	1.99	8.60	35.16
Sandy2	4	1.25	0.37	1.48	5.46
Clay2	4	0.74	0.34	1.44	4.50
Clay1	3	0.45	0.09	1.46	1.16
Clayey	6	0.17	0.03	0.73	0.66
Forest1	4	6.45	1.42	5.82	25.71
Forest2	4	17.57	4.70	13.35	68.22
Forest3	4	8.60	2.45	7.82	36.47
Forest4	4	31.35	8.35	19.64	111.68
Garden1	4	2.00	0.62	2.90	7.36
Garden2	4	4.76	0.99	3.84	20.59
PottingSoil1	5	41.96	11.41	30.38	167.13
Peat	5	81.29	23.18	40.59	288.39
PottingSoil2	3	57.44	13.55	32.55	201.25

Table 12.3. *Details of the Qt FID, Qt CO and CO_2 characteristics obtained during the analysis of around 20 uncontaminated natural soils*

Figure 12.5 illustrates the responses obtained in the pyrolysis phase during cracking of the natural organic matter for forest soil no. 4.

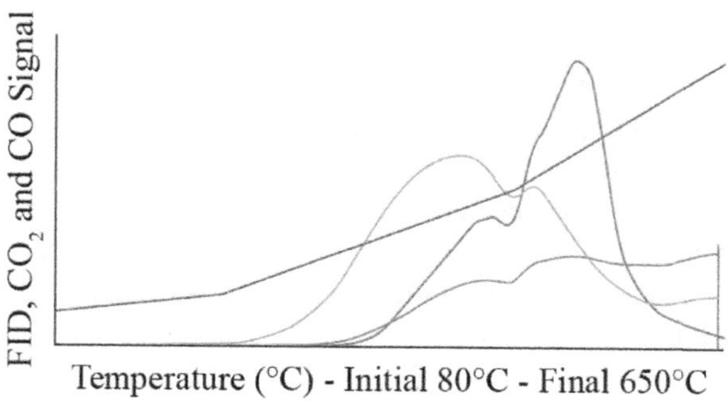

Figure 12.5. *FID pyrogram (pink) and CO/CO_2 signals (blue and green) generated by pyrolysis of forest soil no. 4, rich in humus. For a color version of this figure, see www.iste.co.uk/baudin/rock-eval.zip*

12.3.2.1. *Correlations observed during the pyrolysis phase between the different signals*

For the 20 surface soils, different correlations were studied between the FID, CO and CO_2 signals detected in zones Q2 and Q3 during the pyrolysis phase. Other correlations were then studied between the FID responses in pyrolysis and those detected during oxidation for CO and CO_2 (results not shown).

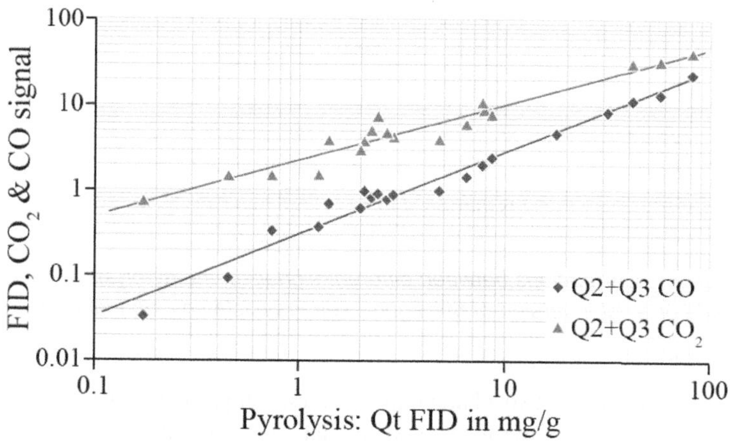

Figure 12.6. *Correlations between signals Qt FID, CO and CO_2 in zone (Q2 + Q3) during pyrolysis. For a color version of this figure, see www.iste.co.uk/baudin/rock-eval.zip*

For the correlations of signals detected during pyrolysis of natural organic matter, Figure 12.6 illustrates those observed between the sum of the FID signals (Qt FID) and the CO and CO_2 signals detected in zones Q2 + Q3. As shown in Table 12.4, the correlation coefficients signify the existence of a close proportionality between the generation of CO/CO_2 and the FID trace of hydrocarbon compounds of natural organic matter. The FID signals generated by cracking of natural organic matter are proportionally linear to the sum of the CO and CO_2 signals detected in zones Q2 + Q3.

12.3.2.2. *Correlations observed between the pyrolysis phase and the oxidation phase*

Concerning the correlations of the signals detected during the pyrolysis and oxidation phase, Figure 12.7 presents the relations observed for the 20 surface soils between the sum of the CO + CO_2 responses observed in the oxidation phase and the FID, CO or CO_2 responses detected in the pyrolysis phase. In all three cases, there is visible proportionality.

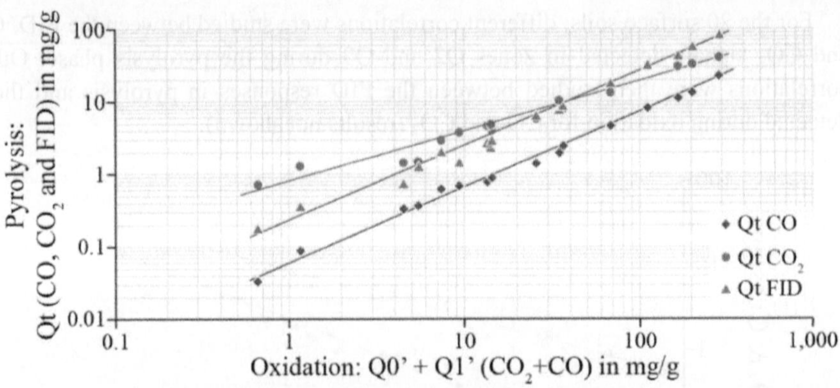

Figure 12.7. *Correlations between different signals of the pyrolysis and oxidation phases. For a color version of this figure, see www.iste.co.uk/baudin/rock-eval.zip*

These results allow two characteristic relationships of the behavior of natural organic matter during pyrolysis and oxidation to be defined. As previously mentioned, some of the samples analyzed contain carbonates, some of which degrade as soon as the pyrolysis phase is complete, mainly releasing CO_2. This is the case for the following carbonates:

– Nahcolite ($NaHCO_3$) with a degradation temperature of ~250°C;

– Magnesite ($MgCO_3$) with a degradation temperature of ~450°C;

– Siderite ($FeCO_3$) with a degradation temperature of ~580°C.

The presence of these carbonates can therefore disturb the relationships presented above between the different signals, which led to the analysis of different mineral matrices in order to define the relationships presented above. A summary of the results and the characteristic equations of the relationships observed are shown in Table 12.4.

	Without intersection at (0;0)		With intersection at (0;0)	
Relationship type	Equation	R^2	Relationship type	Equation
	Equation	R^2	Equation	R^2
Qt FID/Qt CO	$Y = 0.2684X + 0.278$	0.99	$Y = 0.2689X$	0.99
Qt FID/Qt CO_2	$Y = 0.5108X + 3.066$	0.96	$Y = 0.5763X$	0.91
Q0' + Q1' (CO + CO_2)/QT FID	$Y = 0.2815X + 1.0275$	0.99	$Y = 0.2753X$	0.99
Q0' + Q1' (CO + CO_2)/Qt CO	$Y = 0.0758X + 0.2788$	0.99	$Y = 0.0741X$	0.99

Table 12.4. *Summary of the relationships observed*

Thanks to the various equations above, it is therefore possible to evaluate the overall proportion of the FID signal generated by a variety of natural organic matter. The majority of compounds of natural organic matter are detected in pyrolysis zones Q2/Q3; therefore, it was retained for the Pollut-Eval® method to work with the averages of the relative values (R2 and R3) of the signals detected in these two zones. This choice enabled better distribution of the interfering proportion of the FID signals when heavy pollutants are vaporized or cracked in these temperature ranges.

The final equations retained in the method correlate the pyrolysis FID response detected in Q2 + Q3 with $Q2CO_2$ or Q3CO. These data are derived from immature Type III organic matter.

Relationship 1: $(Q_2 + Q_3)\ \text{FID} = Q_2CO_2/0.3562$

Relationship 2: $(Q_2 + Q_3)\ \text{FID} = Q_3CO/0.1139$

Following this study and thanks to these relationships, it was therefore possible in the course of a diagnosis of soils polluted by hydrocarbons, to evaluate (and not to quantify) the proportion of the interfering FID signal, due to natural organic matter. Zone Q2 represents 20 % of the overall response, and zone Q3 represents 80 %. This result is significant because the limit between Q2 and Q3 (350°C) corresponds to the transition temperature zone in the vaporization and cracking of pollutant hydrocarbons. The quantity of natural organic matter having cracked in these two zones is expressed, via the previously calibrated FID detector, as "hydrocarbon equivalent". The calculator can thus distribute the parasitic response of the organic matter between these two zones, and the values obtained are deduced from the measurements carried out on the polluted matrices. Nevertheless, it is important to remain cautious because the evaluation of a hydrocarbon equivalent is marred by the error related to the respective carbon richness of hydrocarbons and natural organic matter. During numerous case studies, extraction from the soil using a cyclohexane/acetone 85/15 mixture enabled validation of the pyrolysis profiles of the unpolluted natural organic matter and that contaminated by the hydrocarbon petroleum cut.

13

Analysis of Carbonates

François BAUDIN[1], Nicolas BOUTON[2] and Adrien WATTRIPONT[2]

[1] *ISTeP, Sorbonne Université, Paris, France*

[2] *Vinci Technologies, Nanterre, France*

13.1. Reminder of the principles of mineral carbon analysis using the Rock-Eval® method

Carbon can be present in geological materials in organic or inorganic forms, and most often these two forms are mixed in soils, sediments and sedimentary rocks. In the early versions of Rock-Eval®, the devices were set in such a way as to avoid possible pollution of organic effluents by the inorganic carbon fraction of the samples. Thus, the oxidation phase did not exceed 600°C and was limited to measuring the CO_2 released at this temperature for 7 minutes to prevent the thermal cracking of calcite ($CaCO_3$), which starts at around 580–600°C, from overly disturbing the analysis (Espitalié et al. 1985a). For the same reason, the CO_2 released during the pyrolysis phase was trapped up to 390°C (or 400°C depending on the device) to avoid possible contamination of organic CO_2 by thermal cracking of siderite ($FeCO_3$), which starts at around 400°C.

With Rock-Eval® 6 and the continuous detection of CO and CO_2 during the pyrolysis and oxidation phases, it then becomes possible to attribute an organic or mineral origin to carbon dioxide and oxide (Figure 13.1). Thus, the S3 peak, which until then corresponded to the CO_2 produced at a temperature below 400°C, gains surface area with the addition of an S3' peak, corresponding to the CO_2 released in the interval 400–650°C. This S3' peak is considered by Lafargue et al. (1998a) and Behar

---

For a color version of all figures in this chapter, see www.iste.co.uk/baudin/rock-eval.zip.

The Rock-Eval Method,
coordinated by François BAUDIN. © ISTE Ltd 2023.

et al. (2001) as being of exclusively inorganic origin. The CO produced during the pyrolysis phase, which was not considered in earlier versions of Rock-Eval®, is, for its part, separated into an S3CO peak and an S3'CO peak. The separation between these two peaks is generally situated at around 550°C but can fluctuate between 450°C and 600°C (Behar et al. 2001), depending on the position of the minimum between the two peaks. The S3CO peak has organic matter as its source and the S3'CO peak is counted as coming half from organic matter and half from mineral carbon, due to the Boudouard reactions (see the explanation of these reactions in sections 2.4.5 and 3.2.2). During the oxidation phase, CO is exclusively produced by organic matter (S4CO peak), while CO_2 is separated into two peaks: S4CO_2 corresponding to CO_2 of organic origin and S5 to that of mineral origin. The separation between S4CO_2 and S5 is generally situated between 550°C and 720°C according to Behar et al. (2001). In practice, the temperature at which the S4CO peak disappears marks the separation between residual organic carbon and mineral carbon. Logically, the percentage of mineral carbon, called MinC, is therefore the sum of a pyroMinC and an oxiMinC detected and quantified during the pyrolysis and oxidation phases, respectively, and whose bases of calculation are recalled in Table 13.1.

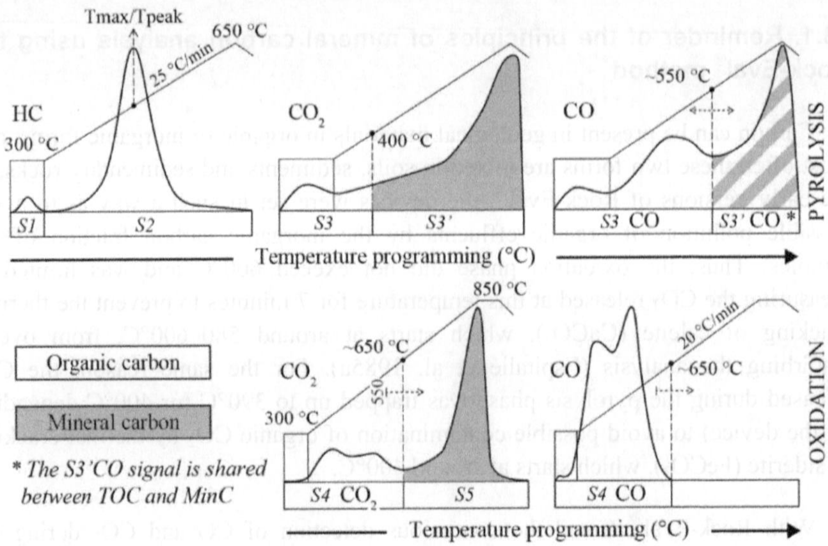

Figure 13.1. *Thermograms obtained during Rock-Eval® analysis in Bulk Rock Basic cycle mode of a sample of sedimentary rock containing organic matter and carbonate (the most common case) and separation into different peaks: S1, S2, ... S5 of these different thermograms. The bluish sections correspond to the effluents (CO_2 and CO) resulting from the thermal decomposition of the carbonate fraction. All thermograms are normalized to their maximum values, i.e. the surfaces/heights are not comparable from one thermogram to another. Modified from Behar et al. (2001)*

pyroMinC	% weight	$\frac{\left[S3'\times\frac{12}{44}\right]+\left[\left(\frac{S3'CO}{2}\right)\times\frac{12}{28}\right]}{10}$	Pyrolysis Mineral Carbon
oxiMinC	% weight	$\frac{S5\times\frac{12}{44}}{10}$	Oxidation Mineral Carbon
MinC	% weight	PyroMinC + OxiMinC	Mineral Carbon

Table 13.1. *Calculations of the % by weight of mineral carbon (MinC) resulting from the decomposition of the carbonate fraction during pyrolysis (pyroMinC) and oxidation (oxiMinC) cycles during a Rock-Eval® Basic cycle analysis (Behar et al. 2001)*

13.2. Kinetic effects

The thermal decomposition of calcium carbonate following the reaction:

$$CaCO_3 \rightarrow CaO + CO_2 \qquad [13.1]$$

has been known since the second half of the 18th century. It is implemented on a large scale in the cement industry for the production of quicklime or calcium oxide (CaO). This highly endothermic decomposition reaction occurs around 800°C according to the literature, but it is easy to verify with Rock-Eval® that the thermal cracking rate of a calcium carbonate, and therefore the decomposition temperature, very much depends on the mass of the sample that has been introduced into the crucible (Pillot et al. 2014; Figure 13.2). Thus, the S5 Tpeak varies between 750°C and 850°C when 2 and 50 mg, respectively, of pure calcite are analyzed in oxidation alone. Thermogravimetric analysis (TGA) under conditions similar to the Rock-Eval® oxidation phase (sweeping by 100 $mL.min^{-1}$ N_2/O_2 80:20 with a thermal gradient of 20°C min^{-1}) shows exactly the same difference of 100°C (between 720°C and 820°C) for the same variation in weight.

This effect of the variation of the decomposition Tpeak of the carbonate peak as a function of the weight of the sample can be explained by the kinetics of the calcite decomposition reaction. In the case of a chemical reaction

$$R(eagent) \rightarrow P(roduct).$$

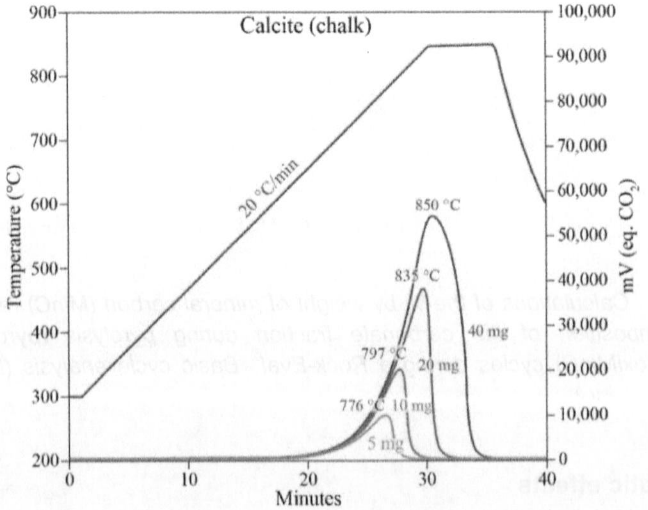

Figure 13.2. *Effect of the weight of a chalk sample on CO_2 production during its oxidation in a Rock-Eval® 6 (modified from Pillot et al. (2014)). Production is all the earlier, the lower the weight of the sample, which is marked by an increase in Tpeak of nearly 80°C, passing from 5 to 40 mg*

The rate of production of *P*, v_P, is equal to the opposite of the rate of consumption of *R*, v_R:

$$v_P = -v_R = \frac{d[P]}{dt} = -\frac{d[R]}{dt}, \quad [13.2]$$

with [R] as the concentration of *R*, which decreases as the reaction progresses, and [P] as the concentration of *P*, which increases with time. According to Arrhenius and Boltzmann, this rate is written as follows:

$$\frac{d[P]}{dt} = A\, e^{\frac{-Ea}{RT}}\, [R]_t \quad [13.3]$$

with:

– *t* as the time, in seconds;

– *A*, in s^{-1}, or Hz, which is the Arrhenius factor, or pre-exponential factor, corresponding, for unimolecular reactions, to typical vibrational frequencies of the molecular bonds[1];

1. This pre-exponential factor, *A*, is often considered as a constant.

– T as the temperature, in K;

– R as the perfect gas constant, 8.314 $J.mol^{-1}.K^{-1}$, which is also equal to k_b x N_A, (Boltzmann's constant times Avogadro's number);

– $[R]_t$ as the concentration of reagent remaining to be transformed at each time, t;

– energy, in two forms: the chemical activation energy, necessary to initiate the reaction, Ea, in $J.mol^{-1}$, and RT, which is the thermal energy given to the reaction system.

The numerical resolution of this differential equation gives a Gaussian curve, which is the usual shape of the S2 curve. This reaction is called order-1:

$$\frac{d[P]}{dt} = A\, e^{\frac{-Ea}{RT}}\, [R]_t^{\,1} \qquad [13.4]$$

If a chemical reaction involved two molecules of R to produce a molecule of P:

$$R + R \rightarrow P$$

The solution of the Arrhenius equation would be as follows:

$$\frac{d[P]}{dt} = A\, e^{\frac{-Ea}{RT}}\, [R]_t\, [R]_t \qquad [13.5]$$

or

$$\frac{d[P]}{dt} = A\, e^{\frac{-Ea}{RT}}\, [R]_t^{\,2} \qquad [13.6]$$

This equation is then said to be order-2, with reference to the term $[R]^2$.

In fact, there are reactions of order n, where n can be any number and sometimes even negative. The decomposition kinetics of solids is often described as zero order. In this case, the equation is:

$$\frac{d[P]}{dt} = A\, e^{\frac{-Ea}{RT}}\, [R]_t^{\,0} \qquad [13.7]$$

If the term of the quantity of reagent remaining to be transformed disappears, then the equation is no longer differential and ultimately becomes:

$$\frac{d[P]}{dt} = A\, e^{\frac{-Ea}{RT}} \qquad [13.8]$$

The reaction takes place only under the effect of time and temperature. It stops when the reagent is consumed. The shape of the curve is no longer Gaussian, but purely exponential, and suddenly drops to zero when the reagent is finally consumed. Therefore, the more reagent there is to consume, the more time and temperature the reaction takes. According to Roduit (2000), the decomposition reaction of calcite is of order 0.177, i.e. an order not quite equal to zero, but very far from 1.

The cracking kinetics of chemically different carbonates should therefore be studied separately. For example, the reaction mechanism of thermal decomposition of siderite under inert atmosphere is different from that of calcite. For siderite, according to Lafargue et al. (1998a), we have:

$$FeCO_3 \rightarrow FeO + CO_2 \quad [13.9]$$

$$3\,FeO + CO_2 \rightarrow Fe_3O_4 + CO \quad [13.10]$$

Here, two reaction rates are superimposed as the species produced react with one another to form CO, CO_2 and Fe_3O_4 in fine.

Therefore, not all carbonate forms exhibit a shift in their decomposition temperature as a function of the sample weight (Pillot et al. 2014). While this effect is very pronounced for calcite, aragonite, dolomite, cerussite, ankerite, strontianite, magnesite and huntite, it is not very pronounced for siderite, hydromagnesite, azurite, malachite and rhodochrosite, as shown in Figure 13.3.

As a first approximation, the kinetics of thermal decomposition of carbonates depends strongly on the cations associated with the carbonate ion (CO_3^{2-}), and more specifically on their electronegativity. The lower the electronegativity of the cation, the higher the decomposition temperature or, in other words, the higher the energy binding the cation to the carbonate anion. Thus, for Ca, Mn, Fe and Cu (four elements from period 4 of the periodic table), the electronegativity passes, respectively, from 1 to 1.55, then to 1.83, and lastly, to 1.9 (in the Pauling scale), hence a decrease in the decomposition temperature going from calcite to rhodochrosite, then siderite, and lastly, azurite (Figure 13.3). Moreover, strontianite needs to be analyzed with Rock-Eval® 7 in order to be able to observe its decomposition peak (well above 850°C) because strontium is more electronegative than calcium.

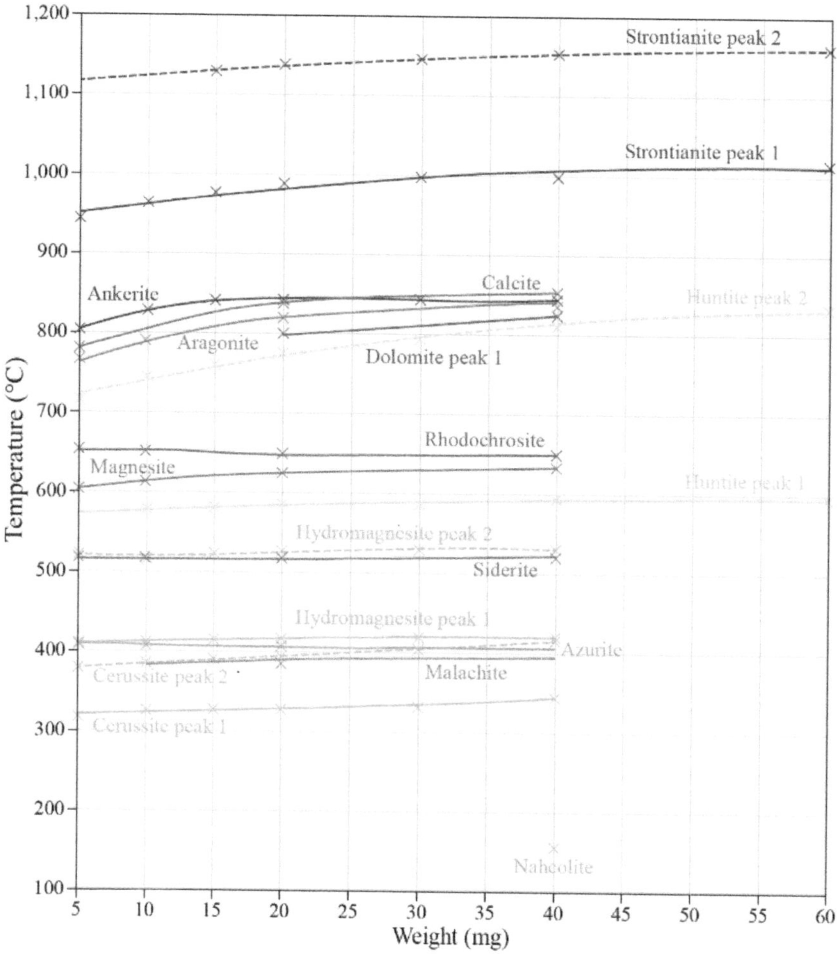

Figure 13.3. *Effect of sample weight on Tpeak, i.e. the temperature of the CO_2 production maximum(s) during oxidation of different carbonates using the Rock-Eval® method. Data for aragonite, azurite, dolomite, magnesite, malachite, rhodochrosite and siderite from Pillot et al. (2014)*

Moreover, the lower the cation electronegativity, the more pronounced the effect of weight on the decomposition temperature. This is explained by a cracking kinetics tending toward zero as the electronegativity decreases. The numerical resolutions of order-1 kinetics do not show a change of curve as a function of the initial weight of the reagent in the crucible, whereas the more the reaction order tends toward 0, the more pronounced the effect of the weight on Tpeak, and the more concave the shape of the peak on the final temperature plateau (Figure 13.4), as seen on a Rock-Eval® (Figure 13.2) or TGA. The final shape of the curve is therefore a visual indication of percentage decomposition.

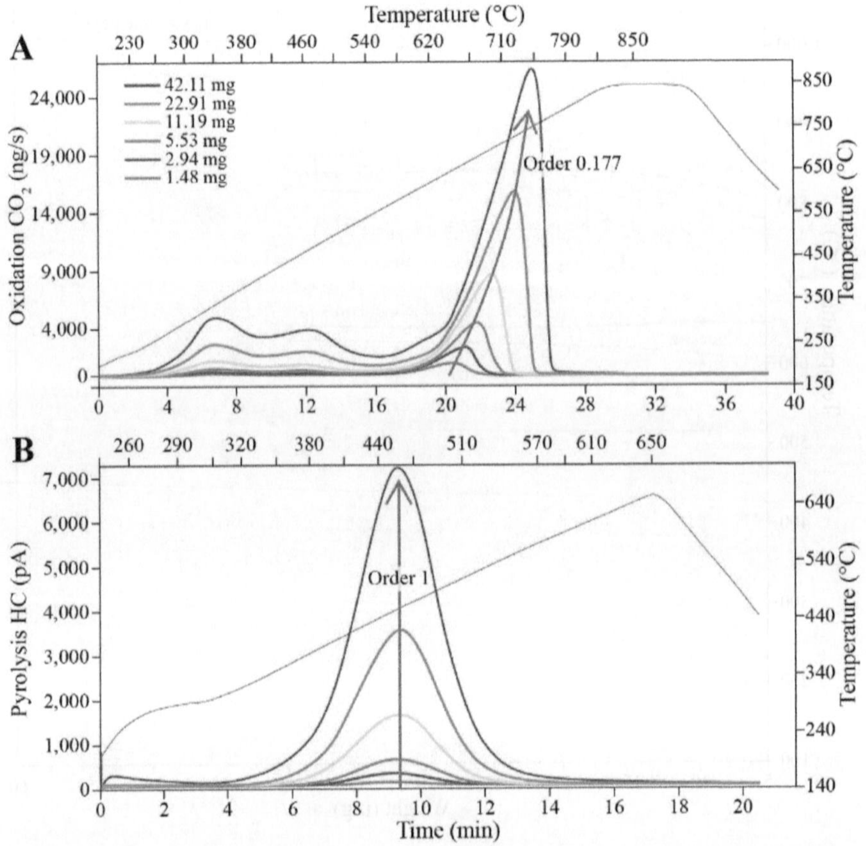

Figure 13.4.*(A) Rock-Eval CO_2 signal of increasing weights of the IFPEN reference sample 160,000: the peak on the right-hand side is a sign of calcite decomposition, while the peaks on the left-hand side are a sign of the combustion of the residual carbon; (B) Rock-Eval FID signal of increasing weights of the same sample: the peak area increases in an evenly distributed manner toward low and high temperatures, but Tpeak values remain constant, which is an indication of order-1 decomposition of organic matter reaction*

The heating rate also has a strong influence on the cracking temperature of calcite and aragonite. Thus, 40 mg of calcite heated in a Rock-Eval® crucible in the oxidation phase alone with rising temperature ramps of 2, 5, 10, 20 and 30°C. min^{-1}

shows CO_2 release peaks, the temperature of which varies from 680°C to 830°C (Figure 13.5), which is very close to that obtained in TGA (690°C–810°C). Since kinetics are a linear function of time, and an exponential (logarithmic) function of temperature, the peaks differ in appearance depending on whether the *x*-axis is annotated as time (Figure 13.5a) or as temperature (Figure 13.5b). The slower the temperature rise, the longer the calcite decomposition reaction, and the lower the decomposition temperature, and since the effect of time is linear, and the effect of temperature is logarithmic, according to Arrhenius' equation, time becomes the limiting factor when the temperature drops. For this reason, it is recommended not to significantly modify the oxidation cycle of the Rock-Eval® method, whether for the analysis of sedimentary rocks, sediments or soils.

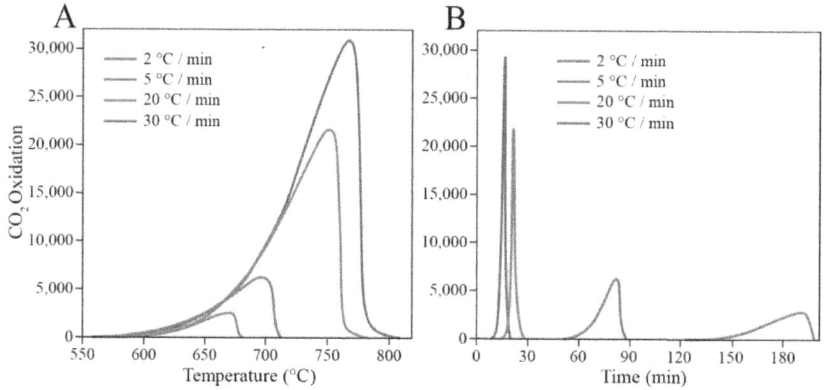

Figure 13.5. *Highlighting of a kinetic effect during cracking of a pure calcite. (A) ~40 mg are oxidized between 300 and 850°C according to temperature gradients from 2 to 30°C.min^{-1}. It can be seen that S5 Tpeak shifts from 670°C to 820°C when the gradients increase. (B) Same analyses but with the x-axis in time (min) and not as a function of temperature (in °C)*

Finally, it remains to be explained why some carbonates crack with a single peak of CO_2 release while others show two or even three peaks (Figure 13.3). It is easy to understand that for mixed carbonates such as dolomite and huntite, thermal cracking between the carbonate ion and Mg^{2+} and Ca^{2+} to which it is bound determines a double peak[2]. The $MgCO_3$ bond cracks at a lower temperature than that of $CaCO_3$ due to an electronegativity passing from 1.31 for Mg to 1 for Ca. For hydrated carbonate forms (such as hydromagnesite, $Mg_5(CO_3)_4(OH)_2$ 4 H_2O), dehydration

2. On the contrary, we do not observe several peaks for ankerite, whereas theoretically, the elements Ca, Fe, Mg and Mn can be present. The fact that Ca largely dominates over the other elements likely explains this singularity.

leads to a first destabilization of the mineral with the release of CO_2, followed by a second later when there is thermal cracking between Mg and CO_3. Carbonate forms that associate a single cation (Ca, Mg, Cu, Mn, Fe, etc.) generally present a single peak of CO_2 release. However, cerussite ($PbCO_3$) and strontianite ($SrCO_3$) are exceptions to this rule because their thermal decomposition occurs in stages. For cerussite, the reactions are as follows:

$$2PbCO_3 \rightarrow PbO{,}PbCO_3 + CO_2 \text{ around } 330°C \quad [13.11]$$

$$3(PbO{,}PbCO_3) \rightarrow 2(2PbO{,}PbCO_3) + CO_2 \text{ around } 390°C \quad [13.12]$$

$$2(2PbO{,}PbCO_3) \rightarrow 3PbO + CO_2 \text{ around } 410°C \quad [13.13]$$

The last two reactions, at similar temperatures, are merged in the same second peak during a Rock-Eval® analysis (Figure 13.3).

For strontianite ($SrCO_3$), the reason is less clear, and there are two possible explanations. The first is crystallographic in the case of pure strontianite. During heating, a change occurs at about 1,000°C with passage from the orthorhombic system to the hexagonal system with a first release of CO_2 and total decarbonatation then occurs at about 1,150°C (Figure 13.3). The alternative explanation is chemical. Since the majority of strontianites are solid solutions of $CaCO_3 - SrCO_3$, both peaks may be related to a release associated with each of the two cations in the case of non-pure strontianite.

13.3. Artifacts

As mentioned in the previous section, calcite and aragonite crack earlier, i.e. at a lower temperature, when analyzed alone and in small quantities (a few mg) in the crucible. This effect can also be expected to appear when these carbonates are present in a small proportion in a sample where they are then mixed with other minerals and with organic matter. This can be seen on certain thermograms of the oxidation phase where the S5 peak is shifted toward temperatures below 850°C. Note that the S5 Tpeak can drop to 775°C for calcite, or even 750°C for aragonite, when these minerals account for less than 5% by weight in the sample. This slight shift has sometimes been misinterpreted as resulting from the poor crystallization state of calcite or aragonite (Pillot et al. 2014; Baudin et al. 2015). Indeed, an X-ray diffraction analysis of marno-limestone samples of a series where the S5 Tpeak varies from 770 to 850°C reveals no calcite crystallization defects. It should be noted that this shift in the S5 Tpeak has no effect on the calculation of MinC or

TOC, as even with an S5 peak being expressed a little sooner, the $S4CO_2$ peak stands out and it is easy to distinguish them from each other.

The situation is different if the decomposition of calcite or aragonite occurs significantly earlier (with a decomposition start temperature << 650°C) and the S5 peak then interferes with the $S4CO_2$ peak (Figure 13.6). This case is rare but occurs when analyzing certain recent salt-rich sediments. Wattripont et al. (2019) clearly described the phenomenon that leads to such an overlap of the $S4CO_2$ and S5 peaks. This phenomenon occurs when a marine sediment containing both carbonates and clay minerals is oven-dried or freeze-dried with its pore water rich in dissolved salts. The powder obtained will therefore contain $CaCO_3$ and NaCl (plus other salts, in lesser quantities), combined with clay minerals and possibly organic matter. When clay minerals are hydrated, which is the case with the family of smectites, or hydroxylated, as with kaolinite, water or OH^- ions are released between 550 and 750°C. These molecules then react with NaCl to give hydrochloric acid vapors (HClv), which has the effect of chemically attacking the carbonates, thus releasing CO_2 at a much lower temperature than during its thermal decomposition (Figure 13.7).

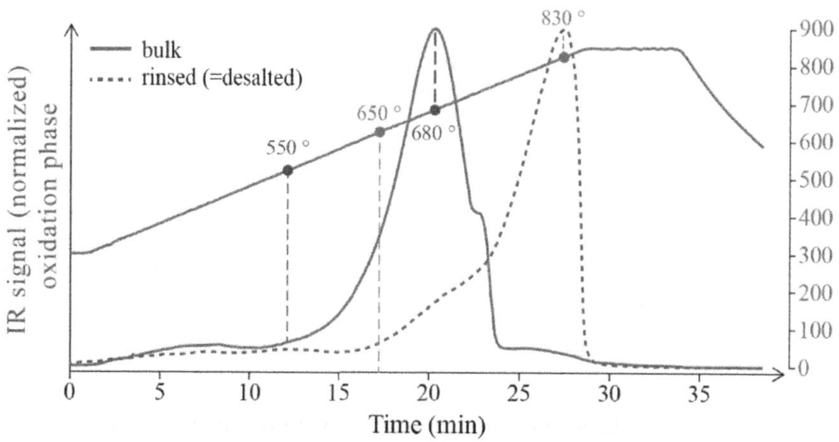

Figure 13.6. *Comparison of the oxidation thermograms of clay-carbonate (~30% $CaCO_3$), and organic hemipelagic mud from the Timor Sea. The analysis was performed on the bulk sediment and after washing with distilled water to remove salts present in the pore water. It should be noted that the release of CO_2 from the carbonate fraction occurs earlier (S5 Tpeak lower; 680°C compared to 830°C) when the sediment is bulk. S5 then overlaps $S4CO_2$; the separation between organic and carbonate-based CO_2 then becomes more difficult. Modified from Wattripont et al. (2019)*

The consequence is an underestimation of the S4CO_2 peak because its separation from the S5 peak is positioned much earlier, around 550°C rather than around 650°C during the oxidation phase. Since the area of the S4CO_2 peak is smaller, the percentage of residual organic carbon is reduced and, conversely, the percentage of mineral carbon is increased. The % TOC calculated on this basis is lesser, which also leads to an increase in the values of the hydrogen and oxygen indices (Figure 13.8).

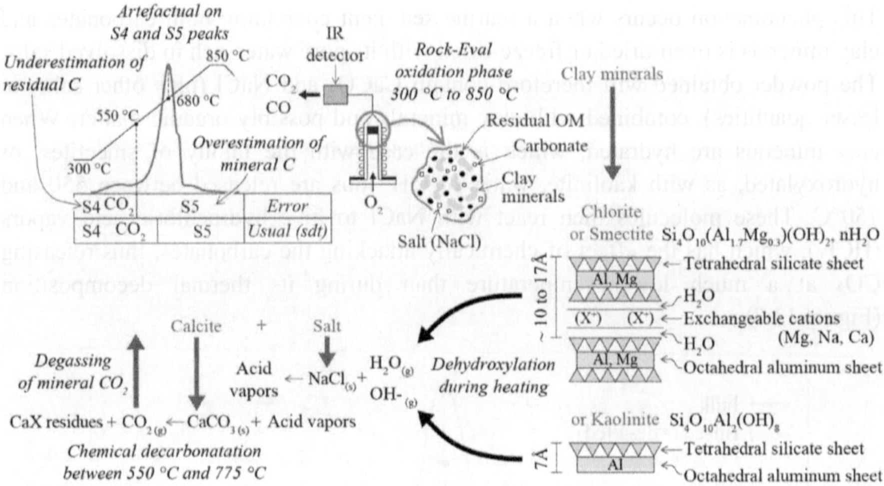

Figure 13.7. *Explanation of the origin of the S5 peak shift during the oxidation of recent salt-rich clay-carbonate sediments. A reaction between NaCl and the water released by the dehydration/dehydroxylation of clays produces acid vapors at temperatures between 550°C and 775°C, which causes dissolution of the calcite and not its thermal cracking, which usually occurs at higher temperatures. From Wattripont et al. (2019)*

Most often, rinsing the sediment once or twice with distilled water leads to the removal of dissolved salts and the return of the S5 peak to its usual position, with an S5 Tpeak close to 850°C. In the absence of hydrated or hydroxylated clay minerals in the marine sediment, these reactions do not take place. Wattripont et al. (2019) verified this on pure carbonate muds dried with its salt pore water. Although two molecules involved in the reaction (NaCl and $CaCO_3$) are present in the sample, since no mineral releases water at high temperature, the acid vapors do not form and there is no acid attack of the carbonate.

It is therefore essential to rinse recent sediment samples deposited in highly saline waters before Rock-Eval® analysis to avoid these artifacts and guard against any misinterpretation. Furthermore, as seen in section 4.3, the presence of salt could generate parasitic ionizations during pyrolysis, thus disturbing the FID signal. This is another good reason to remove salt from recent sediments. Finally, if any more reason was needed, it is evident that the acid vapors strongly corrode the crucibles, pistons and ovens of the device.

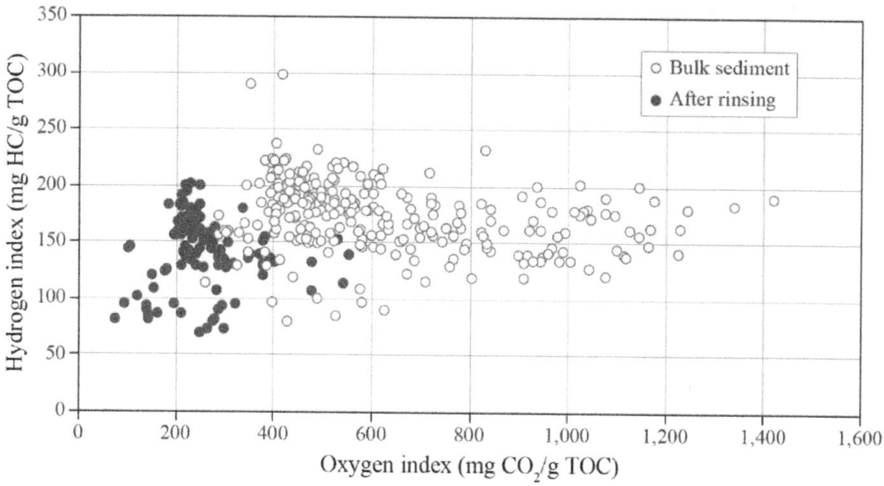

Figure 13.8. *Underestimating the organic carbon for recent unrinsed marine samples leads, mathematically, to a decrease in %TOC and thus to an increase in the hydrogen and oxygen indices, since %TOC is in the denominator in the calculation of these two indices. In the example of the sediments of the Timor Sea, the increase in the OI is much greater than that in the HI. Modified from Wattripont et al. (2019)*

13.4. Discrimination and quantification of different carbonate mineral species using the Rock-Eval® method

Pillot et al. (2014) proposed the use of a Rock-Eval® 6 in oxidation mode alone to identify the most common carbonate species (Table 13.2). The shape of the peaks during thermal destabilization and the maximum CO_2 production temperature (S5 Tpeak) appear to be sufficiently different from one another to allow six to seven types of carbonate to be differentiated (Figure 13.9). According to this study, when two carbonate species are mixed in a sample, the quantification of each peak gives the respective proportions of each species; this was verified for synthetic magnesite/calcite and natural magnesite/dolomite mixtures. It should nevertheless be highlighted that these types of mixtures are fairly rare in nature. It would be interesting to see whether this method is also applicable to calcite/dolomite mixtures, which occur more frequently in nature and whose destabilization peaks overlap much more than those of magnesite with calcite or dolomite.

Mineral species	Formula	Mol. mass	Theoretical %C
Ankerite	$Ca(Fe,Mg,Mn)(CO_3)_2$	206	11.7
Aragonite *	$CaCO_3$	100	12.0
Azurite *	$Cu_3(CO_3)_2(OH)_2$	344	7.0
Calcite *	$CaCO_3$	100	12.0
Cerussite	$PbCO_3$	267	4.5
Dolomite *	CaMg(CO3)2	184	13.0
Huntite	$Mg_3Ca\ (CO_3)_4$	353	13.6
Hydromagnesite *	$Mg_5(CO_3)_4(OH)_2\ 4\ H_2O$	468	10.3
Magnesite *	$MgCO_3$	84	14.3
Malachite *	$Cu_2CO_3(OH)_2$	221	5.4
Nahcolite	$NaHCO_3$	84	14.3
Calcium oxalate	CaC_2O_4	128	18.8
Rhodochrosite *	$MnCO_3$	115	10.4
Siderite *	$FeCO_3$	116	10.3
Strontianite	$SrCO_3$	148	8.1

Table 13.2. *Chemical formula, molecular mass and % by weight of carbon of the carbonates mentioned in this chapter. *Species studied by Pillot et al. (2014)*

If we add the ankerite, azurite, cerussite, hydromagnesite, huntite and strontianite peaks to the seven species represented in Figure 13.9, we can see that several peaks overlap, which makes them difficult to differentiate (Figure 13.10). Moreover, if the effect of the weight of the carbonate species has an influence on the position of

Tpeak (see section 13.2), it becomes especially difficult to distinguish, let alone hope to quantify, a mixture of carbonate species.

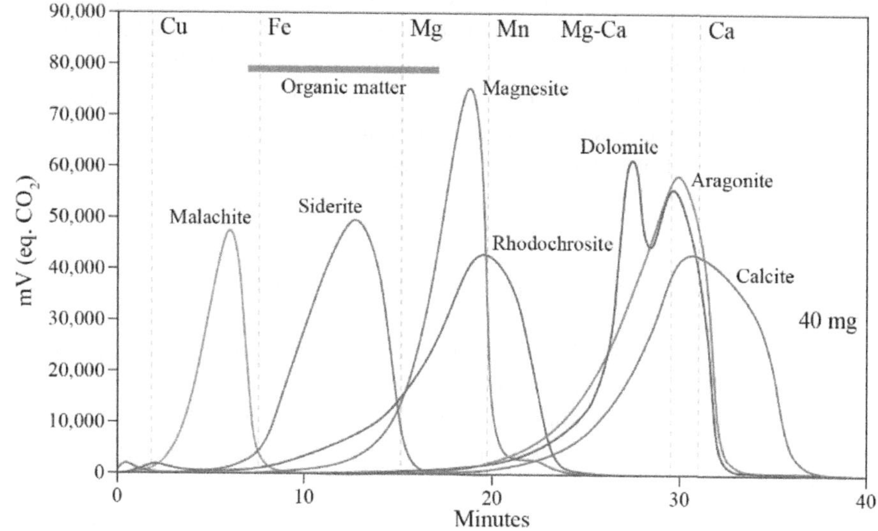

Figure 13.9. *Rock-Eval® S5 peak distribution for seven carbonate species analyzed in oxidation alone with a sample of 40 mg. Modified from Pillot et al. (2014)*

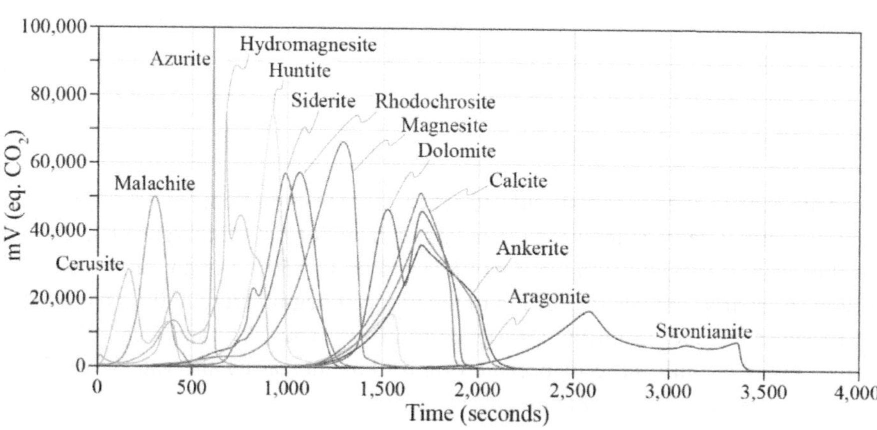

Figure 13.10. *Rock-Eval® S5 peak distribution for 13 carbonate species analyzed in oxidation alone with a sample of 40 mg. Strontianite was analyzed with an RE7S to reach temperatures > 900°C, at which it cracked*

Furthermore, hydromagnesite, azurite and siderite are found to crack within the temperature range where CO_2 of organic origin is released. An additional difficulty is therefore to be expected if the sample contains organic matter along with these types of carbonate. Finally, the approach proposed by Pillot et al., eliminating the pyrolysis phase, is not suitable for a routine analysis where the Bulk Rock method is used to quantify an organic fraction and a carbonate fraction simultaneously. That being said, this approach certainly remains of interest for the cement industry, and it has had the merit of opening up a field of research on carbonate analysis using the Rock-Eval® method.

13.5. Corrections in case of the presence of siderite

As recalled in section 13.1, the CO_2 released during the pyrolysis phase was trapped up to 400°C in the first versions of Rock-Eval® to avoid it potentially being contaminated by thermal cracking of siderite ($FeCO_3$), which starts at approximately 400°C. With the advent of Rock-Eval® 6 and the continuous measurement of CO_2 and CO throughout the whole pyrolysis phase, this question is not put back on the agenda since the definition of S3 remains unchanged. Nevertheless, it is clear that in sediments and sedimentary rocks rich in siderite, part of the CO_2 and CO produced by its thermal decomposition interferes with these same gases produced by pyrolysis of organic matter (Figure 13.11). Under an inert atmosphere, the thermal decomposition of siderite produces both CO_2 and CO following the reactions already mentioned in section 13.2:

$$FeCO_3 \rightarrow FeO + CO_2 \qquad [13.9]$$

$$3FeO + CO_2 \rightarrow Fe_3O_4 + CO \qquad [13.10]$$

This second reaction is stimulated when siderite is present in large quantities, and its thermal degradation produces a large amount of FeO (Lafargue et al. 1998a). In this case, it is obvious that S3CO (detected up to 550°C) is not only organic in origin. Similarly, the decomposition of siderite can begin before 400°C, particularly in recent lacustrine sediments (Sebag et al. 2018). Thus, CO and CO_2 of sideritic origin interfere with those of organic origin, thus increasing the values of S3 and therefore the OI and TOC (Figure 13.11).

Sebag et al. (2018) proposed a correction method[3] when sediments contain only siderite as carbonate and when the siderite disturbs the signal enough to have a visible effect on the HI (Figure 13.12). The first stage consists of verifying that the MinC values are related to the siderite content. This can be achieved by verifying that the MinC values and the XRD intensities of siderite are significantly and positively correlated. In a second stage, equations should be defined to correct the S3 and S3' signals entering into the TOC and OI calculations, i.e. the surface area of the S3CO_2 and S3CO thermograms before 400°C and 550°C, respectively. The third stage is the calculation of the zones of CO_2 (and CO) peaks associated with siderite pyrolysis. Lastly, the fourth stage evaluates the proportion of the CO_2 peak involved in the calculation of TOC and the OI. Once these relations have been validated, it is possible to deduce the proportion of CO_2 and of CO of sideritic origin. Finally, using the bases of calculations of Behar et al. (2001), it is possible to recalculate the Rock-Eval® parameters (TOC, HI and OI) corrected for the effects of siderite cracking. Comparison of the original and corrected parameters confirms that the effect of correction is minimal for low-MinC samples, whereas it can be significant for high-MinC samples, particularly with respect to the OI (Figure 13.13).

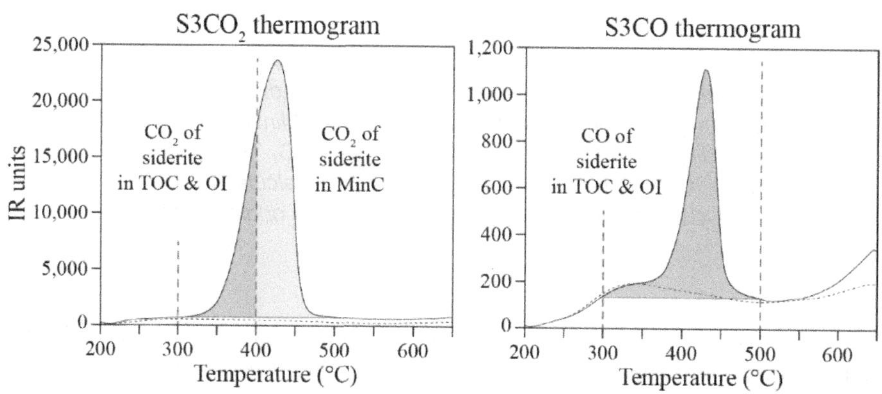

Figure 13.11. *Impact of siderite pyrolysis on the calculation of standard Rock-Eval® parameters: MinC, TOC and OI. The dotted line is that of a siderite-free sample and serves as a baseline for the calculation of the surfaces. Modified from Sebag et al. (2018)*

3. This method requires pyrolysis and oxidation of both CO and CO_2 to be quantified by the same IR detector, which is the case in the Standard Rock-Eval® 6. In the case where a Rock-Eval® 6 Turbo is used, it should be ensured that KCO and KCO_2 are identical or at least close for the IR cell detecting the pyrolysis effluents and the oxidation effluents.

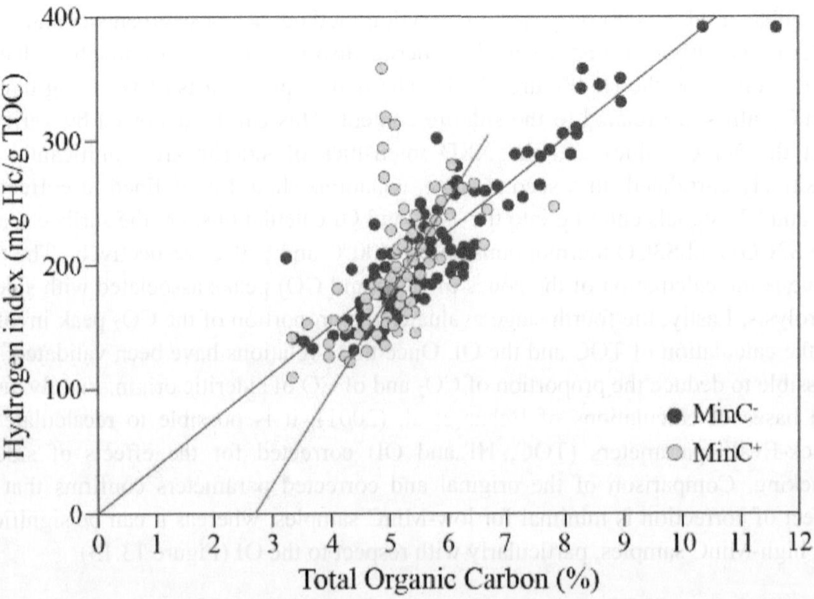

Figure 13.12. *Sediment distribution of Lake Baromi (Cameroon) in an HI=f(TOC) diagram. Samples containing less than 0.4% MinC are distributed along a straight line passing through the origin, while those with a % MinC > 0.4% are distributed along a straight line intercepting the x-axis at 2.5% TOC. This 2.5% TOC does not correspond to a hydrogen-free inert carbon, but is a calculation artifact due to the interference of CO_2 of sideritic origin with that of organic origin. Modified from Sebag et al. (2018)*

Ordoñez et al. (2019) studied recent sideritic sediments in a lake in Indonesia (Lake Towuti) and applied a four-term (siderite, labile Corg, refractory Corg and resistant Corg) end-member analysis (EMA) mixing model to correct Rock-Eval® thermograms affected by the presence of siderite (Figure 13.14). The use of EMA allowed them to improve TOC, MinC and OI calculations to be able to discuss the evolution of the environmental conditions in Lake Towuti. These corrective methods are therefore well adapted when the series contains only siderite. However, they are ineffective if siderite is mixed with another type of carbonate, such as calcite or aragonite. Indeed, it then becomes impossible to distinguish the MinC originating from siderite from that coming from another carbonate species on the basis of the Rock-Eval® thermograms alone. The correction proposed by Sebag et al. (2018) is therefore no longer possible. The same applies if we have mixtures of hydromagnesite or huntite (cracking wholly or partly at relatively low temperature, Figures 13.3 and 13.10) with aragonite. Such mixtures are encountered in recent sediments of certain hypersaline and hyperalkaline lakes. The same difficulties are encountered in soils containing calcium oxalates. Indeed, the latter crack at a very low temperature (around 170°C, Figure 13.3) and release CO_2 that is indistinguishable from that of organic origin.

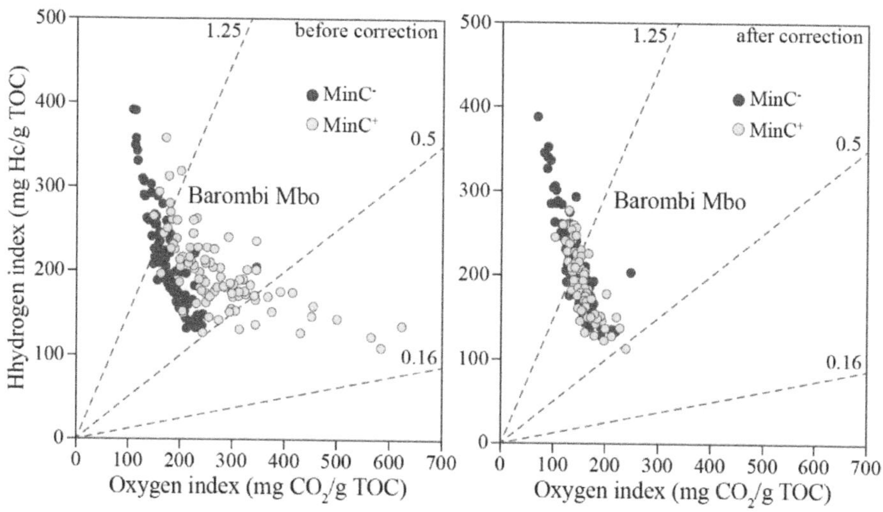

Figure 13.13. *Comparison of the position of Lake Baromi sediment samples in a van Krevelen pseudo-diagram [HI=f(OI)] before and after correction of the effect of early siderite cracking. Modified from Sebag et al. (2018)*

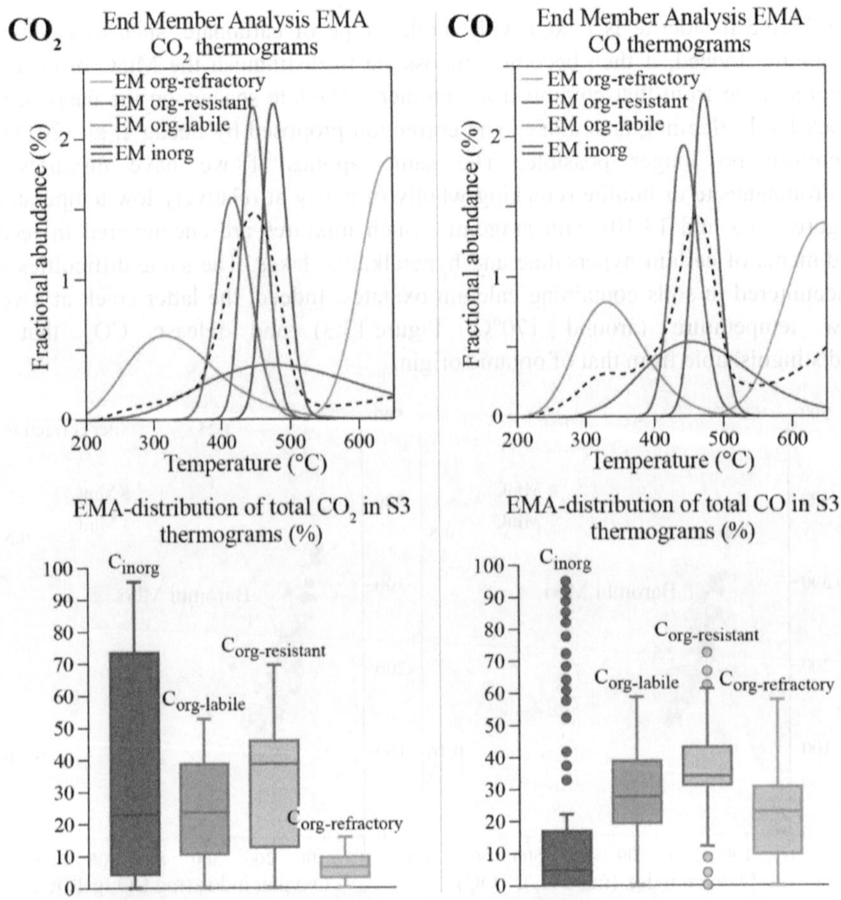

Figure 13.14. *Results of mixing model analysis performed on CO_2 and CO thermograms (dotted lines) to identify and quantify carbon phases and distinguish them from the inorganic phases (according to Ordoñez et al. (2019))*

13.6. From Rock-Eval® MinC to carbonate percentage

The Rock-Eval® method has of course been compared to other calcimetric measurement methods (Bernard calcimeter or Mélières manocalcimeter, elemental analysis, intensity of X-ray diffraction peaks, etc.), and it turns out to be just as effective in quantifying mineral carbon (Lafargue et al. 1998a; Behar et al. 2001; Figure 13.15). This is all the more remarkable because MinC is derived from the sum of CO_2 and CO measured during the pyrolysis phase (pyroMinC) and CO_2 measured during the oxidation phase (oxiMinC), and because there is an approximation of the effect of the Boudouard reaction on the % mineral carbon associated with S3'CO. It is nevertheless advisable to ensure from time to time that the device is well calibrated by analyzing a pure calcite and verifying that the MinC value is very close to 12%, i.e. 100% of $CaCO_3$.

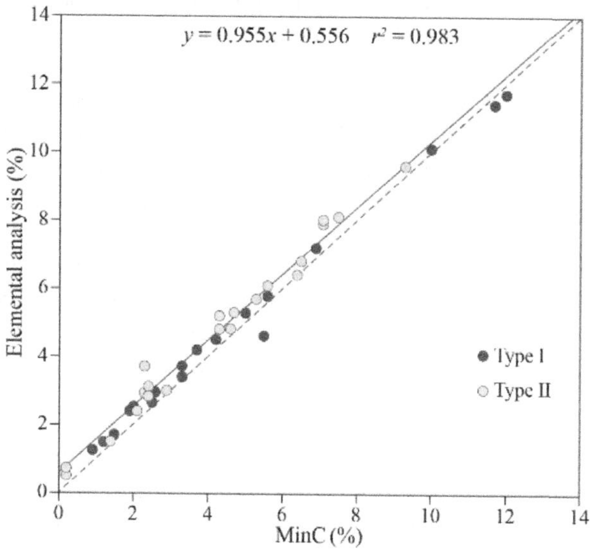

Figure 13.15. *Comparison of mineral carbon measurements determined by elemental analysis with those determined using the Rock-Eval® method. Modified from Behar et al. (2001)*

Certain researchers prefer to work with percentage carbonate rather than with percentage mineral carbon, which requires the type of carbonate present in the samples to be known. The carbonates occurring most frequently in the sedimentary series are calcite, aragonite and dolomite, siderite being a subordinate form. The molecular mass of calcite or its polymorph, aragonite, is 100, i.e. 12% MinC. Therefore, the value of MinC simply needs to be multiplied by 8.33 (100÷12) to have a calcium carbonate percentage equivalent in a sample that contains only these forms of carbonate. For pure dolomite, MinC is 13%. It is therefore necessary to multiply MinC by 14.15 (184÷13) to have a dolomite percentage equivalent in a sample. For pure siderite, MinC is 10.3%, and it is necessary to multiply by 11.26 (116÷10.3) to have a siderite percentage equivalent in a sample.

Jiang et al. (2017) proposed a statistical relationship between MinC and the % carbonate in clayey source rocks and reservoir rocks based on the analysis of nearly 900 Canadian samples ranging in age from Ordovician to Cretaceous. By comparing the % carbonate determined by X-ray diffraction with the MinC determined by Rock-Eval®, they propose the following relationship: % Carbonate = 7.976% MinC with $R^2 = 0.974$. This value of 7.976 is derived from a calculation (linear regression) on the average composition of the 890 samples studied by these authors, in which they considered that MinC was a mixture of calcite + dolomite + siderite + rhodochrosite. The value of 7.976 therefore has no universal value and is simply related to the study dataset. For a carbonate series where calcite dominates, the only correct relationship is % Calcite = 8.33%MinC with $R^2 = 0.989$, while for a dolomitic series, the relationship is % Dolomite = 14.15%MinC. In the case of a mixture of different types of carbonate, the relationship by Jiang et al. (2017) can be adopted as a first approximation, but it is preferable to check (or another method) the nature and proportion of the different carbonates present using XRD and to determine the relationship between %carbonate and %MinC for each case study.

13.7. TOC and MinC relationships in sediments and sedimentary rocks

Let us set aside the methodological aspects of the previous parts of this chapter, and use a few examples to highlight the benefits of Rock-Eval® in providing TOC and MinC content for the same sample.

If we exclude evaporative environments, many sediments and sedimentary rocks are composed of a mixture in varying proportions of carbonate, siliciclastic minerals (quartz and/or clays) and organic matter. Variations in one of these fractions during sediment deposition therefore contribute to the dilution or concentration of the other two fractions. The resulting changes are easily expressed in binary diagrams (Figure 13.16) showing the relationship between the $CaCO_3$ (or MinC) and TOC contents, parameters conventionally provided by the Rock-Eval® method. In the case of sedimentation where the carbonate flow varies cyclically, the carbonate content is of course anticorrelated with that of TOC, although the TOC flow also fluctuates (Figure 13.16, case A). The slope of the regression line between $CaCO_3$ and TOC is inclined to a lesser or greater extent depending on the flow of organic matter. As a result, the lithologies that result from such variations can range from white bioturbated limestone to well-laminated black shale.

If the fluctuations are induced by siliciclastic contributions (Figure 13.16, case B), the percentages of $CaCO_3$ and TOC are positively correlated. Finally, if the variations are related to fluctuations in the quantities of organic matter, the carbonate content is only very partially affected since organic matter is generally a minor constituent. The regression line between $CaCO_3$ and TOC is virtually parallel to the *y*-axis in a binary diagram, whereas the TOC percentages of sediments can vary rapidly up to an order of magnitude (Figure 13.16, case C).

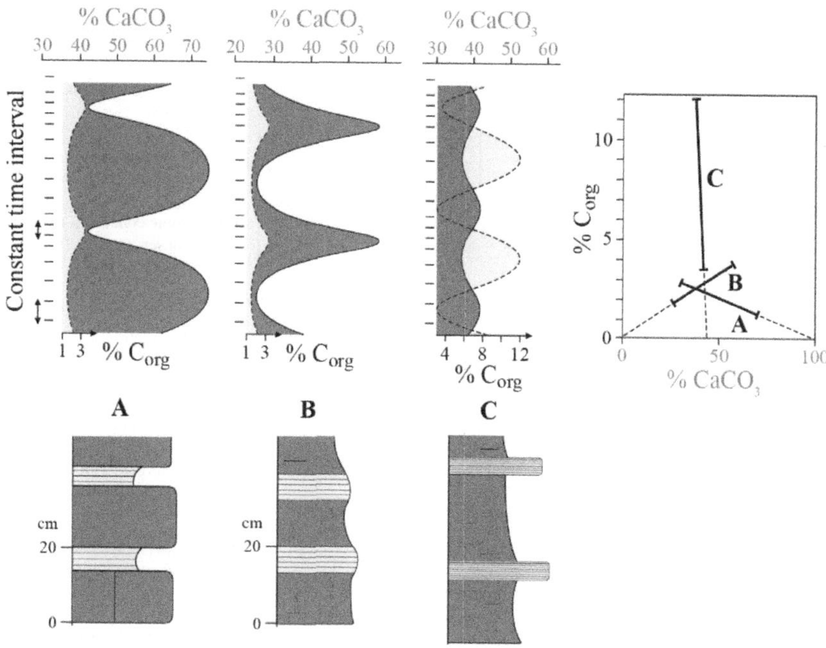

Figure 13.16. *Different types of relationships between %TOC and %CaCO3 as a function of the variations in the relative carbonate, siliciclastic or organic matter contents in the case of sedimentation under allocyclic influence. The cyclicities are expressed by the curves of the variations in carbonate and organic carbon (in %) for alternations where the carbonate content is average (centered around 50%). The changes in sedimentation rate are represented to the left of the curves by time intervals of variable thickness (according to Ricken (1993) in Baudin et al. (2017c))*

This type of approach was conducted on Frasnian facies of the Canadian Rocky Mountains at the scale of a transgressive/regressive cycle (van Buchem et al. 1996). The changes in facies in the basin adjacent to the platform, seen through TOC versus $CaCO_3$ diagrams, make it possible to distinguish the different phases of construction of the platform and characterize the associated organic facies in the basin (Figure 13.17).

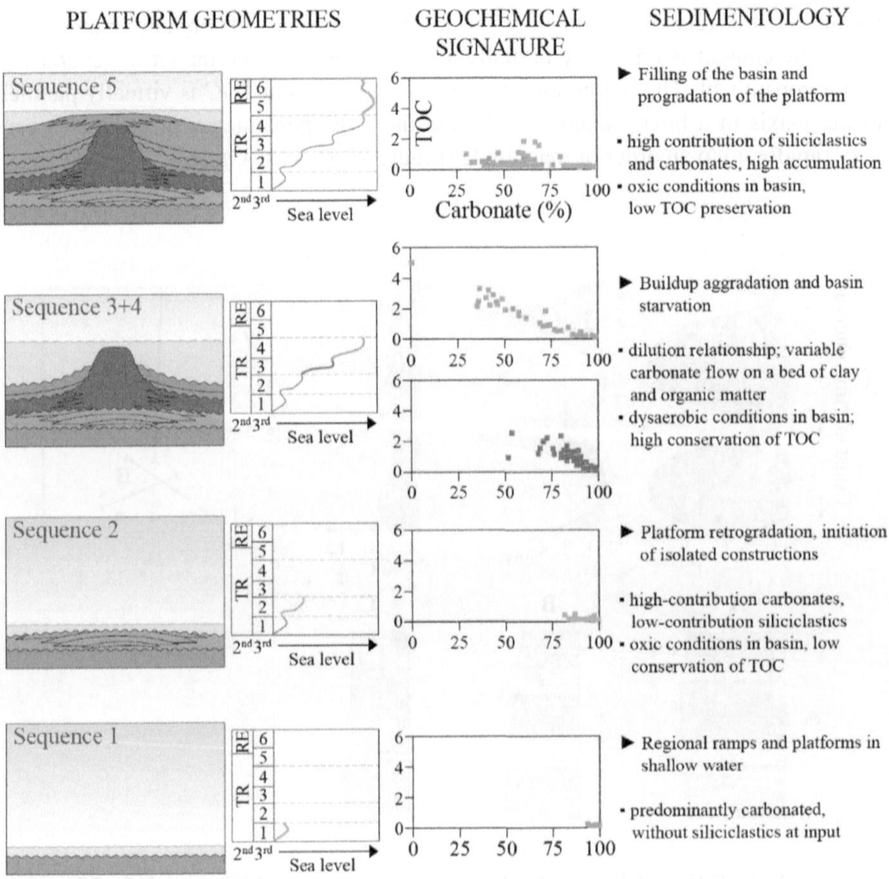

Figure 13.17. *Relationships between the geometries of a Frasnian platform in the Canadian Rocky Mountains and the geochemical characteristics ($CaCO_3$ and TOC) of the deposits of the adjacent basin at the scale of a transgressive/regressive cycle. Modified from van Buchem et al. (1996)*

13.8. Concluding remarks

While the Rock-Eval® method was first developed for the analysis of organic matter, the continuous measurement of CO_2 and CO during the pyrolysis and oxidation phases from Rock-Eval® 6 onward allowed the addition of carbonate analysis through the MinC parameter. This is what makes the method interesting since in less than an hour and in a single sample, it gives access not only to the organic carbon concentration and the mineral carbon concentration of the sample, but also to the total carbon concentration. Rock-Eval® is thus an instrument for the precise and rapid quantification of carbonates. If the analyst is attentive to the shape of the CO_2 peaks, they can even derive information on the nature of the carbonates. Finally, the temperature programming of Rock-Eval® enables the analysis of the kinetics of carbonate cracking, a field that has been overlooked until now, and that is only waiting to be explored by academic researchers or those in the cement industry.

14

What's Next for Rock-Eval®?

François BAUDIN[1] and Nicolas BOUTON[2]

[1] *ISTeP, Sorbonne Université, Paris, France*

[2] *Vinci Technologies, Nanterre, France*

This story, broken down throughout the chapters of this book, was made possible because the Rock-Eval® tool was developed by passionate, visionary people, for people who were just as much so. It is impossible to name them all, but the original impulse was provided by a team formed around Jean Espitalié at the *Institut Français du Pétrole* (IFP), considered by all to be the inventor of Rock-Eval®. They were able to find the right partners to transform their ideas into an instrument ultimately marketed by Vinci Technologies, thanks to a number of colleagues, especially Jean Espitalié and Roland Antonas. This success is of course mainly due to the users in laboratories all over the world, who recognized the benefit in acquiring instruments to perform measurements on their samples, and to gain an ever-finer understanding of how the data obtained with Rock-Eval® could be useful to them, either for research or commercial purposes. This story is just waiting to be continued.

In the oil industry, the list of publications citing Rock-Eval® is increasing year by year (see Chapter 1), and the number of patents filed, both instrumental and application-related, is not diminishing. The instrument continues to evolve, to adapt to the needs of researchers and geochemistry laboratories. For example, patent no. WO/2020/157389 (Bouton and Espitalié 2020) should, in the near future, enable the measurement of hydrocarbon gases, as well as the H_2S resulting from primary cracking of kerogen, and therefore provide access to multi-compositional kinetics of primary cracking: gaseous hydrocarbons on the one hand (methane and butane) and oil on the other hand.

While it is always difficult to predict the future, it is worth noting that in the founding paper by Espitalié et al. (1977a), the last sentence of the conclusion referred to the use of Rock-Eval® for soil characterization. Nearly 45 years after this prediction, Rock-Eval® analysis of soil carbon is reaching maturity. In the meantime, the world has changed and our view of the world has changed. Chris Bowler, head of Tara-Océan, reminds us that since the industrial revolution, humanity has disrupted the carbon cycle by burning as much carbon every year as it took nature 1 million years to bury and transform into coal, gas and oil.

With the announced shutdown of thermal engines in Europe in 2035 as part of the fight against global warming, and the capping of oil production, hydrocarbon needs are likely to decrease in the coming decades. After having participated in the development of the oil industry, and so having participated in putting even more CO_2 back into the atmosphere, Rock-Eval® saw how it could play a useful role in the fight against global warming. Perhaps one day it will even be mentioned in an IPCC report as a tool to better quantify the carbon sinks that are soils? Indeed, it can be noted that Rock-Eval® analysis of organic matter in oil source rocks has gradually migrated to recent sediments, then to soils. It may well be about to prove useful for tomorrow's agriculture, by contributing to the fight against global warming, and also by securing the source of our food for future generations. This is the reason behind the 4p1000 Initiative, founded in Paris in 2015 at COP 21, with Vinci Technologies joining forces.

Renewable energies based on methanation (biogas), or on second-generation biofuels, biomass recovery processes, fermenters, etc., could benefit from Rock-Eval® analysis. Rock-Eval® analysis could thus provide information on incoming flows, to anticipate process parameters, as well as on outgoing flows (digestates), to measure energy efficiency and the remaining value of products.

Indeed, the list of publications shows the dissemination of the data acquired with Rock-Eval® across ever more varied fields. After fossil fuels, and the environment (with the detection of micro-plastic pollution; Romero-Sarmiento et al. 2022), it is possible to foresee applications in biology. Indeed, since carbon is the key element of chemistry, materials and plastics, it is conceivable that Rock-Eval® will be found in all types of analytical laboratories, including for medical or ecotoxicological analyses.

In recent years, cement manufacturers and catalyst producers have shown an interest in Rock-Eval®. This instrument, which is a hybrid between a chromatographic, thermal and elemental analyzer, is evolving into an analytical platform on which new fields of application can be tested in the coming years (Figure 14.1).

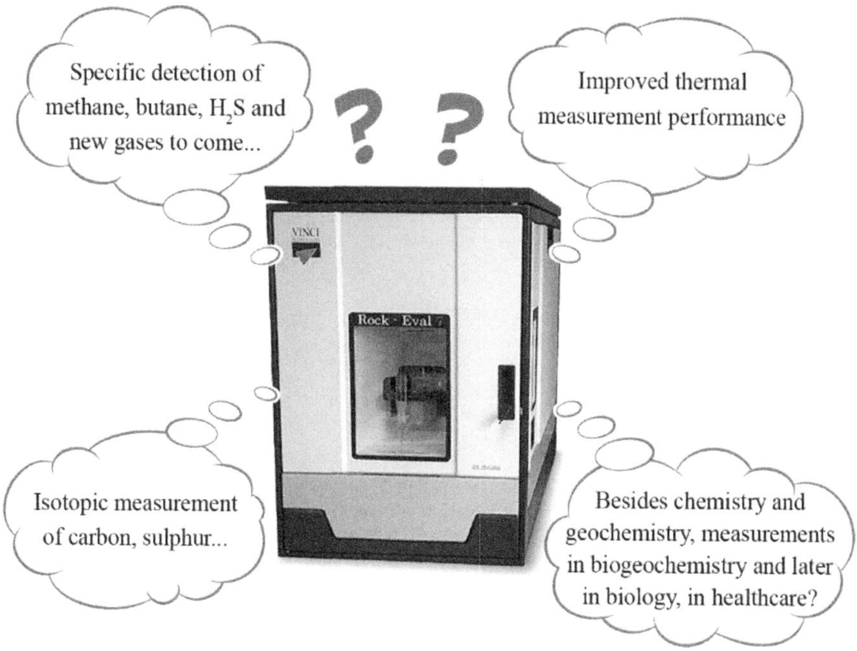

Figure 14.1. *Ideas for possible applications of the Rock-Eval® method. For a color version of this figure, see www.iste.co.uk/baudin/rock-eval.zip*

Rock-Eval® has already been tested in combination with isotope-ratio mass spectrometers. Analyses of $\delta^{13}C$ in the S1 and S2 parameters have already been carried out (Dörner et al. 2019). Work on the measurement and interpretation of sulfur isotopes coupled with Rock-Eval® 7S is currently being published (Cohen-Sadon et al. 2021), with these couplings of Rock-Eval® analysis and measurements with isotopic mass spectrometers just waiting to be pursued further.

In addition to carbon, Rock-Eval® has opened up to other elements, with hydrogen, via the HI, and oxygen, via the OI, as well as with sulfur, and soon with other elements, in new patents that are currently pending. The instrumental development work is very long, and multidisciplinary work is slowly ensuring the insertion of the techniques into other areas of research. Let us hope for a second edition of this book a few years from now, to open up new chapters.

References

Aarnes, I., Svensen, H., Polteau, S., Planke, S. (2011). Contact metamorphic devolatilization of shales in the Karoo Basin, South Africa, and the effects of multiple sill intrusions. *Chemical Geology*, 281(3–4), 181–194.

Aboussou, A. (2018). New Rock-Eval method for pyritic and organic sulphur quantification: Application to study organic matter preservation in Jurassic sediments. Thesis, Sorbonne Université.

Agirrezabala, L.M., Permanyer, A., Suárez-Ruiz, I., Dorronsoro, C. (2014). Contact metamorphism of organic-rich mudstones and carbon release around a magmatic sill in the basque-cantabrian basin, western Pyrenees. *Organic Geochemistry*, 69, 26–35.

Ahad, J.M.E., Jautzy, J.J., Cumming, B.F., Das, B., Laird, K.R., Sanei, H. (2015). Sources of polycyclic aromatic hydrocarbons (PAHs) to northwestern Saskatchewan lakes east of the Athabasca oil sands. *Organic Geochemistry*, 80, 35–45.

Aizhenshtat, Z., Krein, E.B., Vairavamurthy, M.A., Goldstein, T.P. (1995). Role of sulfur in the transformations of sedimentary organic matter: A mechanistic overview. In *Geochemical Transformations of Sedimentary Sulfur*, Vairavamurthy, M.A. and Schoonen, M.A.A. (eds). American Chemical Society, Washington.

Albrecht, R., Sebag, D., Verrecchia, E.P. (2015). Organic matter decomposition: Bridging the gap between Rock–Eval pyrolysis and chemical characterization (CPMAS ^{13}C NMR). *Biogeochemistry*, 122, 101–111.

Amrani, A. and Aizhenshtat, Z. (2004). Mechanisms of sulfur introduction chemically controlled: $\delta^{34}S$ imprint. *Organic Geochemistry*, 35, 1319–1336.

Andriulo, A., Mary, B., Guérif, J. (1999). Modelling soil carbon dynamics with various cropping sequences on the rolling pampas. *Agronomie*, 19, 365–377.

Ariztegui, D., Chondrogianni, C., Lami, A., Guilizzoni, P., Lafargue, E. (2001). Lacustrine organic matter and the Holocene paleoenvironmental record of Lake Albano (Central Italy). *Journal of Paleolimnology*, 26(3), 283–292.

Aroa, A., Dutta, S., Gogoi, B., Banerjee, S. (2017). The effects of igneous dike intrusion on organic geochemistry of black shale and its implications: Late Jurassic Jhuran formation, India. *International Journal of Coal Geology*, 178, 84–99.

Bajard, M., Poulenard, J., Sabatier, P., Develle, A.L., Giguet-Covex, C., Jacob, J., Crouzet, C., David, F., Pignol, C., Arnaud, F. (2017). Progressive and regressive soil evolution phases in the Anthropocene. *Catena*, 150, 39–52.

Balesdent, J. (1996). The significance of organic separates to carbon dynamics and its modelling in some cultivated soils. *European Journal of Soil Science*, 47, 485–493.

Balesdent, J., Mariotti, A., Guillet, B. (1987). Natural ^{13}C abundance as a tracer for studies of soil organic matter dynamics. *Soil Biology and Biochemistry*, 19, 25–30.

Balesdent, J., Basile-Doelsch, I., Chadoeuf, J., Cornu, S., Derrien, D., Fekiacova, Z., Hatté, C. (2018). Atmosphere–soil carbon transfer as a function of soil depth, *Nature*, 559, 599–602.

Bao, Z., Zhao, Z., Guha, J. (2005). Organic geochemistry of sedimentary rock-hosted disseminated gold deposits in southwestern Guizhou Province, China. *Acta Geologica Sinica*, 79(1), 120–133.

Barker, C. (1974a). Pyrolysis techniques for source rock evaluation. *American Association of Petroleum Geologists Bulletin*, 58, 2349–2361.

Barker, C. (1974b). Programmed-temperature pyrolysis of vitrinite of various rank. *Fuel*, 53, 176–177.

Barker, C. and Pawlewicz, M.J. (1994). Calculation of vitrinite reflectance from thermal histories and peak temperatures. A comparison of methods. In *Vitrinite Reflectance as a Maturity Parameter*, Mukhopadhyay, P.K. and Dow W.G. (eds). Americal Chemical Society, Washington.

Barré, P., Eglin, T., Christensen, B.T., Ciais, P., Houot, S., Kätterer, T., van Oort, F., Peylin, P., Poulton, P.R., Romanenkov, V. et al. (2010). Quantifying and isolating stable soil organic carbon using long-term bare fallow experiments. *Biogeosciences*, 7, 3839–3850.

Barré, P., Plante, A.F., Cécillon, L., Lutfalla, S., Baudin, F., Bernard, S., Christensen, B.T., Eglin, T., Fernandez, J.M., Houot, S. et al. (2016). The energetic and chemical signatures of persistent soil organic matter. *Biogeochemistry*, 130, 1–12.

Barré, P., Quénéa, K., Vidal, A., Cécillon, L., Christensen, B.T., Kätterer, T., Macdonald, A., Petit, L., Plante, A.F., van Oort, F. et al. (2018). Microbial and plant-derived compounds both contribute to persistent soil organic carbon in temperate soils. *Biogeochemistry*, 140, 81–92.

Baudin, F., Disnar, J.R., Martinez, P., Dennielou, B. (2010). Distribution of the organic matter in the channel-levees systems of the Congo mud-rich deep sea fan (West Africa). Implication for deep offshore petroleum source rocks and global carbon cycle. *Marine and Petroleum Geology*, 27, 995–1010.

Baudin, F., Disnar, J.R., Aboussou, A., Savignac, F. (2015). Guidelines for Rock–Eval analysis of recent marine sediments. *Organic Geochemistry*, 86, 71–80.

Baudin, F., Martinez, P., Dennielou, B., Charlier, K., Marsset, T., Droz, L., Rabouille, C. (2017a). Organic carbon accumulation in modern sediments of the Angola basin influenced by the Congo deep-sea fan. *Deep Sea Research Part II: Topical Studies in Oceanography*, 142, 64–74.

Baudin, F., Stetten, E., Schnyder, J., Charlier, K., Martinez, P., Dennielou, B., Droz, L. (2017b). Origin and distribution of the organic matter in the distal lobe of the Congo deep-sea fan – A Rock-Eval survey. *Deep Sea Research Part II: Topical Studies in Oceanography*, 142, 75–90.

Baudin, F., Tribovillard, N., Trichet J. (2017c). *Géologie de la matière organique*, 2nd edition. EDP Sciences, Paris.

Baudin, F., Dennielou, B., Rabouille, C. (2020). Routing of terrestrial organic matter from the Congo River to the ultimate sink in the abyss: A mass balance approach. *Geologica Belgica*, 23(1–2), 41–52.

Behar, F. and Vandenbroucke, M. (1996). Experimental determination of rate constants of the n-C_{25} thermal cracking at 120, 140, and 800 bar: Implication for high-pressure/high-temperature prospects. *Energy and Fuels*, 10, 932–940.

Behar, F., Leblond, C., Saint-Paul, C. (1989). Analyse quantitative des effluents de pyrolyse en milieu ouvert et fermé. *Revue d'IFP*, 44, 387–397.

Behar, F., Vandenbroucke, M., Tang, Y., Marquis, F., Espitalié, J. (1997). Thermal cracking of kerogen in open and closed systems: Determination of kinetic parameters and stoichiometric coefficients for oil and gas generation. *Organic Geochemistry*, 26(5/6), 321–339.

Behar, F., Budzinski, H., Vandenbroucke, M., Tang, Y. (1999). Methane generation from oil cracking: Kinetics of 9-methylphenanthrene cracking and comparison with other pure compounds and oil fractions. *Energy and Fuels*, 13, 471–481.

Behar, F., Beaumont, V., De Penteado, B. (2001). Rock-Eval 6 technology: Performances and developments. *Oil & Gas Science and Technology*, 56, 111–134.

Behar, F., Lorant, F., Budzinski, H., Desavis, E. (2002). Thermal stability of alkylaromatics in natural systems: Kinetics of thermal decomposition of dodecylbenzene. *Energy and Fuels*, 16, 831–841.

Behar, F., Lewan, M.D., Lorant, F., Vandenbroucke, M. (2003). Comparison of artificial maturation of lignite in hydrous and non-hydrous conditions. *Organic Geochemistry*, 34, 575–600.

Behar, F., Lorant, F., Lewan, M. (2008). Role of NSO compounds during primary cracking of a type II kerogen and a type III lignite. *Organic Geochemistry*, 39, 1–22.

Behar, F., Roy, S., Jarvie, D. (2010). Artificial maturation of a type I kerogen in closed system: Mass balances and kinetic modelling. *Organic Geochemistry*, 41(11), 1235–1247.

Behar, F., Delhaye-Prat, V., Garel, S. (2020). Detrital input quantification in lacustrine petroleum systems: An example of the pre-salt source rocks from the Lower Congo Basin (Congo). *The Depositional Record*, 7, 147–171.

Ben Slimane, A., Raclot, D., Evrard, O., Sanaa, M., Lefèvre, I., Ahmadi, M., Tounsi, M., Rumpel, C., Mammou, A.B., Le Bissonnais, Y. (2013). Fingerprinting sediment sources in the outlet reservoir of a hilly cultivated catchment in Tunisia. *Journal of Soils and Sediments*, 13, 801–815.

Benoit, Y., Dalmazzone, C., Audibert-Hayet, A. (2002). Pollut-Eval: An innovative tool for direct hydrocarbon characterization in oil base mud & cuttings. *SPE Annual Technical Conference and Exhibition*, San Antonio, Texas, September, SPE 77472.

Berner, R.A. (1984). Sedimentary pyrite formation: An update. *Geochimica et Cosmochimica Acta*, 48, 605–615.

Bertrand, P., Brocero, S., Lallier-Vergès, E., Tribovillard, N., Bonifay, E. (1992). Lacustrine organic sedimentation and paleoclimates at medium latitudes during the last Pleistocene: Example of the Lac du Bouchet, Haute Loire, France (preliminary results). *Bulletin de la Société géologique de France*, 163(4), 427–433.

Biester, H., Knorr, K.H., Schellekens, J., Basler, A., Hermanns, Y.M. (2014). Comparison of different methods to determine the degree of peat decomposition in peat bogs. *Biogeosciences*, 11, 2691–2707.

Bjorøy, M., Jensen, H.K.B., Connan, J., Hall, K., Ferriday, A.L. (1993). Geoelf Sulphur Analyser (GSA), a new instrument for rapid determination of organic and inorganic sulphur in sediments and coals. *Organic Geochemistry, 16th International Organic Geochemistry Meeting*. Falch Hurtigtrykk, Oslo, 773–778.

Blair, N.E. and Aller, R.C. (2012). The fate of terrestrial organic carbon in marine environment. *Annual Review of Marine Science*, 4, 401–423.

Blanke, R. (2004). Sédimentation organique profonde associée au système d'upwelling du courant du Benguela (Atlantique Sud-Est). Facteurs de contrôle à l'échelle de la marge au cours du Quaternaire. Thesis, Université Pierre et Marie Curie.

Bodineau, G., Volland, N., Cousin, I., Disnar, J.R., Balbino, L., Brossard, M., Bruand, A. (2002). Dynamique de la structure et caractéristiques des constituants organiques dans les ferralsols de la région du Cerrado (Brésil). *Livre des résumés 19ème Réunion des Sciences de la Terre*, RST 2002, Nantes, 68.

Bonnetti, C., Cuney, M., Michels, R., Truche, L., Malartre, F., Liu, X., Yang, J. (2015). The multiple roles of sulfate-reducing bacteria and Fe-Ti oxides in the genesis of the Bayinwula Roll Front-Type Uranium deposit, Erlian Basin, NE China. *Economic Geology*, 110, 1059–1081.

Bordenave, M., Combaz, A., Giraud, A. (1966). Influence de l'origine des matières organiques et de leur degré d'évolution sur les produits de pyrolyse du kérogène. In *Advances in Organic Geochemistry*, Hobson G.D and Speers G.C. (eds). Pergamon, London.

Bottinelli, N., Kaupenjohann, M., Märten, M., Jouquet, P., Soucémarianadin, L., Baudin, F., Tran, T.M., Rumpel, C. (2020). Age matters: Fate of soil organic matter during ageing of earthworm casts produced by the anecic earthworm Amynthas khami. *Soil Biology and Biochemistry*, 148, 107906.

Bourdon, S., Laggoun-Défarge, F., Disnar, J.R., Maman, O., Guillet, B., Derenne, S., Largeau, C. (2000). Organic matter sources and early diagenetic degradation in a tropical peaty marsh (Tritrivakely, Madagascar). Implications for environmental reconstruction during the Sub-Atlantic. *Organic Geochemistry*, 31(5), 421–438.

Boussafir, M. and Lallier-Vergès, E. (1997). Accumulation of organic matter in the Kimmeridge Clay Formation (KCF): An update fossilization model for marine petroleum source-rocks. *Marine and Petroleum Geology*, 14(1), 75–83.

Boussafir, M., Gelin, F., Lallier-Vergès, E., Derenne, S., Bertrand, P., Largeau, C. (1995). Electron microscopy and pyrolysis of kerogens from the Kimmeridge Clay Formation, UK: Source organisms, preservation processes, and origin of microcycle. *Geochimica et Cosmochimica Acta*, 59, 3731–3747.

Boussafir, M., Sifeddine, A., Jacob, J., Foudi, M., Cordeiro, R.C., Albuquerque, A.L.S., Abrao, J.J., Turcq, B. (2012). Petrographical and geochemical study of modern lacustrine sedimentary organic matter (Lagoa do Caçò, Maranão, Brazil): Relationship between early diagenesis, organic sedimentation and lacustrine filling. *Organic Geochemistry*, 47, 88–98.

Bouton, N. and Espitalié, J. (2020). Procédé et dispositif pour l'analyse en continu des hydrocarbures gazeux et du H_2S dans des échantillons de produits pétroliers. Rock-Eval® multi-gas patent, WO/2020/157389.

Braun, R.L. and Burnham, A.K. (1991). PMOD: A flexible model of oil and gas generation, cracking, and expulsion. *Organic Geochemistry*, 19, 161–172.

Brazeau, M.L., Poulain, A.J., Paterson, A.M., Keller, W.B., Sanei, H., Blais, J.M. (2013). Recent changes in mercury deposition and primary productivity inferred from sediments of lakes from the Hudson Bay Lowlands, Ontario, Canada. *Environmental Pollution*, 173, 52–60.

Bréhéret, J.G. and Herbin, J.P. (1987). Matière organique : validité des données d'affleurement montrées sur un exemple crétacé moyen dans le bassin vocontien (SE de la France) ; corrélation entre données de surface et de forage. *Comptes rendus de l'Académie des Sciences*, 305, 199–202.

Burdige, D.J. (2007). Preservation of organic matter in marine sediments: Controls, mechanisms, and an imbalance in sediment organic carbon budget? *Chemical Reviews*, 107, 467–485.

Burnham, A.K. and Braun, R.L. (1990). Development of a detailed model of petroleum formation, destruction, and expulsion from lacustrine and marine source rocks. *Organic Geochemistry*, 16, 27–39.

Cadeau, P., Romero-Sarmiento, M.F., Sissmann, O., Beaumont, V. (2020). On-line recovery system coupled to a Rock-Eval® device: An analytical methodology for characterization of liquid and solid samples. *Organic Geochemistry*, 144, 104014.

Calkins, W.H. (1994). The chemical forms of sulfur in coal: A review. *Fuel*, 73(4), 475–484.

Canfield, D., Raiswell, R., Westrich, J., Reaves, C., Berner, R. (1986). The use of chromium reduction in the analysis of reduced inorganic sulfur in sediments and shales. *Chemical Geology*, 54(1), 149–155.

Carpentier, B., Arab, H., Pluchery, E., Chautru, J.M. (2007). Tar mats and residual oil distribution in a giant oil field offshore Abu Dhabi. *Journal of Petroleum Science and Engineering*, 58, 472–490.

Carrie, J., Sanei, H., Goodarzi, F., Stern, G., Wang, F. (2009). Characterization of organic matter in surface sediments of the Mackenzie River Basin, Canada. *International Journal of Coal Geology*, 77, 416–423.

Carrie, J., Sanei, H., Stern, G. (2012). Standardisation of Rock–Eval pyrolysis for the analysis of recent sediments and soils. *Organic Geochemistry*, 46, 38–53.

Carvajal-Ortiz, H. and Gentzis, T. (2015). Critical considerations when assessing hydrocarbon plays using Rock-Eval pyrolysis and organic petrology data: Data quality revisited. *International Journal of Coal Geology*, 152, 113–122.

Carvajal-Ortiz, H., Gentzis, T., Ostadhassan, M. (2021). Sulfur differentiation in organic-rich shales and carbonates via open-system programmed pyrolysis and oxidation: Insights into fluid souring and H_2S production in the Bakken Shale, United State. *Energy and Fuels*, 35(15), 12030–12044.

Cécillon, L., Soucémarianadin, L., Berthelot, A., Duverger, M., De Boisseson, J.M., Gosselin, F., Guenet, B., Barthès, B., Danieli, S.D., Abiven, S. et al. (2017). piCaSo – Pilotage sylvicole et contrôle pédologique des stocks de carbone des sols forestiers. Report, ADEME.

Cécillon, L., Baudin, F., Chenu, C., Houot, S., Jolivet, R., Kätterer, T., Lutfalla, S., Macdonald, A., van Oort, V., Plante, A.F. et al. (2018). A model based on Rock-Eval thermal analysis to quantify the size of the centennially persistent organic carbon pool in temperate soils. *Soil Biology and Biochemistry*, 117, 117–124.

Cécillon, L., Baudin, F., Chenu, C., Christensen, B.T., Franko, U., Houot, S., Kanari, E., Kätterer, T., Merbach, I., van Oort, F. et al. (2021). Partitioning soil organic carbon into its centennially stable and active fractions with statistical models based on Rock-Eval® thermal analysis (PARTYsocv2.0 and PARTYsocv2.0eu). *Geoscientific Model Development*, 14, 3879–3898.

Cerri, C., Feller, C., Balesdent, J., Victoria, R., Plencassagne, A. (1985). Application du traçage isotopique naturel en ^{13}C, à l'étude de la dynamique de la matière organique dans les sols. *Comptes rendus de l'Académie des Sciences*, 423–428.

Chassé, M., Luftalla, S., Cécillon, L., Baudin, F., Abiven, S., Chenu, C., Barré, P. (2021). Long-term bare fallow soil fractions reveal thermo-chemical properties controlling soil organic carbon dynamics. *Biogeosciences*, 18, 1703–1718.

Chassiot, L., Miras, Y., Chapron, E., Develle, A.L., Arnaud, F., Motelica-Heino, M., Di-Giovanni, C. (2018). A 7000-year environmental history and soil erosion record inferred from the deep sediments of Lake Pavin (Massif Central, France). *Palaeogeography Palaeoclimatology Palaeoecology*, 497, 218–233.

Chin, Y. and Gschwend, P.M. (1991). The abundance, distribution, and configuration of porewater organic colloids in recent sediments. *Geochim. Cosmochim. Acta*, 55, 1309–1317.

Ciais, P., Borges, A.V., Abril, G., Meybeck, M., Folberth, G., Hauglustaine, D., Janssens, I.A. (2008). The impact of lateral carbon fluxes on the European carbon balance. *Biogeosciences*, 5, 1259–1271.

Clayton, J.L. and Bostick, N.H. (1986). Temperature effects on kerogen and on molecular isotopic composition of organic matter in Pierre Shales near an ignous dike. *Organic Geochemistry*, 10, 135–143.

Clivot, H., Mouny, J.C., Duparque, A., Dinh, J.L., Denoroy, P., Houot, S., Vertès, F., Trochard, R., Bouthier, A., Sagot, S. et al. (2019). Modeling soil organic carbon evolution in long-term arable experiments with AMG model. *Environmental Modelling & Software*, 118, 99–113.

Cohen-Sadon, H., Amrani, A., Feinstein, S., Rosenberg, Y.O. (2021a). Rapid evaluation of organic and pyritic sulfur content in sedimentary rocks by Rock-Eval 7S analyzer. *Conference Proceedings, 30th International Meeting on Organic Geochemistry (IMOG 2021)*, September, 1–3.

Cohen-Sadon, H., Rosenberg, Y.O., Amrani, A. (2021b). Rapid evaluation of $\delta^{34}S$ values in sedimentary rocks by Rock-Eval 7S analyser coupled to a MC-ICPMS. *Conference Proceedings, 30th International Meeting on Organic Geochemistry (IMOG 2021)*, September, 1–3.

Collet, C., Akroume, E., Bureau, F., Cécillon, L., Deleuze, C., Derrien, D., Richter, C., Berthelot, A. (2019). Rapport du projet CAPSOL : dynamique du CArbone et de la croissance après Préparation du SOL dans les plantations forestières. Report, ADEME.

Cooper, H.V., Vane, C.H., Evers, S., Aplin, P., Girkin, N.T., Sjögersten, S. (2019). From peat swamp forest to oil palm plantations: The stability of tropical peatland carbon. *Geoderma*, 342, 109–117.

Cooper, H.V., Sjögersten, S., Lark, R.M., Girkin, N.T., Vane, C.H., Calonego, J.C., Rosolem, C., Mooney, S.J. (2021). Long-term zero-tillage enhances the protection of soil carbon in tropical agriculture. *European Journal of Soil Science*, 72(6), 2477–2492.

Copard, Y. (2002). Altération diagénétique et post-diagénétique (thermicité, oxydation) des charbons carbonifères du Massif Central français (Saint-Etienne, Graissessac et autres lieux). Thesis, Université d'Orléans.

Copard, Y., Disnar, J.R., Becq-Giraudon, J.F., Boussafir, M. (2000). Evidence and effects of fluid circulation on organic matter in intramontane coalfields (Massif Central, France). *International Journal of Coal Geology*, 44, 49–68.

Copard, Y., Disnar, J.R., Becq-Giraudon, J.F. (2002). Erroneous maturity assessment given by Tmax and HI Rock-Eval parameters on highly mature weathered coals. *International Journal of Coal Geology*, 49, 57–65.

Copard, Y., Disnar, J.R., Becq-Giraudon, J.F., Laggoun-Défarge, F. (2004). Erroneous coal maturity assessment caused by low temperature oxidation. *International Journal of Coal Geology*, 58, 171–180.

Copard, Y., Di-Giovanni, C., Martaud, T., Albéric, P., Ollivier, J.E. (2006). Using Rock-Eval 6 pyrolysis for tracking fossil organic carbon in modern environments: Implications for the roles of erosion and weathering. *Earth Surface Processes and Landforms*, 31, 135–153.

Copard, Y., Eyrolle-Boyer, F., Radakovitch, O., Poirel, A., Raimbault, P., Gairoard, S., Di-Giovanni, C. (2018). Badlands as a hot spot of petrogenic contribution of the riverine particulate organic carbon (Gulf of Lion, NW. Mediterranean Sea). *Earth Surface Processes and Landforms*, 43, 2495–2509.

Czirbus, N., Nyilas, T., Raucsik, B., Hetényi, M. (2016). Investigation of the effect of soil mineral composition on soil organic matter stability. *Soil & Water Research*, 11, 147–154.

Dahl, J., Hallberg, R., Kaplan, I.R. (1988). The effects of radioactive decay of uranium on elemental and isotopic ratios of Alum Shale kerogen. *Applied Geochemistry*, 3, 583–589.

Dartiguelongue, C., Behar, F., Budzinski, H., Scacchi, G., Marquaire, P.M. (2006). Thermal stability of dibenzothiophene in closed system pyrolysis: Experimental study and kinetic modelling. *Organic Geochemistry*, 37, 98–116.

Delarue, F., Laggoun-Défarge, F., Disnar, J.R., Lottier, N., Gogo, S. (2011). Organic matter sources and decay assessment in a Sphagnum-dominated peatland (Le Forbonnet, Jura Mountains, France): Impact of moisture conditions. *Biogeochemistry*, 106(1), 39–52.

Delarue, F., Disnar, J.R., Copard, Y., Gogo, S., Jacob, J., Beny, J.M., Laggoun-Defarge, F. (2013). Can Rock-Eval pyrolysis assess the biogeochemical composition of organic matter during peatification? *Organic Geochemistry*, 61, 66–72.

Derenne, S. and Quénéa, K. (2015). Analytical pyrolysis as a tool to probe soil organic matter. *Journal of Analytical and Applied Pyrolysis*, 111, 108–120.

Dewar, R.A. (1961). A flame ionization detector. A theoretical approach. *Journal of Chromatography*, 6, 312–323.

Di-Giovanni, C., Disnar, J.R., Bichet, V., Campy, M., Guillet, B. (1998). Geochemical characterization of soil organic matter and variability of a postglacial detrital organic supply (Chaillexon Lake, France). *Earth Surface Processes and Landforms*, 23, 1057–1069.

Di-Giovanni, C., Disnar, J.R., Bakyono, J.P., Kéravis, D., Millet, F., Olivier, J.E. (2000). Application de l'étude de la matière organique à l'analyse de l'érosion : exemple du bassin versant du Moulin, dans les terres noires des Alpes-de-Haute-Provence (France). *Comptes rendus de l'Académie des Sciences*, 331, 7–14.

Disnar, J.R. and Sureau, J.F. (1990). Organic matter in ore genesis: Progress and perspectives. *Advanced in Organic Geochemistry*, 16(1–3), 577–590.

Disnar, J.R., Marquis, F., Espitalié, J., Barsony, I., Drouet, S., Giot, D. (1997). Géochimie organique et reconstitution de l'histoire thermique et tectono-sédimentaire de la marge ardéchoise (programme GPF; France). *Bulletin de la Société Géologique de France*, 168(1), 73–81.

Disnar, J.R., Guillet, B., Keravis, D., Di-Giovanni, C., Sebag, D. (2003). Soil organic matter (SOM) characterization by Rock-Eval pyrolysis: Scope and limitations. *Organic Geochemistry*, 34, 327–343.

Disnar, J.R., Jacob, J., Morched-Issa, M., Lottier, N., Arnaud, F. (2008). Assessment of peat quality by molecular and bulk geochemical analysis: Application to the Holocene record of the Chautagne marsh (Haute Savoie, France). *Chemical Geology*, 254, 101–112.

Disnar, J.R., Stefanova, M., Bréhéret, J.G., Macaire, J.J. (2011). Microbial mat development and dolomite formation under pre-evaporitic conditions during the Atlantic in a temperate area: The Sarliève Lake (French Massif Central). *Organic Geochemistry*, 42(9), 1089–1098.

Dominé, F. (1989). Kinetics of hexane pyrolysis at very high pressure 1. Experimental study. *Energy and Fuels*, 3, 89–96.

Dörner, M., Berner, U., Erdmann, M., Barth, T. (2019). Carbon isotopic analysis of reactive organic matter using a new pyrolysis-cryotrapping-isotope ratio mass spectrometry method: The isotope variation of organic matter within the S1 and S2 peaks of Rock-Eval. *Organic Geochemistry*, 136, 103886.

Douglas, A.G., Eglinton, G., Henderson, W. (1970). Thermal alternation of the organic matter in sediments. *Advances in Organic Geochemistry Proceedings, Third International Congress on Organic Geochemistry*, London, 26–28.

Downey, M., Garvin, J., Lagomarsino, R., Nicklin, D. (2011). Quick look determination of oil-in-place in oil shale resource plays. Search and discovery article #40764. *American Association of Petroleum Geologists Annual Convention and Exhibition*, Houston, Texas.

Duan, D., Ran, Y., Cheng, H., Chen, J., Wan, G. (2014). Contamination trends of trace metals and coupling with algal productivity in sediment cores in Pearl River Delta, South China. *Chemosphere*, 103, 35–43.

Duan, D., Huang, Y., Cheng, H., Ran, Y. (2015). Relationship of polycyclic aromatic hydrocarbons with algae-derived organic matter in sediment cores from a subtropical region. *Journal of Geophysical Research: Biogeosciences*, 120(11), 2243–2255.

Duan, D., Zhang, D., Yang, Y., Wang, J., Chen, J., Ran, Y. (2017). Source, composition, and environmental implication of neutral carbohydrates in sediment cores of subtropical reservoirs, South China. *Biogeosciences*, 14(18), 4009–4022.

Ducreux, J., Lafargue, E., Marquis, F., Pillot, D., Bocard, C. (1997). Use of Rock-Eval method for the evaluation of soils contaminated by hydrocarbons. *Analusis*, 25(9–10), 40–44.

Durand, B. (1980). *Kerogen, Insoluble Organic Matter from Sedimentary Rocks*. Editions Technip, Paris.

Durand, B. and Espitalié, J. (1973). Etude de la matière organique au cours de l'enfouissement des sédiments. *Comptes rendus de l'Académie des Sciences*, 216, 2253–2256.

Durand, B. and Nicaise, G. (1980). Procedure for kerogen isolation. In *Kerogen, Insoluble Organic Matter from Sedimentary Rocks*, Durand, B. (ed.). Editions Technip, Paris.

Durand, B., Espitalié, J., Nicaise, G., Combaz, A. (1972). Étude de la matière organique insoluble (kérogène) des argiles du Toarcien du Bassin de Paris I – Étude par les procédés optiques, analyse élémentaire, étude en microscopie et diffraction électroniques. *Oil & Gas Science and Technology*, 27, 865–884.

Eglin, T., Ciais, P., Piao, S.L., Barré, P., Bellassen, V., Cadule, P., Chenu, C., Gasser, T., Koven, C., Reichstein, M. et al. (2010). Historical and future perspectives of global soil carbon response to climate and land-use changes. *Tellus B: Chemical and Physical Meteorology*, 62, 700–718.

Elie, M., Techer, I., Trotignon, L., Khoury, H., Salameh, E., Vandamme, D., Boulvais, P., Fourcade, S. (2007). Cementation of kerogen-rich marls by alkaline fluids released during weathering of thermally metamorphosed marly sediments. Part II: Organic matter evolution, magnetic susceptibility and metals (Ti, Cr, Fe) at the Khushaym Matruk natural analogue (Central Jordan). *Applied Geochemistry*, 22(7), 1311–1328.

Emmings, J.F., Davies, S.J., Vane, C.H., Leng, M.J., Moss-Hayes, V., Stephenson, M.H., Jenkin, G.R.T. (2017). Stream and slope weathering effects on organic-rich mudstone geochemistry and implications for hydrocarbon source rock assessment: A Bowland Shale case study. *Chemical Geology*, 471, 74–91.

Espitalié, J. (1984). Analyseur d'indices (huile et gaz) sur les chantiers de forage. In *Nouvelles technologies pour l'exploration et l'exploitation des ressources de pétrole et de gaz*, volume 2, de Bauw, R. (ed.). Editions Technip pour la Commission des communautés européennes, Luxembourg.

Espitalié, J. and Durand, B. (1974). Méthode et appareil permettant l'évaluation rapide du potentiel de production pétrolière de sédiments géologiques sur la base de faibles prélèvements. Rock-Eval® 1 patent, FR_2 227.

Espitalié, J. and Marquis, F. (1997). Méthode et dispositif pour déterminer des caractéristiques pétrolières de sédiments géologiques. Rock-Eval® 6 patent, FR_2 739 694_A1, US-5958777_A.

Espitalié, J. and Pages, L. (1981). Méthode rapide d'analyse géochimique des déblais en suivi de forage. *Pétrole et techniques*, 283, 99–103.

Espitalié, J., Laporte, J.F., Madec, M., Marquis, F., Leplat, P., Paulet, J. (1977a). Méthode rapide de caractérisation des roches mères, de leur potentiel pétrolier et de leur degré d'évolution. *Oil & Gas Science and Technology*, 32, 23–42.

Espitalié, J., Laporte, J.L., Madec, M., Marquis, F. (1977b). Méthode et dispositif pour déterminer des caractéristiques pétrolières de sédiments géologiques sur la base de faibles prélèvements. Rock-Eval® 1 patent, FR_2 339 173_A1.

Espitalié, J., Laporte, J.L., Madec, M., Marquis, F. (1978). Méthode et dispositif pour déterminer des caractéristiques pétrolières de sédiments géologiques sur la base de faibles prélèvements. Rock-Eval® 1 patent, FR_2 370 979_A2.

Espitalié, J., Madec, M., Leplat, P., Paulet, J. (1979). Méthode et dispositif pour déterminer notamment la quantité de carbone organique contenue dans un échantillon. Rock-Eval® 2 patent, FR_2 472 754_A1, US-4519983_A.

Espitalié, J., Laporte, J.L., Madec, M., Marquis, F. (1980a). Méthode et dispositif pour déterminer des caractéristiques pétrolières de sédiments géologiques sur la base de faibles prélèvements, Rock-Eval® 3-OSA patent, FR_2 335 038_A2.

Espitalié, J., Madec, M., Tissot, B. (1980b). Role of mineral matrix in kerogen pyrolysis: Influence on petroleum generation and migration. *American Association of Petroleum Geologists*, 64, 59–66.

Espitalié, J., Senga Makadi, K., Trichet, J. (1984a). Role of mineral matrix during kerogen pyrolysis. *Organic Geochemistry*, 6, 365–382.

Espitalié, J., Marquis, F., Barsony, I. (1984b). Geochemical logging. In *Analytical Pyrolysis, Techniques and Applications*, Voorhees, K.J. (ed.). Butterworth, Boston.

Espitalié, J., Deroo, G., Marquis, F. (1985a). La pyrolyse Rock-Eval et ses applications – Première partie. *Oil & Gas Science and Technology*, 40(5), 563–579.

Espitalié, J., Deroo, G., Marquis, F. (1985b). La pyrolyse Rock-Eval et ses applications – Deuxième partie. *Oil & Gas Science and Technology*, 40(6), 755–784.

Espitalié, J., Deroo, G., Marquis, F. (1986). La pyrolyse Rock-Eval et ses applications – Troisième partie. *Oil & Gas Science and Technology*, 41(1), 73–89.

Espitalié, J., Antonas, R., Lamoureux-Var, V., Letort, G., Pillot, D., Beaumont, V., Haeseler, F. (2010). Méthode et dispositif pour déterminer des caractéristiques pétrolières de sédiments géologiques sur la base de faibles prélèvements. Rock-Eval® 7 sulfur patent, FR_2 937 737, WO_2010049609_A1.

Feller, C., Brossard, M., Chen, Y., Landa, E.R., Trichet, J. (2010). Selected pioneering works on humus in soils and sediments during the 20th century: A retrospective look from the International Humic Substances Society view. *Physics and Chemistry of the Earth, Parts ABC, Studies from the History of Soil Science and Geology*, 35, 903–912.

Fernandez, P., Vilanova, R.M., Grimalt, J.O. (1999). Sediment fluxes of polycyclic aromatic hydrocarbons in European high altitude mountain lakes. *Environmental Science & Technology*, 33, 3716–3722.

Fitzgerald, D. and van Krevelen, D.W. (1959). Chemical structure and properties of coal: The kinetics of coal carbonization. *Fuel*, 38, 17–37.

Forbes, P., Landais, P., Bertrand, P., Brosse, E., Espitalié, J., Yahaya, M. (1988). Chemical transformation of type-III organic matter associated with the akouta uranium deposit (Niger). *Chemical Geology*, 71, 267–282.

Forsman P. and Hunt, J.M. (1958). Insoluble organic matter (kerogen) in sedimentary rocks. *Geochimica et Cosmochimica Acta*, 15, 170–182.

Franko, U. and Merbach, I. (2017). Modelling soil organic matter dynamics on a bare fallow Chernozem soil in Central Germany. *Geoderma*, 303, 93–98.

Freudenthal, T., Wagner, T., Wenzhöfer, F., Zabel, M., Wefer, G. (2001). Early diagenesis of organic matter from sediments of the eastern subtropical Atlantic: Evidence from stable nitrogen and carbon isotopes. *Geochimica et Cosmochimica Acta*, 65(11), 1795–1808.

Fujisaki, K., Perrin, A.S., Boussafir, M., Gogo, S., Sarrazin, M., Brossard, M. (2015). Decomposition kinetics and organic geochemistry of woody debris in a ferralsol in a humid tropical climate. *European Journal of Soil Science*, 66, 876–885.

Fusetti, L., Behar, F., Bounaceur, R., Marquaire, P.M., Grice, K., Derenne, S. (2010). New insights into secondary gas generation from oil thermal cracking: Methylated monoaromatics. A kinetic approach using 1,2,4-trimethylbenzene. Part I: A free-radical mechanism. *Organic Geochemistry*, 41, 146–167.

Gardes, T., Portret-Koltalo, F., Debret, M., Humbert, K., Levaillant, R., Simon, M., Copard, Y. (2020). Temporal trends, sources and relationships with sediment characteristics of polycyclic aromatic hydrocarbons (PAHs) and polychlorinated biphenyls (PCBs) in the major tributary of the Seine estuary, France. *Applied Geochemistry*, 122, 104749.

Gascón Díez, E., Corella, J.P., Adatte, T., Thevenon, F., Loizeau, J.L. (2017). High-resolution reconstruction of the 20th century history of trace metals, major elements, and organic matter in sediments in a contaminated area of Lake Geneva, Switzerland. *Applied Geochemistry*, 78, 1–11.

Gennari, G., Tamburini, F., Ariztegui, D., Hajdas, I., Spezzaferri, S. (2009). Geochemical evidence for high-resolution variations during deposition of the Holocene S1 sapropel on the cretan ridge, eastern mediterranean. *Palaeogeography, Palaeoclimatology, Palaeoecology*, 273, 239–248.

George, S.C. (1992). Effect of igneous intrusion on the organic geochemistry of a siltstone and an oil shale horizon in the Midland Valley of Scotland. *Organic Geochemistry*, 18(5), 705–723.

Georgiev, S., Stein, H.J., Hannah, J.L., Weiss, H.M., Bingen, B., Xu, G., Rein, E., Hatlø, V., Løseth, H., Nali, M. et al. (2012). Chemical signals for oxidative weathering predict Re-Os isochroneity in black shales, East Greenland. *Chemical Geology*, 324–325, 108–121.

Giguet-Covex, C., Arnaud, F., Poulenard, J., Disnar, J.R., Delhon, C., Francus, P., David, F., Enters, D., Rey, P.J., Delannoy, J.J. (2011). Changes in erosion patterns during the Holocene in a currently treeless subalpine catchment inferred from lake sediment geochemistry (Lake Anterne, 2063 m a.s.l., NW French Alps): The role of climate and human activities. *The Holocene*, 21(4), 651–665.

Gillespie, A.W., Sanei, H., Diochon, A., Ellert, B.H., Regier, T.Z., Chevrier, D., Dynes, J.J., Tarnocai, C., Gregorich, E.G. (2014). Perennially and annually frozen soil carbon differ in their susceptibility to decomposition: Analysis of Subarctic earth hummocks by bioassay, XANES and pyrolysis. *Soil Biology and Biochemistry*, 68, 106–116.

Girkin, N.T., Vane, C.H., Cooper, H.V., Moss-Hayes, V., Craigon, J., Turner, B.L., Ostle, N., Sjögersten, S. (2019). Spatial variability of organic matter properties determines methane fluxes in a tropical forested peatland. *Biogeochemistry*, 142(2), 231–245.

Gogo, S., Laggoun-Defarge, F., Delarue, F., Lottier, N. (2011). Invasion of a *Sphagnum*-peatland by *Betula* spp. and *Molinia caerulea* impacts organic matter biochemistry. Implications for carbon and nutrient cycling. *Biogeochemistry*, 106, 53–69.

Graz, Y., Di-Giovanni, C., Copard, Y., Laggoun-Défarge, F., Boussafir, M., Lallier-Vergès, E., Baillif, P., Perdereau, L., Simonneau, A. (2010). Quantitative palynofacies analysis as a new tool to study transfers of fossil organic matter in recent terrestrial environments. *International Journal of Coal Geology*, 84, 49–62.

Graz, Y., Di-Giovanni, C., Copard, Y., Elie, M., Faure, P., Laggoun-Défarge, F., Leveque, J., Michels, R., Ollivier, J.E. (2011). Fossil organic matter occurrence in modern environments: Optical, geochemical and isotopical evidences. *Applied Geochemistry*, 26, 1302–1314.

Graz, Y., Di-Giovanni, C., Copard, Y., Mathys, N., Cras, A., Marc, V. (2012). Annual fossil organic carbon delivery due to mechanical and chemical weathering of marly badlands areas. *Earth Surface Processes Landforms*, 37, 1263–1271.

Gregorich, E.G., Gillespie, A.W., Beare, M.H., Curtin, D., Sanei, H., Yanni, S.F. (2015). Evaluating biodegradability of soil organic matter by its thermal stability and chemical composition. *Soil Biology and Biochemistry*, 91, 182–191.

Hao, F., Li, S., Dong, W., Hu, Z., Huang, B. (1998). Abnormal organic-matter maturation in the Yinggehai Basin, South China Sea: Implications for hydrocarbon expulsion and fluid migration from overpressured systems. *Journal of Petroleum Geology*, 21(4), 427–444.

Hao, F., Li, S., Gong, Z., Yang, J. (2000). Thermal regime, interreservoir compositional heterogeneities, and reservoir-filling history of the Dongfang Gas Field, Yinggehai Basin, South China Sea: Evidence for episodic fluid injections in overpressured basins? *American Association of Petroleum Geologists Bulletin*, 84(5), 607–626.

Hare, A.A., Kuzyk Z.Z.A., Macdonald, R.W., Sanei, H., Barber, D., Stern, G.A., Wang, F. (2014). Characterization of sedimentary organic matter in recent marine sediments from Hudson Bay, Canada, by Rock-Eval pyrolysis. *Organic Geochemistry*, 68, 52–60.

Harouna, M., Disnar, J.R., Martinez, L., Trichet, J. (1993). Discrepancies between different organic maturity indicators in a coal series affected by an abnormal thermal event (Visean, Niger). *Chemical Geology*, 106, 397–413.

Hasinger, O., Spangenberg, J.E., Millière, L., Bindschedler, S., Cailleau, G., Verrecchia, E.P. (2015). Carbon dioxide in scree slope deposits: A pathway from atmosphere to pedogenic carbonate. *Geoderma*, 247–248, 129–139.

Hazra, B., Dutta, S., Kumar, S. (2017). TOC calculation of organic matter rich sediments using Rock-Eval pyrolysis: Critical consideration and insights. *International Journal of Coal Geology*, 169, 106–115.

Hetényi, M. and Nyilas, T. (2014). Soil organic matter characterization using S3 and S4 signals from Rock-Eval pyrolysis. *Pedosphere*, 24, 563–574.

Hetényi, M., Nyilas, T., Tóth, T.M. (2005). Stepwise Rock-Eval pyrolysis as a tool for typing heterogeneous organic matter in soils. *Journal of Analytical and Applied Pyrolysis*, 74, 45–54.

Hetényi, M., Nyilas, T., Sajgó, C., Brukner-Wein, A. (2006). Heterogeneous organic matter from the surface horizon of a temperate zone marsh. *Organic Geochemistry*, 37, 1931–1942.

Holm, T. (1999). Aspects of the mechanism of the flame ionization detector. *Journal of Chromatography*, 842, 221–227.

Holtvoeth, J., Wagner, T., Horsfield, B., Schubert, C., Wand, U. (2001). Late Quaternary supply of terrigenous organic matter to the Congo Deep Sea Fan (ODP site 1075): Implications for equatorial African paleoclimate. *Geo-Marine Letters*, 21, 23–33.

Holtvoeth, J., Wagner, T., Horsfield, B., Schubert, C. (2003). Organic matter in river influenced continental margin sediments: The land-ocean and climate linkage at the Late Quaternary Congo fan (ODP Site 1075). *Geochemistry Geophysics Geosystem*, 4(12), 1109.

Holtvoeth, J., Kolonic, S., Wagner, T. (2005). Soil organic matter as an important contributor to late Quaternary sediments of the tropical West African continental margin. *Geochimica et Cosmochimica Acta*, 69, 2031–2041.

Horsfield, B. and Douglas, A.G. (1980). The influence of minerals on the pyrolysis of kerogens. *Geochimica et Cosmochimica Acta*, 44(8), 1119–1131.

Hoş-Çebi, F. and Korkmaz, S. (2015). Organic geochemistry of Ağaçbaşı Yayla peat deposits, Köprübaşı/Trabzon, NE Turkey. *International Journal of Coal Geology*, 146, 155–165.

Hudson, S.M., Hillam, S.A., Barker, J., Nelson, S.T., Rey, K.A. (2019). Pyrolysis of modern wetland sediment: Extracting climate records from fens in the Uinta Mountains and Fish Lake Plateau, Utah, USA. *Boreas*, 48(3), 810–824.

Hutcheon, I. (1998). The potential role of pyrite oxidation in corrosion and reservoir souring. *Journal of Canadian Petroleum Technology*, 37(1), 27–31.

Inan, S., Yalçin, M.N., Mann, U. (1998). Expulsion of oil from petroleum source rocks: Inferences from pyrolysis of samples of unconventional grain size. *Organic Geochemistry*, 29(1–3), 45–61.

Jacob, J., Disnar, J.R., Boussafir, M., Sifeddine, A., Turcq, B., Spadano Albuquerque, A.L. (2004). Major environmental changes recorded by lacustrine sedimentary organic matter since the last glacial maximum near the equator (Lagoa do Caçó, NE Brazil). *Palaeogeography, Palaeoclimatology, Palaeoecology*, 205(3–4), 183–197.

Jarvie, D.M. (2012). Shale resource systems for oil and gas. Part 2: Shale-oil resource systems. Shale reservoirs – Giant resources for the 21st century. *American Association of Petroleum Geologists Memoir*, 97, 89–119.

Jarvie, D.M. (2014). Components and processes affecting producibility and commerciality of shale resource systems. *Geologica Acta*, 12(4), 307–325.

Jenkinson, D.S. (1990). The turnover of organic carbon and nitrogen in soil. *Philosophical Transactions of the Royal Society B*, 329, 361–368.

Jensen, H.K.B., Connan, J., Brorøy, M., Hall, K., Wold, S. (1998). Geoelf sulphur analyser: Quantification of thermally extractable and pyrolysable organic and mineral sulphur in source rocks. *Organic Geochemistry*, 28(1–2), 87–110.

Jiang, C., Chen, Z., Mort, A., Milovic, M., Robinson, R., Stewart, R., Lavoie, D. (2016). Hydrocarbon evaporative loss from shale core samples as revealed by Rock-Eval and thermal desorption – Gas chromatography analysis: Its geochemical and geological implications. *Marine and Petroleum Geology*, 70, 294–303.

Jørgensen, B.B. (1982). Mineralization of organic matter in the sea-bed: The role of sulphate reduction. *Nature*, 296, 643–645.

Jørgensen, B.B. (1983). Processes at the sediment-water interface. In *The Major Biogeochemical Cycles and their Interactions*, Bolin, B. and Cook, R.B. (eds). Wiley, New York.

Joseph, D. and Oberlin, A. (1983). Oxidation of carbonaceous matter – I elemental analysis (C, H, O) and IR spectrometry. *Carbon*, 21(6), 559–564.

van Kaam-Peters, H.M.E., Schouten, S., Köster, J., Sinninghe Damsté, J.S. (1998). Controls on the molecular and carbon isotopic composition of organic matter deposited in a Kimmeridgian euxinic shelf sea: Evidence for preservation of carbohydrates through sulfurization. *Geochimica et Cosmochimica Acta*, 62, 3259–3283.

Kanari, E. (2018). Estimer la persistance du carbone organique des sols par Rock-Eval. Master's thesis, University of Berlin.

Kanari, E., Cécillon, L., Baudin, F., Clivot, H., Ferchaud, F., Houot, S., Levavasseur, F., Mary, B., Soucémarianadin, L., Chenu, C. et al. (2022). A robust initialization method for accurate soil organic carbon simulations. *Biogeosciences*, 19, 375–387.

Keesstra, S.D., Bouma, J., Wallinga, J., Tittonell, P., Smith, P., Cerdà, A., Montanarella, L., Quinton, J.N., Pachepsky, Y., Van der Putten, W.H. et al. (2016). The significance of soils and soil science towards realization of the United Nations Sustainable Development Goals. *Soil*, 2, 111–128.

Kemp, A.C., Horton, B.P., Nikitina, D., Vane, C.H., Potapova, M., Weber-Bruya, E., Culver, S.J., Repkina, T., Hill, D.F. (2017). The distribution and utility of sea-level indicators in Eurasian sub-Arctic salt marshes (White Sea, Russia). *Boreas*, 46(3), 562–584.

Khedim, N., Cécillon, L., Poulenard, J., Barré, P., Baudin, F., Marta, S., Rabatel, A., Dentant, C., Cauvy-Fraunié, S., Anthelme, F. et al. (2021). Topsoil organic matter build-up in glacier forelands around the world. *Global Change Biology*, 27, 1662–1677.

Koch, A., McBratney, A., Adams, M., Field, D., Hill, R., Crawford, J., Minasny, B., Lal, R., Abbott, L., O'Donnell, A. et al. (2013). Soil security: Solving the global soil crisis. *Global Policy*, 4, 434–441.

Kögel-Knabner, I. (2000). Analytical approaches for characterizing soil organic matter. *Organic Geochemistry*, 31, 609–625.

Kögel-Knabner, I. (2002). The macromolecular organic composition of plant and microbial residues as inputs to soil organic matter. *Soil Biology and Biochemistry*, 34, 139–162.

Kowalewski, I., Lamoureux-Var, V., Huc, A.Y. (2010). Thermal enhanced oil recovery: New field of application for the "in-situ" reaction of the formation of H_2S and organo-sulfur compounds by hydrous thermal alteration. *12nd ALAGO Conference*, Montevideo.

van Krevelen, D.W. (1981). *Coal: Typology, Chemistry, Physics, Constitution*, 2nd edition. Elsevier, Amsterdam.

Kuske, S., Horsfield, B., Jweda, J., Michael, G.E., Song, Y. (2019). Geochemical factors controlling the phase behavior of Eagle Ford Shale petroleum fluids. *American Association of Petroleum Geologists Bulletin*, 103(4), 835–870.

Lacey, J.H., Leng, M.J., Vane, C.H., Radbourne, A.D., Yang, H., Ryves, D.B. (2018). Assessing human impact on Rostherne Mere, UK, using the geochemistry of organic matter. *Anthropocene*, 21, 52–65.

Lafargue, E., Marquis, F., Pillot, D. (1998a). Rock-Eval 6 applications in hydrocarbon exploration, production, and soil contamination studies. *Revue de l'Institut Francais du Petrole*, 53, 421–437.

Lafargue, E., Ducreux, J., Marquis, F., Pillot, D. (1998b). Méthode et dispositif d'évaluation d'une caractéristique de pollution d'un échantillon de sol. Pollut-Eval® patent, FR_ 2 753 271_A1, US_8046497_A.

Lallier-Vergès, E., Sifeddine, A., de Beaulieu, J.L., Reille, M., Tribovillard, N., Bertrand, P., Mongenot, T., Thouveny, N., Disnar, J.R., Guillet, B. (1993). Sensibilité de la sédimentation organique aux variations climatiques du Tardi-Würm et de l'Holocène ; Le Lac du Bouchet (Haute-Loire, France). *Bulletin de la Société Géologique de France*, 164(5), 661–673.

Lamoureux-Var, V. and Barroux, C. (2013). Using geochemistry to address H_2S production risk due to steam injection in oil sands. *13th SPE Heavy Oil Conference*, Calgary, Extended Abstracts, 13HOCC-P-412-SPE.

Lamoureux-Var, V. and Lorant, F. (2005). Experimental evaluation of H_2S yields in reservoir rocks submitted to steam injection. *13th European Symposium on Improved Oil Recovery, AGE*, Budapest, Extended Abstracts, D08.

Lamoureux-Var, V., Kohler, E., Huc, A.Y., Berger, E., Fusetti, L. (2011). Laboratory simulation of H_2S and CO_2 generation in steamed oil sands, Orinoco Belt, Venezuela. *American Association of Petroleum Geologists Hedberg Research Conference Natural Gas Geochemistry: Recent Developments, Applications, and Technologies*, Beijing, Extended Abstracts.

Lamoureux-Var, V., Espitalié, J., Pillot, D., Bouton, N., Garcia, B., Antonas, R., Aboussou, A., Wattripont, A., Ravelojaona, H., Noirez, S. et al. (2019). Rock-Eval 7S: Technology and performance. *29th International Meeting on Organic Geochemistry (IMOG 2019)*, Gothenburg.

Landais, P. (1996). Organic geochemistry of sedimentary uranium ore deposits. *Ore Geology Reviews*, 11, 33–51.

Landais, P., Monthioux, M., Meunier, J.D. (1984). Importance of the oxidation/maturation pair in the evolution of humic coals. *Organic Geochemistry*, 7(3/4), 249–260.

Langford, F.F. and Blanc-Valleron, M.M (1990). Interpreting Rock-Eval pyrolysis data using graphs of pyrolyzable hydrocarbons vs. total organic carbon. *American Association of Petroleum Geologists Bulletin*, 74, 799–804.

Larter, S.R. and Douglas, A.G. (1982). Pyrolysis methods in organic geochemistry: An overview. *Journal of Analytical and Applied Pyrolysis*, 4, 1–19.

Lavrieux, M., Bréhéret, J.G., Disnar, J.R., Jacob, J., Le Milbeau, C., Zocatelli, R. (2012). Preservation of an ancient grassland biomarker signature in a forest soil from the French Massif Central. *Organic Geochemistry*, 51, 1–10.

Lavrieux, M., Disnar, J.R., Chapron, E., Bréhéret, J.G., Jacob, J., Miras, Y., Reyss, J.L., Andrieu-Ponel, V., Arnaud, F. (2013). 6700 yr sedimentary record of climatic and anthropogenic signals in Lake Aydat (French Massif Central). *The Holocene*, 23(9), 1317–1328.

Le Mer, G., Barthod, J., Dignac, M.F., Barré, P., Baudin, F., Rumpel, C. (2020). Inferring the impact of earthworms on the stability of organo-mineral associations, by Rock-Eval thermal analysis and ^{13}C NMR spectroscopy. *Organic Geochemistry*, 144, 104016.

Leininger, J.P., Lorant, F., Minot, C., Behar, F. (2006). Mechanism of 1-methylnaphthalene pyrolysis in a batch reactor and relevance with other methylated polyaromatics. *Energy and Fuels*, 20, 2518–2530.

Leng, M.J., Wagner, B., Boehm, A., Panagiotopoulos, K., Vane, C.H., Snelling, A., Haidon, C., Woodley, E., Vogel, H., Zanchetta, G. et al. (2013). Understanding past climatic and hydrological variability in the Mediterranean from Lake Prespa sediment isotope and geochemical record over the Last Glacial cycle. *Quaternary Science Reviews*, 66, 123–136.

Lewan, M.D. (1993). Laboratory simulation of petroleum formation: Hydrous pyrolysis. In *Organic Geochemistry Principles and Applications*, Engel, M.H. and Macko, S. (eds). Plenum, New York.

Lewan, M.D. (1997). Experiments on the role of water in petroleum formation. *Geochimica et Cosmochimica Acta*, 61, 3691–3723.

Lewan, M.D. (1998). Sulfur radical control on petroleum formation rates. *Nature*, 391, 164–166.

Lewan, M.D. and Ruble, T.E. (2002). Comparison of petroleum generation kinetics by isothermal hydrous and non-isothermal open-system pyrolysis. *Organic Geochemistry*, 33, 1457–1475.

Littke, R., Klussmann, U., Krooss, B., Leythaeuser, D. (1991). Quantification of loss of calcite, pyrite, and organic matter due to weathering of Toarcian black shales and effects on kerogen and bitumen characteristics. *Geochimica et Cosmochimica Acta*, 55, 3369–3378.

Lo, H.B. and Cardott, B.J. (1995). Detection of natural weathering of Upper McAlester coal and Woodford Shale, Oklahoma, U.S.A. *Organic Geochemistry*, 22(1), 73–83.

Lorant, F., Behar, F., Vandenbroucke, M., McKinney, D.E., Tang, Y. (2000). Methane generation from methylated aromatics: Kinetic study and carbon isotope modelling. *Energy and Fuels*, 14, 1143–1155.

Luo, Y., Ahlström, A., Allison, S.D., Batjes, N.H., Brovkin, V., Carvalhais, N., Chappell, A., Ciais, P., Davidson, E.A., Finzi, A. et al. (2016). Toward more realistic projections of soil carbon dynamics by Earth system models. *Global Biogeochemical Cycles*, 30, 40–56.

Lutfalla, S., Chenu, C., Barré, P. (2014). Are chemical oxidation methods relevant to isolate a soil pool of centennial carbon? *Biogeochemistry*, 118, 135–139.

von Lützow, M., Kögel-Knabner, I., Ekschmitt, K., Matzner, E., Guggenberger, G., Marschner, B., Flessa, H. (2001). Stabilization of organic matter in temperate soils: Mechanisms and their relevance under different soil conditions – A review: Mechanisms for organic matter stabilization in soils. *European Journal of Soil Science*, 57, 426–445.

Machel, H.G. (2001). Bacterial and thermochemical sulfate reduction in diagenetic settings – Old and new insights. *Sedimentary Geology*, 140, 143–175.

Madec, M. and Espitalié, J. (1985). Determination of organic sulphur in sedimentary rocks by pyrolysis. *Journal of Analytical and Applied Pyrolysis*, 8, 201–219.

Madec, M., Espitalié, J., Laporte, J.L., Barsony, I. (1978). Méthode et dispositif d'analyse de sédiments géologiques, permettant notamment de déterminer leur teneur en soufre organique. Rock-Eval® Sulfur_1 patent, FR_2 376 414.

Magny, M., Leroux, A., Bichet, V., Gauthier, E., Richard, H., Walter-Simonnet, A.V. (2012). Climate, vegetation and land use as drivers of Holocene sedimentation: A case study from Lake Saint-Point (Jura Mountains, eastern France). *The Holocene*, 23(1), 137–147.

Mahmud, W.M. and Ermila, M. (2020). Case study: Barium sulfate scale removal by dynamic underbalance. *SPE International Conference and Exhibition on Formation Damage Control*, SPE-199232-MS, Lafayette, Louisiana.

Malou, O.P., Sebag, D., Moulin, P., Chevallier, T., Badiane-Ndour, N.Y., Thiam, A., Chapuis-Lardy, L. (2020). The Rock-Eval® signature of soil organic carbon in arenosols of the Senegalese groundnut basin. How do agricultural practices matter? *Agriculture, Ecosystems & Environment*, 301, 107030.

Malounguila-Nganga, D., Giresse, P., Boussafir, M., Miyouna, T. (2017). Late Holocene swampy forest of Loango Bay (Congo). Sedimentary environments and organic matter deposition. *Journal of African Earth Sciences*, 134, 419–434.

Manalt, F., Beck, C., Disnar, J.R., Deconinck, J.F., Recourt, P. (2001). Evolution of clay mineral assemblages and organic matter in the late glacial-Holocene sedimentary infill of Lake Annecy, (northwestern Alps): Paleoenvironmental implications. *Journal of Paleolimnology*, 25, 179–192.

Marchand, C., Lallier-Vergès, E., Baltzer, F. (2003). The composition of sedimentary organic matter in relation to the dynamic features of a mangrove-fringed coast in French Guiana. *Estuarine Coastal and Shelf Sciences*, 56, 119–130.

Marchand, C., Lallier-Vergès, E., Disnar, J.R., Kéravis, D. (2008). Organic carbon sources and transformations in mangrove sediments: A Rock-Eval pyrolysis approach. *Organic Geochemistry*, 39(4), 408–421.

Matteodo, M., Grand, S., Sebag, D., Rowley, M.C., Vittoz, P., Verrecchia, E.P. (2018). Decoupling of topsoil and subsoil controls on organic matter dynamics in the Swiss Alps. *Geoderma*, 330, 41–51.

Mayr, C., Lücke, A., Maidana, N.I., Wille, M., Haberzettl, T., Corbella, H., Ohlendorf, C., Schäbitz, F., Fey, M., Janssen, S. et al. (2009). Isotopic fingerprints on lacustrine organic matter from Laguna Potrok Aike (southern Patagonia, Argentina) reflect environmental changes during the last 16,000 years. *Journal of Paleolimnology*, 42(1), 81–102.

McIver, R.D. (1967). Composition of kerogen – Clue to its role in the origin of petroleum. *7th World Petroleum Congress Proceedings*. Elsevier, Amsterdam, 25–36.

McWilliam, I.G. and Dewar, R.A. (1958). Flame ionization detector for gas chromatography. *Nature*, 181(4611), 760.

Meinhold, G., Howard, J.P., Strogen, D., Kaye, M.D., Abutarruma, Y., Elgadry, M., Thusu, B., Whitham, A.G. (2013). Hydrocarbon source rock potential and elemental composition of lower Silurian subsurface shales of the eastern Murzuq Basin, southern Libya. *Marine and Petroleum Geology*, 48, 224–246.

Melenevskii, V.N., Leonova, G.A., Konyshev, A.S. (2011). The organic matter of the recent sediments of Lake Beloe, West Siberia (from data of pyrolytic studies). *Russian Geology and Geophysics*, 52(6), 583–592.

Melenevskii, V.N., Leonova, G.A., Bobrov, V.A., Kashirtsev, V.A., Krivonogov, S.K. (2015). Transformation of organic matter in the Holocene sediments of Lake Ochki (south Baikal region): Evidence from pyrolysis data. *Geochemistry International*, 53(10), 903–921.

Melenevskii, V.N., Klimin, M.A., Tolstokorov, S.V. (2019). Diagenesis of organic matter in peat: Rock–Eval pyrolysis data. *Geochemistry International*, 57(2), 227–231.

Meyers, P.A. (1997). Organic geochemical proxies of paleoceanographic, paleolimnologic and paleoclimatic processes. *Organic Geochemistry*, 27(5/6), 213–250.

Meyers, P.A. and Lallier-Vergès, E. (1999). Lacustrine sedimentary organic matter records of Late Quaternary paleoclimates. *Journal of Paleolimnology*, 21, 345–372.

Meyers, P.A. and Simoneit, B.R.T. (1999). Effects of extreme heating on the elemental and isotopic compositions of an Upper Cretaceous coal. *Organic Geochemistry*, 30, 299–305.

Millet, L., Vannière, B., Verneaux, V., Magny, M., Disnar, J.R., Laggoun-Défarge, F., Walter-Simonnet, A.V., Bossuet, G., Ortu, E., de Beaulieu, J.L. (2007). Response of littoral chironomid communities and organic matter to late glacial lake level, vegetation and climate changes at Lago dell'Accesa (Tuscany, Italy). *Journal of Paleolimnology*, 38, 525–539.

Mort, A. (2004). Mécanismes de formation et évolution des pyrobitumes dans les réservoirs pétroliers : cas naturels et approches expérimentales. Thesis, Université d'Orléans.

Nádudvari, Á. and Fabiańska, M.J. (2015). Coal-related sources of organic contamination in sediments and water from the Bierawka River (Poland). *International Journal of Coal Geology*, 152, 94–109.

Newell, A.J., Vane, C.H., Sorensen, J.P.R., Moss-Hayes, V., Gooddy, D.C. (2016). Long-term Holocene groundwater fluctuations in a chalk catchment: Evidence from Rock-Eval pyrolysis of riparian peats: Rock-Eval evidence for groundwater fluctuations in a chalk catchment. *Hydrological Processes*, 30(24), 4556–4567.

Noël, H., Garbolino, E., Brauer, A., Lallier-Vergès, E., de Beaulieu, J.L., Disnar, J.R. (2001). Human impact and soil erosion during the last 5000 yrs as recorded in lacustrine sedimentary organic matter at Lac d'Annecy, the French Alps. *Journal of Paleolimnology*, 25, 229–244.

Ordoñez, L., Vogel, H., Sebag, D., Ariztegui, D., Adatte, T., Russell, J.M., Kallmeyer, J., Vuillemin, A., Friese, A., Crowe, S.A. et al. (2019). Empowering conventional Rock-Eval pyrolysis for organic matter characterization of the siderite-rich sediments of Lake Towuti (Indonesia) using End-Member Analysis. *Organic Geochemistry*, 134, 32–44.

Orr, W.L. (1986). Kerogen/asphaltene/sulfur relationships in sulfur-rich Monterey oils. *Organic Geochemistry*, 10, 499–516.

Outridge, P.M. and Sanei, H. (2010). Does organic matter degradation affect the reconstruction of pre-industrial atmospheric mercury deposition rates from peat cores? A test of the hypothesis using a permafrost peat deposit in northern Canada. *International Journal of Coal Geology*, 83(1), 73–81.

Peters, K.E. (n.d.). Guidelines for evaluating petroleum source rock using programmed pyrolysis. *American Association of Petroleum Geologists Bulletin*, 70, 318–329.

Philp, R.P. and DeGarmo, C.D. (2020). Geochemical characterization of the Devonian-Mississippian woodford shale from the McAlister cemetery quarry, Criner Hills Uplift, Ardmore Basin, Oklahoma. *Marine and Petroleum Geology*, 112, 104078.

Pichevin, L., Bertrand, P., Boussafir, M., Disnar, J.R. (2004). Organic matter accumulation and preservation controls in a deep sea modern environment: An example from Namibian slope sediments. *Organic Geochemistry*, 35(5), 543–559.

Pillot, D., Deville, E., Prinzhofer, A. (2014). Identification and quantification of carbonate species using Rock-Eval pyrolysis. *Oil Gas Science Technology – Revue d'IFP*, 69, 341–349.

Poeplau, C., Don, A., Six, J., Kaiser, M., Benbi, D., Chenu, C., Cotrufo, M.F., Derrien, D., Gioacchini, P., Grand, S. et al. (2018). Isolating organic carbon fractions with varying turnover rates in temperate agricultural soils – A comprehensive method comparison. *Soil Biology and Biochemistry*, 125, 10–26.

Poeplau, C., Barré, P., Cécillon, L., Baudin, F., Sigurdsson, B.D. (2019). Changes in the Rock-Eval signature of soil organic carbon upon extreme soil warming and chemical oxidation – A comparison. *Geoderma*, 337, 181–190.

Poot, A., Quik, J.T.K., Veld, H., Koelmans, A.A. (2009). Quantification methods of black carbon: Comparison of Rock-Eval analysis with traditional methods. *Journal of Chromatography A*, 1216, 613–622.

Poot, A., Jonker, M.T.O., Gillissen, F., Koelmans, A.A. (2014). Explaining PAH desorption from sediments using Rock Eval analysis. *Environmental Pollution*, 193, 247–253.

Preux, C., Lamoureux-Var, V., Michel, P., Ayache, S. (2018). Forecasting H_2S production risk in thermal projects for EOR. *First EAGE/IFPEN Conference on Sulfur Risk Management in E&P*, Rueil-Malmaison, Extended Abstracts.

Rabouille, C., Dennielou, B., Baudin, F., Raimonet, M., Droz, L., Khripounoff, A., Martinez, P., Méjanelle, L., Michalopoulos, P., Pastor, L. et al. (2019). Carbon and silica megasink in deep-sea sediments of the Congo terminal lobes. *Quaternary Science Reviews*, 222, 105854.

Rahman, M.W. and Rimmer, S.M. (2014). Effects of rapid thermal alteration on coal: Geochemical and petrographic signatures in the Springfield (No. 5) Coal, Illinois Basin. *International Journal of Coal Geology*, 131, 214–226.

Raiswell, R. and Berner, R.A. (1986). Pyrite and organic matter in Phanerozoic normal marine shales. *Geochimica et Cosmochimica Acta*, 50, 1967–1976.

Rakotoalimanana, D.A., Behar, F., Bounaceur, R., Burklé-Vitzthum, V., Marquaire, P.M. (2016). Thermal cracking of n-butylcyclohexane at high pressure (100 bar) – Part 1: Experimental study. *Journal of Analytical and Applied Pyrolysis*, 117, 1–16.

Rickard, D. and Luther III, G.W. (1997). Kinetics of pyrite formation by the H2S oxidation of iron (II) monosulfide in aqueous solutions between 25 and 125°C: The mechanism. *Geochimica et Cosmochimica Acta*, 61, 135–147.

Rickard, D. and Morse, J.W. (2005). Acid volatile sulfide (AVS). *Marine Chemistry*, 97, 141–197.

Roduit, B. (2000). Computational aspects of kinetic analysis. Part E: The ICTAC Kinetics Project – Numerical techniques and kinetics of solid state processes. *Thermochimica Acta*, 355, 171–180.

Romanens, R., Pellacani, F., Mainga, A., Fynn, R., Vittoz, P., Verrecchia, E.P. (2019). Soil diversity and major soil processes in the Kalahari basin, Botswana. *Geoderma Regional*, 19, e00236.

Romero-Sarmiento, M.F. (2019). A quick analytical approach to estimate both free versus sorbed hydrocarbon contents in liquid-rich source rocks. *American Association of Petroleum Geologists Bulletin*, 103, 2031–2043.

Romero-Sarmiento, M.F., Pillot, D., Letort, G., Lamoureux-Var, V., Beaumont, V., Huc, A.Y., Garcia, B. (2014). New Rock-Eval method for characterization of shale plays. *14th Latin American Congress on Organic Geochemistry (ALAGO)*, Buzios, Rio de Janeiro.

Romero-Sarmiento, M.F., Pillot, D., Letort, G., Lamoureux-Var, V., Beaumont, V., Huc, A.Y., Garcia, B. (2016a). New Rock-Eval method for characterization of unconventional shale resource systems. *Oil & Gas Science and Technology*, 71, 37.

Romero-Sarmiento, M.F., Euzen, T., Rohais, S., Jiang, C., Littke, R. (2016b). Artificial thermal maturation of source rocks at different thermal maturity levels: Application to the Triassic Montney and Doig formations in the Western Canada Sedimentary basin. *Organic Geochemistry*, 97, 148–162.

Romero-Sarmiento, M.F., Ramiro-Ramirez, S., Berthe, G., Fleury, M., Littke, R. (2017a). Geochemical and petrophysical source rock characterization of the Vaca Muerta Formation, Argentina: Implications for unconventional petroleum resource estimations. *International Journal of Coal Geology*, 184, 27–41.

Romero-Sarmiento, M.F., Youssouf, S., Letort, G. (2017b). Procédé pour estimer les hydrocarbures libres dans un échantillon de roche. IFP Energies nouvelles patent 17/59447, October 9.

Romero-Sarmiento, M.F., Cadeau, P., Sissmann, O., Beaumont, V. (2019). Système et procédé pour la détermination de la composition chimique des composés contenus dans un échantillon. IFP Energies nouvelles patent 19/0258, July 4.

Romero-Sarmiento, M.F., Ravelojaona, H., Pillot, D., Rohais, S. (2022). Polymer quantification using the Rock-Eval® device for identification of plastics in sediments. *Science of the Total Environment*, 807(3), 151068.

Rosenberg, Y.O., Meshoulam, A., Said-Ahmad, W., Shawar, L., Dror, G., Reznik, I., Feinstein, S., Amrani, A. (2017). Study of thermal maturation processes of sulfur-rich source rock using compound specific sulfur isotope analysis. *Organic Geochemistry*, 112, 59–74.

Rühlmann, J. (1999). A new approach to estimating the pool of stable organic matter in soil using data from long-term field experiments. *Plant and Soil*, 213, 149–160.

Saenger, A., Cécillon, L., Sebag, D., Brun, J.J. (2013). Soil organic carbon quantity, chemistry and thermal stability in a mountainous landscape: A Rock–Eval pyrolysis survey. *Organic Geochemistry*, 54, 101–114.

Saenger, A., Cécillon, L., Poulenard, J., Bureau, F., De Daniéli, S., Gonzalez, J.M., Brun, J.J. (2015). Surveying the carbon pools of mountain soils: A comparison of physical fractionation and Rock-Eval pyrolysis. *Geoderma*, 241–242, 279–288.

Saint-Germès, M., Baudin, F., Bazhenova, O., Derenne, S., Fadeeva, N., Largeau, C. (2002). Origine et préservation de la matière organique amorphe dans la série de Maykop (Oligocène- Miocène inférieur) du Précaucase et de l'Azerbaïdjan. *Bulletin de la Société Géologique de France*, 173, 5, 423– 436.

Sakho, I., Mesnage, V., Copard, Y., Deloffre, J., Disnar, J.R., Lafite, R., Niang, I., Faye, G. (2015). A cross-section analysis of sedimentary organic matter in a mangrove ecosystem under dry climate conditions, Somone, Senegal. *Journal of African Earth Sciences*, 101, 220–231.

Salmon, V., Derenne, S., Lallier-Vergès, E., Largeau, C., Beaudoin, B. (2000). Protection of organic matter by mineral matrix in a Cenomanian black shale. *Organic Geochemistry*, 31(5), 463–474.

Sanei, H. and Goodarzi, F. (2006). Relationship between organic matter and mercury in recent lake sediment: The physical–geochemical aspects. *Applied Geochemistry*, 21, 1900–1912.

Sanei, H., Stasiuk, L.D., Goodarzi, F. (2005). Petrological changes occurring in organic matter from recent lacustrine sediments during thermal alteration by Rock-Eval pyrolysis. *Organic Geochemistry*, 36, 1190–1203.

Sanei, H., Outridge, P.M., Dallimore, A., Hamilton, P.B. (2012). Mercury–organic matter relationships in pre-pollution sediments of thermokarst lakes from the Mackenzie River Delta, Canada: The role of depositional environment. *Biogeochemistry*, 107(1–3), 149–164.

Sanei, H., Wood, J.M., Ardakani, O.H., Clarkson, C.R., Jiang, C. (2015). Characterization of organic matter fractions in an unconventional tight gas siltstone reservoir. *International Journal of Coal Geology*, 150–151, 296–305.

Schomburg, A., Verrecchia, E.P., Guenat, C., Brunner, P., Sebag, D., Le Bayon, R.C. (2018). Rock-Eval pyrolysis discriminates soil macro-aggregates formed by plants and earthworms. *Soil Biology and Biochemistry*, 117, 117–124.

Schomburg, A., Sebag, D., Turberg, P., Verrecchia, E.P., Guenat, C., Brunner, P., Adatte, T., Schlaepfer, R., Le Bayon, R.C. (2019). Composition and superposition of alluvial deposits drive macro-biological soil engineering and organic matter dynamics in floodplains. *Geoderma*, 355, 113899.

Sebag, D., Disnar, J.R., Guillet, B., Di-Giovanni, C., Verrechia, E.P., Durand, B. (2006). Monitoring organic matter dynamics in soil profile by Rock-Eval pyrolysis: Bulk characterization and quantification of degradation. *European Journal of Soil Science*, 57, 344–355.

Sebag, D., Debret, M., M'voubou, M., Obame, R.M., Ngomanda, A., Oslisly, R., Bentaleb, I., Disnar, J.R., Giresse, P. (2013). Coupled Rock-Eval pyrolysis and spectrophotometry for lacustrine sedimentary dynamics: Application for West Central African rainforests (Kamalété and Nguène lakes, Gabon). *The Holocene*, 23(8), 1173–1183.

Sebag, D., Verrecchia, E.P., Cécillon, L., Adatte, T., Albrecht, R., Aubert, M., Bureau, F., Cailleau, G., Copard, Y., Decaens, T. et al. (2016). Dynamics of soil organic matter based on new Rock-Eval indices. *Geoderma*, 284, 185–203.

Sebag, D., Garcin, Y., Adatte, T., Deschamps, P., Ménot, G., Verrecchia, E.P. (2018). Correction for the siderite effect on Rock-Eval parameters: Application to the sediments of Lake Barombi (southwest Cameroon). *Organic Geochemistry*, 123, 126–135.

Shawar, L., Halevy, I., Said-Ahmad, W., Feinstein, S., Boyko, V., Kamyshny, A., Amrani, A. (2018). Dynamics of pyrite formation and organic matter sulfurization in organic-rich carbonate sediments. *Geochimica et Cosmochimica Acta*, 241, 219–239.

Sifeddine, A., Bertrand, P., Lallier-Vergès, E., Patience, A.J. (1996). Lacustrine organic fluxes and palaeoclimatic variations during the last 15 ky: Lac du Bouchet (Massif Central, France). *Quaternary Science Reviews*, 15, 203–211.

Simonneau, A., Chapron, E., Vannière, B., Wirth, S.B., Gilli, A., Di-Giovanni, C., Anselmetti, F., Desmet, M., Magny, M. (2013). Mass-movement and flood-induced deposits in Lake Pedro, southern Italy: Implications for Holocene palaeohydrology and natural hazards. *Climate of the Past*, 9, 825–840.

Sinninghe Damsté, J.S. and De Leeuw, J.W. (1990). Analysis, structure and geochemical significance of organically-bound sulphur in the geosphere: State of the art and future research. *Organic Geochemistry*, 16(4–6), 1077–1101.

Soucémarianadin, L., Cécillon, L., Chenu, C., Baudin, F., Nicolas, M., Girardin, C., Barré, P. (2018a). Is Rock-Eval 6 thermal analysis a good indicator of soil organic carbon lability? – A method-comparison study in forest soils. *Soil Biology and Biochemistry*, 117, 108–116.

Soucémarianadin, L., Cécillon, L., Guenet, B., Chenu, C., Baudin, F., Nicolas, M., Girardin, C., Barré, P. (2018b). Environmental factors controlling soil organic carbon stability in French forest soils. *Plant Soil*, 426, 267–286.

Soucémarianadin, L., Cécillon, L., Chenu, C., Baudin, F., Nicolas, M., Girardin, C., Delahaie, A., Barré, P. (2019). Heterogeneity of the chemical composition and thermal stability of particulate organic matter in French forest soils. *Geoderma*, 342, 65–74.

Spangenberg, J.E. and Frimmel, H.E. (2001). Basin-internal derivation of hydrocarbons in the Witwatersrand Basin, South Africa: Evidence from bulk and molecular $\delta^{13}C$ data. *Chemical Geology*, 173, 339–355.

Stalder, C., El Kateb A., Vertino, A., Rüggeberg, A., Camozzi, O., Pirkenseer, C.M., Spangenberg, J.E., Hajdas, I., Van Rooij, D., Spezzaferri, S. (2018). Large-scale paleoceanographic variations in the western Mediterranean Sea during the last 34,000 years: From enhanced cold-water coral growth to declining mounds. *Marine Micropaleontology*, 143, 46–62.

Stetten, E., Baudin, F., Reyss, J.-L., Martinez, P., Charlier, K., Schnyder, J., Rabouille, C., Dennielou, B., Coston-Guarini, J., Pruski, A. (2015). Organic matter characterization and distribution in sediments of the terminal lobes of the Congo deep-sea fan: Evidence for the direct influence of the Congo River. *Marine Geology*, 369, 182–195.

Stock, A.T., Littke, R., Schwarzbauer, J., Horsfield, B., Hartkopf-Fröder, C. (2017). Organic geochemistry and petrology of Posidonia Shale (Lower Toarcian, Western Europe) – The evolution from immature oil-prone to overmature dry gas-producing kerogen. *International Journal of Coal Geology*, 176–177, 36–48.

Sun, Y., Püttmann, W., Speczik, S. (1995). Differences in the depositional environment of basal Zechstein in southwest Poland: Implication for base metal mineralization. *Organic Geochemistry*, 23(9), 819–835.

Swindles, G.T., Morris, P.J., Whitney, B., Galloway, J.M., Galka, M., Gallego-Sala, A., Macumber, A.L. (2017). Ecosystem state shifts during long-term development of an Amazonian peatland. *Global Change Biology*, 24, 738–757.

Talbot, M.R. and Livingstone, D.A. (1989). Hydrogen index and carbon isotopes of lacustrine organic matter as lake level indicators. *Palaeogeography, Palaeoclimatology, Palaeoecology*, 70(1–3), 121–137.

Talbot, M.R., Jensen, N.B., Lærdal, T., Filippi, M.L. (2006). Geochemical responses to a major transgression in giant African lakes. *Journal of Paleolimnology*, 35(3), 467–489.

Tarafa, M.E., Hunt, J.M., Ericsson, I. (1983). Effect of hydrocarbon volatility and adsorption on source-rock pyrolysis. *Journal of Geochemical Exploration*, 18(1), 75–85.

Thoumazeau, A., Chevallier, T., Baron, V., Rakotondrazafy, N., Panklang, P., Marichal, R., Kibblewhite, M., Sebag, D., Tivet, F., Bessou, C. et al. (2020). A new in-field indicator to assess the impact of land management on soil carbon dynamics. *Geoderma*, 375, 114496.

Tissot, B.P. (1969). Premières données sur les mécanismes et la cinétique de la formation du pétrole dans les bassins sédimentaires. Simulation d'un schéma réactionnel sur ordinateur. *Oil and Gas Science and Technology*, 24, 470–501.

Tissot, B.P. and Espitalié, J. (1975). L'évolution de la matière organique des sédiments : application d'une simulation mathématique. *Oil and Gas Science and Technology*, 24, 470–501.

Tissot, B.P. and Welte, D.H. (1984). *Petroleum Formation and Occurrence*, 2nd revised and enlarged edition. Springer-Verlag, Berlin, Heidelberg, New York, Tokyo.

Tissot, B.P., Durand, B., Espitalié, J., Combaz, A. (1974). Influence of nature and diagenesis of organic matter in formation of petroleum. *American Association of Petroleum Geologists Bulletin*, 58, 499–506.

Tomic, J., Behar, F., Vandenbroucke, M., Tang, Y. (1995). Artificial maturation of Monterey kerogen (Type II-S) in a closed system and comparison with Type II kerogen: Implications on the fate of sulfur. *Organic Geochemistry*, 23, 647–660.

Trabelsi, K., Espitalié, J., Huc, A.Y. (1994). Characterisation of extra heavy oils and tar deposits by modified pyrolysis methods. *Proceedings of the European Symposium on Heavy Oil Technologies in a Wider Europe*, June 7–8, 30–40.

Tribovillard, N., Riboulleau, A., Lyons, T., Baudin, F. (2004). Enhanced trapping of molybdenum by sulfurized marine organic matter of marine origin in Mesozoic limestones and shales. *Chemical Geology*, 213, 385–401.

Tribovillard, N., Hatem, E., Averbuch, O., Barbecot, F., Bout-Roumazeilles, V., Trentesaux, A. (2015). Iron availability as a dominant control on the primary composition and diagenetic overprint of organic-matter-rich rocks. *Chemical Geology*, 401, 67–82.

Tzovolou, D.N., Benoit, Y., Haeseler, F., Klint, K.E., Tsakiroglou, C.D. (2009). Spatial distribution of jet fuel in the vadoze zone of a heterogeneous and fractured soil. *Science of the Total Environment*, 407, 3044–3054.

Ungerer, P. (1990). State of the art of research in kinetic modelling of oil formation and destruction. *Organic Geochemistry*, 16, 1–25.

Ungerer, P. and Pelet, R. (1987). Extrapolation of the kinetics of oil and gas formation from laboratory experiments to sedimentary basins. *Nature*, 327, 52–54.

Upton, A., Vane, C.H., Girkin, N., Turner, B.L., Sjögersten, S. (2018). Does litter input determine carbon storage and peat organic chemistry in tropical peatlands? *Geoderma*, 326, 76–87.

Uzcategui, D., Lamoureux-Var, V., Berger, E. (2011). H_2S and CO_2 generation mechanisms on a steam injection project on Petrocedeño Field, Orinoco Belt. *2011 World Heavy Oil Congress*, Edmonton, Alberta, Extended Abstracts, WHOC11-112.

Vairavamurthy, M., Orr, W., Manowitz, B. (1995). Geochemical transformations of sedimentary sulfur: An introduction. In *Geochemical Transformations of Sedimentary Sulfur*, Vairavamurthy, M.A. and Schoonen, M.A.A. (eds). American Chemical Society, Washington.

Van Buchem, F.S.P., Eberli, G.P., Whalen M.T., Mountjoy, E.W., Homewood, P. (1996). The basinal geochemical signature and platform margin geometries in the Upper Devonian mixed carbonate-siliciclastic system of western Canada. *Bulletin de la Société Géologique de France*, 167, 685–699.

Van der Putten, N., Hébrard, J.P., Verbruggen, C., Van de Vijver, B., Disnar, J.R., Spassov, S., de Beaulieu, J.L., De Dapper, M., Kéravis, D., Hus, J. et al. (2008). An integrated palaeoenvironmental investigation of a 6200 year old peat sequence from Ile de la Possession, Iles Crozet, sub-Antarctica. *Palaeogeography, Palaeoclimatology, Palaeoecology*, 270, 179–195.

Van Exem, A., Debret, M., Copard, Y., Verpoorter, C., de Wet, G., Lecoq, N., Sorrel, P., Werner, A., Roof, S., Laignel, B. et al. (2019). New source-to-sink approach in an arctic catchment based on hyperspectral core-logging (Lake Linné, Svalbard). *Quaternary Science Reviews*, 203, 128–140.

Vandenbroucke, M. and Largeau, C. (2007). Kerogen origin, evolution and structure. *Organic Geochemistry*, 38(5), 719–833.

Vindušková, O., Sebag, D., Cailleau, G., Brus, J., Frouz, J. (2015). Methodological comparison for quantitative analysis of fossil and recently derived carbon in mine soils with high content of aliphatic kerogen. *Organic Geochemistry*, 89–90, 14–22.

Volland-Tuduri, N. (2005). Nature et mode d'assemblage des constituants minéraux et organiques dans des Ferralsols de la région des Cerrados (Brésil) : évolution après mise en culture. Thesis, Université d'Orléans.

Wade, W.J., Hanor, J.S., Sassen, R. (1988). Controls on H_2S concentration and hydrocarbon destruction in the eastern Smackover trend. *Transactions: Gulf Coast Association of Geological Societies*, 309–320.

Wattripont, A., Baudin, F., de Rafelis, M., Deconinck, J.F. (2019). Specifications for carbonate content quantification in recent marine sediments using Rock-Eval pyrolysis. *Journal of Analytical and Applied Pyrolysis*, 140, 393–403.

Werne, J.P., Lyons, T.W., Hollander, D.J., Schouten, S., Hopmans, E.C., Sinninghe Damsté, J.S. (2008). Investigating pathways of diagenetic organic matter sulfurization using compound-specific sulfur isotope analysis. *Geochimica et Cosmochimica Acta*, 72, 3489–3502.

Wilhelms, A. and Larter, S.R. (1994). Petroleum reservoirs. Part I: Introduction and case studies. *Marine and Petroleum Geology*, 11(4), 418–441.

Wilkes, H., Ramrath, A., Negendank, J.F.W. (1999). Organic geochemical evidence for environmental changes since 34,000 yrs BP from Lago di Mezzano, central Italy. *Journal of Paleolimnology*, 22, 349–365.

Wohlfarth, B., Schwark, L., Bennike, O., Filimonova, L., Tarasov, P., Björkmanj, L., Brunnberg, L., Demidov, I., Possnert, G. (2004). Unstable early-Holocene climatic and environmental conditions in northwestern Russia derived from a multidisciplinary study of a lake-sediment sequence from Pichozero, southeastern Russian Karelia. *The Holocene*, 14(5), 732–746.

Woods, J., Kung, J., Kingston, D., Kotlyar, L., Sparks, B., McCracken, T. (2008). Canadian crudes: A comparative study of SARA fractions from a modified HPLC separation technique. *Oil & Gas Science and Technology*, 63(1), 151–163.

Worden, R.H., Smalley, P.C., Fallick, A.E. (1997). Sulfur cycle in buried evaporates. *Geology*, 25, 643–646.

Worden, R.H., Smalley, P.C., Cross, M.M. (2000). The influences of rock fabric and mineralogy upon thermo-chemical sulfate reduction: Khuff Formation, Abu Dhabi. *Journal of Sedimentary Research*, 70, 1218–1229.

Worden, R.H., Smalley, P.C., Barclay S.A. (2003). H_2S and diagenetic pyrite in North Sea sandstones: Due to TSR or organic sulphur compound cracking? *Journal of Geochemical Exploration*, 78/79, 487–491.

Wu, F., Xu, L., Sun, Y., Liao, H., Zhao, X., Guo, J. (2012). Exploring the relationship between polycyclic aromatic hydrocarbons and sedimentary organic carbon in three Chinese lakes. *Journal of Soils and Sediments*, 12(5), 774–783.

Wu, F., Xu, L., Liao, H., Guo, F., Zhao, X., Giesy, J.P. (2013). Relationship between mercury and organic carbon in sediment cores from Lakes Qinghai and Chenghai, China. *Journal of Soils and Sediments*, 13, 1084–1092.

Yang, S. and Horsfield, B. (2020). Critical review of the uncertainty of Tmax in revealing the thermal maturity of organic matter in sedimentary rocks. *International Journal of Coal Geology*, 225, 103500.

Yang, S., Schulz, H.M., Horsfield, B., Schovsbo, N.H., Noah, M., Panova, E., Rothe, H., Hahne, K. (2018). On the changing petroleum generation properties of Alum Shale over geological time caused by uranium irradiation. *Geochimica et Cosmochimica Acta*, 229, 20–35.

Yu, Z., Liu, K., Liu, L., Qu, X., Yu, M., Zhao, S., Ming, X. (2016). Characterization of Paleogene hydrothermal events and their effects on reservoir properties in the Qikou Sag, eastern China. *Journal of Petroleum Science and Engineering*, 146, 1226–1241.

Zaccone, C., Sanei, H., Outridge, P.M., Miano, T.M. (2011). Studying the humification degree and evolution of peat down a Holocene bog profile (Inuvik, NW Canada): A petrological and chemical perspective. *Organic Geochemistry*, 42, 399–408.

Zegouagh, Y., Derenne, S., Dignac, M.F., Baruiso, E., Mariotti, A., Largeau, C. (2004). Demineralisation of a crop soil by mild hydrofluoric acid treatment. *Journal of Analytical and Applied Pyrolysis*, 71, 119–135.

Zocatelli, R., Lavrieux, M., Disnar, J.R., Le Milbeau, C., Jacob, J., Bréhéret, J.G. (2012a). Free fatty acids in Lake Aydat catchment soils (French Massif Central): Sources, distributions and potential use as sediment biomarkers. *Journal of Soils and Sediments*, 12(5), 734–748.

Zocatelli, R., Turcq, B., Boussafir, M., Cordeiro, R.C., Disnar, J.R., Costa, R.L., Sifeddine, A., Albuquerque, A.L.S., Bernardes, M.C., Jacob, J. (2012b). Late Holocene paleoenvironmental changes in Northeast Brazil recorded by organic matter in lacustrine sediments of Lake Boqueirão. *Palaeogeography, Palaeoclimatology, Palaeoecology*, 363–364, 127–134.

Zocatelli, R., Moreira-Turcq, P., Jacob, J., Cordeiro, R.C., Boussafir, M., Le Milbeau, C., Bernardes, M., Turcq, B. (2016). Holocene land cover dynamics in the curuai floodplain inferred from lacustrine biomarkers. *Palaeogeography, Palaeoclimatology, Palaeoecology*, 443, 237–248.

List of Authors

Jérémie AMMOUIAL
Vinci Technologies
Nanterre
France

Pierre BARRÉ
Géologie ENS
CNRS
Paris
France

François BAUDIN
ISTeP
Sorbonne Université
Paris
France

Françoise BEHAR
TotalEnergies
Courbevoie
France

Yves BENOIT
IFP Energies nouvelles
Rueil-Malmaison
France

Nicolas BOUTON
Vinci Technologies
Nanterre
France

Lauric CÉCILLON
ENS/Université de Rouen Normandie
INRAE
Rouen
France

Yoann COPARD
M2C
Université de Rouen Normandie
Rouen
France

Jean ESPITALIÉ
Vinci Technologies
Nanterre
France

Eva KANARI
Géologie ENS
CNRS
and
Sorbonne Université
Paris
France

Violaine LAMOUREUX-VAR
IFP Energies nouvelles
Rueil-Malmaison
France

Maria-Fernanda ROMERO-SARMIENTO
IFP Energies nouvelles
Rueil-Malmaison
France

Adrien WATTRIPONT
Vinci Technologies
Nanterre
France

Index

E, F, G

H, I

K, L

M, N

O, P

R, S

T, V, W